INTERFACES IN COMPUTER SCIENCE AND OPERATIONS RESEARCH

Advances in Metaheuristics, Optimization, and Stochastic Modeling Technologies

OPERATIONS RESEARCH/COMPUTER SCIENCE INTERFACES SERIES

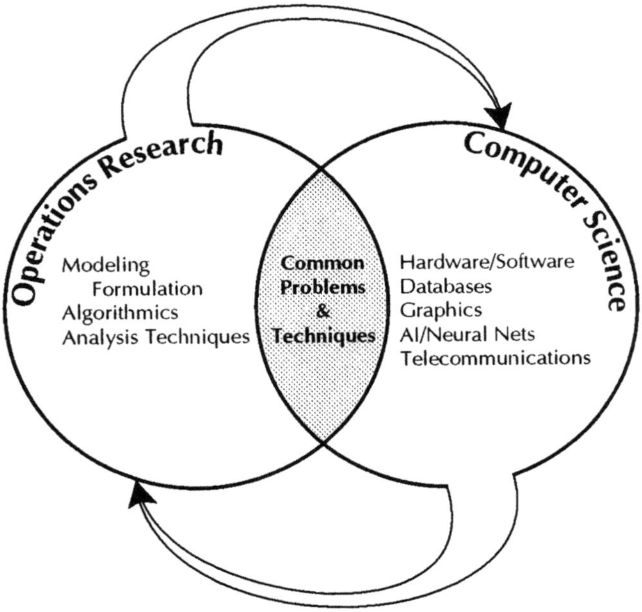

Ramesh Sharda, Series Editor
Conoco/DuPont Chair of Management of Technology
Oklahoma State University
Stillwater, Oklahoma U.S.A.

Other published titles in the series:

Brown, Donald/Scherer, William T.
University of Virginia
Intelligent Scheduling Systems

Barth, Peter
Max-Planck-Institut fur Informatik, Germany
Logic-Based 0-1 Constraint Programming

Jones, Christopher V.
University of Washington
Visualization and Optimization

INTERFACES IN COMPUTER SCIENCE AND OPERATIONS RESEARCH

Advances in Metaheuristics, Optimization, and Stochastic Modeling Technologies

EDITED BY

Richard S. BARR
Southern Methodist University
Dallas, Texas, USA

Richard V. HELGASON
Southern Methodist University
Dallas, Texas, USA

Jeffery L. KENNINGTON
Southern Methodist University
Dallas, Texas, USA

SPRINGER SCIENCE+BUSINESS MEDIA, LLC

Library of Congress Cataloging-in-Publication Data

A C.I.P. Catalogue record for this book is available from the Library of Congress.

ISBN 978-1-4613-6837-3 ISBN 978-1-4615-4102-8 (eBook)
DOI 10.1007/978-1-4615-4102-8

Printed on acid-free paper.

CONTENTS

CONTRIBUTORS

Richard S. Barr
Southern Methodist University
Dallas, Texas, USA

M. Bouzoubaa
INRIA-CERMICS
Sophia-Antipolis Cedex, FRANCE

V. Chandru
Indian Instutute of Science
Bangalore, INDIA

Hai D. Chu
SABRE Decision Technologies
Dallas-Ft. Worth, Texas, USA

Nicolas P. Couellan
University of Oklahoma
Norman, Oklahoma, USA

Steven P. Coy
University of Maryland
College Park, Maryland, USA

Cihan H. Dagli
University of Missouri–Rolla
Rolla, Missouri, USA

Aniruddha Deshpande
Louisana State University
Baton Rouge, Louisana, USA

Eric Gelman
SABRE Decision Technologies
Dallas-Ft. Worth, Texas, USA

Fred Glover
University of Colorado
Boulder, Colorado, USA

Bruce L. Golden
University of Maryland
College Park, Maryland, USA

François Guertin
Université de Montréal
Montréal, Quebec, CANADA

G. Hasle
SINTEF
Oslo, NORWAY

Ellis L. Johnson
Georgia Institute of Technology
Atlanta, Georgia, USA

Joanna Józefowska
Poznań University of Technology
Poznań, POLAND

Jeffery L. Kennington
Southern Methodist University
Dallas, Texas, USA

Boris Kovalerchuk
Louisana State University
Baton Rouge, Louisana, USA

V. G. Kulkarni
University of North Carolina
Chapel Hill, North Carolina, USA

Ohseok Kwon
University of Maryland
College Park, Maryland, USA

Huan Li
NORTEL
Richardson, Texas, USA

Cormac Lucas
Brunel University
Uxbridge, Middlesex, UK

Gautam Mitra
Brunel University
Uxbridge, Middlesex, UK

Riad A. Mohammad
SABRE Decision Technologies
Dallas-Ft. Worth Airport, Texas, USA

Hossein Mousavi
Brunel University
Uxbridge, Middlesex, UK

B. Neveu
INRIO-CERMICS
Sophia-Antipolis Cedex, FRANCE

Soren S. Nielsen
University of Texas at Austin
Austin, Texas, USA

Rema Padman
Carnegie Mellon University
Pittsburgh, Pennsylvania, USA

Jean-Yves Potvin
Université de Montréal
Montréal, Quebec, CANADA

L. C. Puryear
Software Solutions
Research Triangle Park, North Carolina, USA

Stephen F. Roehrig
Carnegie Mellon University
Pittsburgh, Pennsylvania, USA

Thomas F. Siems
Federal Reserve Bank of Dallas
Dallas, Texas, USA

David M. Steiger
University of North Carolina at Greensboro
Greensboro, North Carolina, USA

Theodore B. Trafalis
University of Oklahoma
Norman, Oklahoma, USA

Evangelos Triantaphyllou
Louisana State University
Baton Rouge, Louisana, USA

M. Vidyasagar
Centre for AI and Robotics
Bangalore, INDIA

V. Vinay
Indian Institute of Science
Bangalore, INDIA

Edward A. Wasil
American University
Washington, DC, USA

Dan Zhu
University of Iowa
Iowa City, Iowa, USA

Yixin Zhu
NORTEL
Richardson, Texas, USA

Arkadiusz Zimniak
Poznań University of Technology
Poznań, POLAND

PREFACE

The disciplines of computer science and operations research (OR) have been linked since their origins, each contributing to the dramatic advances of the other. This work explores the connections between these key technologies: how high-performance computing methods have led to advances in OR deployment, and how OR has contributed to the design and development of advanced systems. The collected writings—from researchers and practitioners in Computer Science, Operations Research, Management Science, and Artificial Intelligence—were among those delivered at the Fifth INFORMS Computer Science Technical Section Conference in Dallas, Texas, January 8–10, 1996.

The articles advance both theory and practice. Presented are new approaches to complex problems based on: metaheuristics (neural networks, genetic algorithms, and Tabu Search), optimization and mathematical programming, stochastic methods, constraint programming, and logical analysis. These advanced methodologies are applied to new applications in such areas as: telecommunications network design, financial engineering, manufacturing, project management, and forecasting, airline and machine scheduling, vehicle routing, modeling and decision support systems.

Featured is a remarkable paper by keynote speaker Fred Glover, creator of the Tabu Search family of metaheuristics. In it he develops the principles of memory-based heuristic methods, contrasts them with the popular genetic algorithms and simulated annealing, provides a sweeping survey of application vignettes, and points to promising avenues for future research.

The Editors
Richard S. Barr
Richard V. Helgason
Jeffery L. Kennington

Fifth INFORMS Computer Science Technical Section Conference

General Chair

Richard S. Barr

Advisory Committee

Osman Balci
Narayan Bhat
Saul Gass
Harvey Greenberg

Program Committee

Jeff Kennington, *Co-Chair*
Richard V. Helgason, *Co-Chair*
Robert Bixby
Andrew Boyd
Thomas Cook
Robert Cooper
Fred Glover
Bruce Golden
Terry Harrison
Betty Hickman
Karla Hoffman
John Hooker
Chris Jones
Rex Kinkaid
Irvin Lustig
Richard McBride
Gautam Mitra
John Mulvey
Ken Nygard
James Orlin
Catherine Roucairol
Asim Roy
William Scherer
David Shanno
Ramesh Sharda
Ariela Sofer
William R. Stewart
John Tomlin
Steven Wallach
Edward Wasil
Hossam Zaki
Stavros Zenios

Exhibits Chairs

Bjarni Kristjansson
Harlan Crowder

Referees for This Volume

Narayan Bhat
Weidong Chen
Laura Davis
Robert Fourer
Bruce Golden
W. M. Gray
Harvey Greenberg
John Hooker
James Kelly
Georgia-Ann Klutke
Theodore D. Klastorin
Ali Khotanzhad
Richard McBride
John Ramberg
Robert Russell
William Scherer
Ramesh Sharda
Hossam Zaki

1

TABU SEARCH AND ADAPTIVE MEMORY PROGRAMMING — ADVANCES, APPLICATIONS AND CHALLENGES

Fred Glover

College of Business, CB 419
University of Colorado
Boulder, CO 80309
email: Fred.Glover@colorado.edu

ABSTRACT

Tabu search (TS) has provided advances for solving difficult optimization problems in many domains. At the same time, fundamental TS strategies are often not applied as effectively as they might be, and their underlying rationale is often not completely understood. We examine basic concepts and principles of tabu search, emphasizing those that have sometimes led to applying the label "adaptive memory programming" to this class of methods.

The goal of this paper is to focus on key themes that are given inadequate attention in many treatments of tabu search. We also examine basic TS strategies that provide useful alternatives to procedures often associated with "evolutionary" or "genetic" algorithms. Specific tabu search applications are also summarized to provide a clearer understanding of settings where the method is being used. Finally, we include an Appendix that identifies the elements of tabu search that are most neglected in implementations, and that can significantly improve its performance.

1 INTRODUCTION

Metaheuristic procedures have benefitted from numerous advances in recent years. Developments in new implementations of tabu search (TS) have been especially productive. Most striking are the advances enabling difficult practical problems to be handled with greater effectiveness than previously envisioned possible. At the same time, adaptive memory strategies of tabu search are becoming incorporated into other methods, both significantly modifying the operation of these methods and changing the perspectives on which they are based.

This paper focuses on three main themes. First, in Section 1 we examine the adaptive memory rationale that underlies tabu search, and which has sometimes given this approach the label *adaptive memory programming*. We discuss key precepts that underlie the method and issues that are raised for exploiting them more fully. Next, in Section 2 we examine comparisons and contrasts between specific strategies of tabu search, particularly those based on the *scatter* search model, and currently popular strategies based on models of nature, as represented by approaches that are beginning to be embedded in genetic algorithms (GAs). We show how classical GA procedures and their more modern "evolutionary" counterparts can be improved by strategies for combining solutions that are made available by alternative frameworks. A fuller application of these frameworks that incorporates adaptive memory designs of TS offers a useful alternative to GA based approaches. In Section 3 we present a collection of tabu search vignettes, which give brief descriptions of selected tabu search applications and their outcomes. These vignettes identify a diverse range of settings where tabu search has made useful contributions, and suggest the form of additional applications where similar successes may be anticipated.

Each of these sections is self-contained and independent of the others. Our intent is not to present the detailed workings of tabu search, but to disclose fundamental perspectives and principles, especially those that are often overlooked. To reinforce this theme, we include an Appendix that provides a catalog of some of the most commonly neglected features of tabu search, emphasizing those that can significantly improve the quality of results obtained.

2 TABU SEARCH AND ADAPTIVE MEMORY PROGRAMMING

Tabu search has become the focus of numerous comparative studies and practical applications in recent years. Fruitful discoveries about preferred strategies for solving difficult optimization problems have surfaced as a result.

However, sometimes the nature and implications of these discoveries have not been made entirely clear. The reason for this ambiguity is that tabu search has been presented with two faces in the literature, causing it to be viewed as two different methods — one simpler and one more advanced. The simpler method incorporates a restricted portion of the TS design, and is sometimes used in preliminary analyses to test the performance of a limited subset of its components — usually involving only short term memory. The more advanced method embodies a broader framework that includes longer term memory, with associated intensification and diversification strategies. This second approach, due to its focus on exploiting a collection of strategic memory components, is sometimes referred to as *Adaptive Memory Programming* (AMP).

In fact, in both of its forms (though more strongly on the second) the thrust that gives tabu search its distinctive character is the systematic use of adaptive memory — in contrast to the design of "memoryless" approaches like simulated annealing and genetic algorithms, or "rigid memory" approaches like branch and bound and its various AI cousins such as A* search.

More encompassing forms of tabu search that qualify for the AMP label often prove considerably more effective than the primitive forms. In many situations, attempts to rely predominantly on restricted short term memory is an evident handicap — one that invites comparison to attempts to solve a problem by disengaging a part of the brain. Nevertheless, simplified TS approaches are sometimes surprisingly successful. Since they are also frequently quite easy to implement, these approaches will undoubtedly continue to appear in the literature. However, it is important to be aware that they can be strongly dominated by more complete TS methods. Relevant considerations underlying these differences are sketched in the Appendix.

The remainder of this section will focus on certain issues that underly the orientation of TS/AMP approaches, and on questions that are raised by these issues. (Those who are less interested in "precepts and perspectives" can skip to later sections.)

2.1 Is Memory Really a Good Idea?

Memory would seem unquestionably to be an integral component of any search that deserves to be called "intelligent." Yet, surprisingly, some of the methods widely hailed as innovations in artificial intelligence — as applied to optimization — are largely devoid of memory. The avoidance of memory, or more particularly adaptive memory, is not as unreasonable as might be imagined. Adaptive memory characteristically introduces too many degrees of freedom to be treated in "theorem and proof" developments. Researchers who prefer to restrict consideration to processes (and behavior) that can be characterized by rigorous proofs, which are the lifeblood of academic publishing, must focus their efforts in other directions.

Yet there are more subtle and valid reasons to be wary about the use of memory. Malleable forms of memory entail certain dangers — potential pitfalls that go hand in hand with the ability to provide valuable strategic opportunities. These dangers are the price to be paid for the evolution of "intelligent" mechanisms, including biological mechanisms embodied in a brain.[1]

From an evolutionary standpoint, the emergence of memory may be viewed as posing a challenge comparable to the emergence of oxygen, whose corrosive properties (as evolutionary biologists are fond of telling us) caused considerable destruction until organisms adapted to take advantage of them. Analogous perils may well have been created by the emergence of memory, though today we only see the outcomes that survived and flourished. Characteristically, the blind alleys of poorly designed physical structure are conspicuously imprinted on the fossil record. But the blind alleys of poorly regulated mental adjustment — which may have affected survival in far subtler ways — remain invisible to us.[2]

It is noteworthy, however, that we are memory users whose evolutionary line has survived. Since we tend to endow our problem solving schemes with features that reflect our own disposition, such schemes tend to be protected (at least to a degree) from dangers otherwise presented by adaptive memory. Even so, hastily contrived uses of memory can lead to conspicuously undesirable outcomes.

Accordingly, in order to solve complex problems more effectively, TS/AMP approaches seek to uncover the potential gains of adaptive memory without being caught in the traps of ill-considered memory designs. This leads to a quest for "integrating principles," by which alternative forms of memory are appropriately combined with effective strategies for exploiting them. A novel finding is that such principles are sometimes sufficiently potent to yield effective problem solving behavior with negligible reliance on memory. Over a wide

[1] This perspective invites a reconsideration of popular themes: e.g., if an increase in mental capacities creates such attendant risks, the act of acquiring a "knowledge of good and evil," as in the Garden of Eden story, is an understandable basis for expulsion from a simpler (and safer) existence.

[2] Or perhaps they are more visible than we suspect. Just as we maintain vestiges of physical structure now left behind — a spine still not perfectly adapted to upright walking and joints still not perfectly articulated for intricate grasping — we no doubt maintain vestiges of mental patterns now obsolete, ineffective neural organizations that confine the range of our perceptions and reactions.

range of problem settings, however, strategic use of memory can make dramatic differences in the ability to solve problems.

2.2 Points of Departure.

A starting premise of tabu search is that intelligent inhibition plays a critical role in making effective use of memory. This may be conceived as a reflection of an analogous supposition that appropriate forms of inhibition and restraint correspondingly are essential to survival, although we may not always think of such elements as survival tools. (The usual tenet of our culture is that inhibition represents something which must be overcome, rather than something which can provide important advantages when properly utilized.)

The connotation of the "tabu" term in tabu search carries an implication, as it does in other domains, of rules that are contextual and subject to change. This type of variability can range from narrowly confined interaction to highly complex coordination. The potential intricacy of managing such variation understandably may pose an obstacle to rules that are too rigidly constrained by the quest for mathematical precision, but there is a reverse danger of seeking to handle complexity by the expedient of simplistic rules, particularly those that rely heavily on randomization as a substitute for identifying strategic relationships.

Currently it is fashionable to base the design of search mechanisms on a level of organization represented by primitive organisms. But we may legitimately wonder whether intelligent behavior can be adequately encompassed within physical or biological processes that are distant precursors of the forms of organization embodied in our own brains. If there is value in having the capabilities we call human, then it seems questionable to aspire to mimic something less.

There is of course no reason to limit consideration to forms of intelligence that match our own. Evolution presumably may have honed our skills to handle problems that have typically presented themselves in our surroundings. Our prowess may be less impressive for problems confronted in other settings, including problems created as a result of our own technology. A leading goal of TS/AMP research is to identify memory and strategy combinations that have merit in a wide range of contexts, not restricted to those we have commonly encountered by the accident of history. If this pursuit may yield insights into different types of intelligence, that would be a welcome bonus. Conceivably, by this perspective, the field of memory-based search methods may have something useful to contribute to the field of cognitive behavior. Up to now, the complementary realms of search and psychology have been largely isolated from each other. As findings about the connections between adaptive memory processes and improved problem solving become systematized, however, this situation may change.

2.3 Elements of Adaptive Memory.

Adaptive memory involves an attribute-based focus, and depends intimately on the elements of recency, frequency, influence and logic. This simple catalog disguises a surprising range of alternatives — as becomes apparent when the four basic elements are considered in combination, and differentiated for different attribute classes over varying regions and spans of time. The notion of influence, for example, characteristically refers to changes in quality, structure, feasibility and regionality. The logic classification likewise is not limited to a single dimension, but invites distinctions between "sequential logic" and "event driven logic," whose alternative forms appropriately give rise to different kinds of memory structures.

A number of key questions arise about the nature and interrelation of these elements, which have important implications for designing search methods. A brief listing of some of these questions follows.

1. Which types of solution attributes can be most effectively exploited by adaptive memory? (What is the impact of different exploitation strategies on selecting neighborhoods for conducting the search?)

2. What types of functions are useful for generating new attributes as combinations of others? (What implications do such functions have for *vocabulary building* methods in tabu search? (See Section 2.))

3. What are relevant measures of influence, as reflected in attribute changes caused by moving from one solution to another? (How can these measures assist in isolating characteristics of past trajectories that are relevant for designing current ones?)

4. What thresholds should be established to identify levels of recency, frequency and influence? (What role should these levels play in determining tabu restrictions and aspiration criteria? How can thresholds be used to provide penalties and inducements for selecting particular moves, and for changing the phase of search?)

5. How should probabilities be designed to take advantage of information provided by evaluative measures and thresholds? (Which search functions and domains should be governed by probabilistic variation and which should preferably be treated deterministically?)

6. How may memory be applied most effectively to coordinate the use of compound neighborhoods? (What forms of memory are most useful for ejection chain strategies, as a basis for concatenating component moves into more elaborate alternatives?)

7. What clustering and pattern classification approaches are best suited to take advantage of the search history? (How can these approaches be coordinated to improve intensification and diversification strategies in tabu search?)

8. Which special adaptations of memory and learning give best results for highly context-specific problems? (Conversely, which "generic" forms are most effective over wide ranges of problems whose structure is not predicated in advance?)

Basic considerations and research directions associated with these questions are examined in the following sections, and in the discussion of commonly neglected aspects of tabu search in the Appendix.

3 MODELS OF NATURE — BEYOND "GENETIC" METAPHORS

One of the most misunderstood aspects of tabu search is the connection between a subset of its strategies and certain approaches embodied in genetic algorithms (GAs). TS researchers have tended sometimes to overlook the part of the adaptive memory focus that is associated with strategies for combining sets of elite solutions. Complementing this, GA researchers have been largely unaware that such a collection of strategies outside their domain exists. This has quite possibly been due to the influence of the "genetic metaphor," which on the one hand has helped to launch a number of useful problem solving ideas, and on the other hand has also sometimes obscured fertile connections to ideas that come from different foundations.

To understand the relevant ties, it is useful to go back in time to examine the origins of the GA framework and of an associated set of notions that became embodied in TS strategies. We will first sketch the original genetic algorithm design, as characterized in Holland (1975). Our description is purposely somewhat loose, to be able to include approaches more general than the specific proposals that accompanied the introduction of GAs. Many variations and changes have come about over the years, as we subsequently observe.

Genetic Algorithm Template

1. Begin with a population of binary vectors.
2. Operate repeatedly on the current generation of vectors, for a selected number of steps, choosing two "parent vectors" at random. Then mate the parents by exchanging certain of their components to produce offspring. (The exchange, called "crossover," was originally designed to reflect the process by which chromosomes exchange components in genetic mating and, in common with the step of selecting parents themselves, was organized to rely heavily on randomization. In addition, a "mutation" operation is occasionally allowed to flip bits at random.)
3. Apply a measure of fitness to decide which offspring survive to become parents for the next generation. When the selected number of matings has been performed for the current generation, return to the start of Step 2 to initiate the mating of the resulting new set of parents.
4. Carry out the mating-and-survival operation of Steps 2 and 3 until the population becomes stable or until a chosen number of iterations has elapsed.

A somewhat different model for combining elements of a population comes from a class of relaxation strategies in mathematical optimization known as surrogate constraint methods. The goal of these approaches is to generate new constraints that capture information not contained in the original problem constraints taken independently, but which is implied by their union. While this may seem somewhat removed from the concerns that motivated the development of genetic algorithms, we will see that some unexpected connections emerge.

The information-capturing focus of the surrogate constraint framework has the aim of developing improved methods for solving difficult optimization problems by means of (a) providing better criteria for choice rules to guide a search for improved solutions, (b) inferring new bounds (constraints with special structures) to limit the space of solutions examined. (For background, see Glover

(1965, 1968, 1975), Greenberg and Pierskalla (1970, 1973), Karwan and Rardin (1976, 1979), and Freville and Plateau (1986, 1993).) Based on these objectives, the generation of new constraints proceeds as follows.

Surrogate Constraint Template

1. Begin with an initial set of problem constraints (chosen to characterize all or a special part of the feasible region for the problem considered).
2. Create a measure of the relative influence of the constraints as basis for combining subsets to generate new constraints. The new (surrogate) constraints, are created from nonnegative linear combinations of other constraints, together with cutting planes inferred from such combinations. (The goal is to determine surrogate constraints that are most effective for guiding the solution process.)
3. Change the way the constraints are combined, based on the problem constraints that are not satisfied by trial solutions generated relative to the surrogate constraints, accounting for the degree to which different source constraints are violated. Then process the resulting new surrogate constraints to introduce additional inferred constraints obtained from bounds and cutting planes. (Weaker surrogate constraints and source constraints that are determined to be redundant are discarded.)
4. Progressively update and modify the surrogate constraints to take advantage of different current stages of the solution method and different regions of the solution space. Repeat the process as long as the solution method for the original problem continues to iterate.

A natural first impression is that the surrogate constraint design is quite unrelated to the GA design, stemming from the fact that the concept of combining constraints seems inherently different from the concept of combining vectors. However, this difference is not as great as it may initially appear. In many types of problem formulations, including those where surrogate constraints were first introduced, constraints are summarized by vectors. More particularly, over time, as the surrogate constraint approach became embedded in both exact and heuristic methods, variations led to the creation of a "primal counterpart" called *scatter search*. The scatter search approach combines solution vectors by rules patterned after those that govern the generation of new constraints, and specifically inherits the strategy of exploiting linear combinations and inference (Glover (1977)).[3]

Accordingly, following the principles that underlie the surrogate constraint design, the scatter search process is organized to (1) capture information not contained separately in the original vectors, (2) take advantage of auxiliary heuristic solution methods (to evaluate the combinations produced and to actively generate new vectors), (3) make dedicated use of strategy instead of randomization to carry out component steps.

We sketch the scatter search approach in its original form and identify some novel connections and contrasts to GA methods. Then we examine extensions that additionally take advantage of the memory-based designs of tabu search.

Scatter Search Procedure.

1. Apply heuristic processes to generate a starting set of solution vectors (trial points). Designate a subset of the best vectors to be reference points. (Subsequent iterations of this step, transferring from Step 3 below, incorporate advanced starting solutions and best solutions from previous history as candidates for the reference points.)
2. Form linear combinations of subsets of the current reference points to create new points. The linear combinations are:
   (a) chosen to produce points both inside and outside the convex region spanned by the reference points.
   (b) modified by generalized rounding processes to yield integer values for integer-constrained vector components.
3. Extract a collection of the best points generated in Step 2 to be used as starting points for a new application of the heuristic processes of Step 1. Repeat these steps until reaching a specified iteration limit.

Two particular features of the scatter search proposal, which will be elaborated later, deserve mention. The use of clustering strategies is suggested for selecting subsets of points in Step 2, which allows different blends of intensification and

[3]The motivation for this development goes beyond the goal of producing a primal analog for the surrogate constraint approach. In situations where surrogate constraint relaxations yield a duality gap, the natural response is to combine elite solutions that "rim" this gap as a basis for exploiting information that may be contained in their union.

diversification by generating new points "within clusters" and "across clusters." Also, solutions selected in Step 3 as starting points for re-applying heuristic processes are not required to be feasible, since heuristics proposed to accompany scatter search include those capable of starting from an infeasible solution.

In sum, scatter search is founded on the following premises.

(P1) Useful information about the form (or location) of optimal solutions is typically contained in a (sufficiently diverse) collection of elite solutions[4].

(P2) When solutions are "combined" as a strategy for exploiting such information, it is important to provide for combinations that can extrapolate beyond the regions spanned by the solutions considered, and further to incorporate heuristic processes to map combined solutions into new points.

(P3) Taking account of multiple solutions simultaneously, as a foundation for creating combinations, enhances the opportunity to exploit information contained in the union of elite solutions.

The fact that the heuristic processes of scatter search (as referred to in *(P2)*) are not restricted to a single uniform design, but represent a varied collection of procedures, affords strategic possibilities[5] whose implications are examined in the Appendix.

The table on the following page traces the links between the conceptions underlying scatter search and conceptions that have been introduced over time as amendments to the GA framework.

These innovations in the GA domain, which have subsequently been incorporated in a wide range of studies, are variously considered to be advances or heresies according to whether they are viewed from liberal or traditional

[4] Useful information may also be contained in bad solutions. However, such solutions are usually much more numerous and varied than good ones, and consequently there is less advantage in trying to make use of them. On the other hand, valuable information can be contained in trajectories from bad solutions to good solutions (or from good solutions to other good solutions), and such trajectory-based information is one of the elements that tabu search seeks to exploit.

[5] This theme also shares a link with the original surrogate constraint proposal, where heuristics for surrogate relaxations are introduced to improve the application of exact solution methods. In combination, the heuristics are used to generate strengthened surrogate constraints and, iteratively applied, to generate trial solutions for integer programming problems.

Scatter Search Features (1977) Incorporated into Non-Traditional GA Approaches

- Introduction of "flexible crossover operations." (Scatter search combinations include all possibilities generated by the early GA crossover operations, and also include all possibilities embedded in the more advanced "uniform" and "Bernoulli" crossovers (Ackley (1987), Spears and DeJong (1991)). Path relinking descendants of scatter search provide further possibilities, noted subsequently.)

- Use of heuristic methods to improve solutions generated from processes for combining vectors (Muhlenbein et al. (1988), Ulder et al. (1991)), (Whitley, Gordon and Mathias (1994)).

- Exploitation of vector representations that are not restricted to binary representations (Davis (1989), Eschelman and Schaffer (1992)).

- Introduction of special cases of linear combinations for operating on continuous vectors (Davis (1989), Wright (1990), Bäck et al. (1991), Michalewicz and Janikow (1991)).

- Use of combinations of more than two parents simultaneously to produce offspring (Eiben et al. (1994), Mühlenbein and Voight (1996)).

- Introduction of strategies that subdivide the population into different groupings (Mühlenbein and Schlierkamp-Voosen (1994)).

perspectives. Significantly, their origins are somewhat diffuse, rather than integrated within a single framework[6]. It is clear that a number of the elements of the scatter search approach remain outside of the changes brought about by these proposals. A simple example is the approach of introducing adaptive rounding processes for mapping fractional components into integers. There also has conspicuously been no GA counterpart to the use of clustering to create strategic groupings of points, nor (as a result) to the notion of combining points according to distinctions between membership in different clusters. (The closest approximation to this has been the use of "island populations" that evolve separately, but without concern for analyzing or subdividing populations based on inference and clustering. The relevance of such matters, and of related conditional analyses, also is discussed in the Appendix.)

The most important distinction, however, is the link between scatter search and the theme of exploiting history. The prescriptions for combining solutions within scatter search are part of a larger design for taking advantage of information about characteristics of previously generated solutions to guide current search. In retrospect, it is perhaps not surprising that such a design should share an intimate association with the surrogate constraint framework, with its emphasis on extracting and coordinating information across different solution phases. This orientation, which takes account of elements such as the recency, frequency and quality of particular value assignments, has become the foundation of notions incorporated within tabu search. (The same reference on surrogate constraint strategies that is the starting point for scatter search is also often cited as a source of early TS conceptions.) By this means, the link between tabu search and so-called "evolutionary" approaches becomes apparent[7].

---

[6] The "press" for the GA approach suggests by contrast that it is not subject to such variation, but represents a manifestation of immutable natural law. An amusing quote from the January 16, 1996 issue of the Wall Street Journal is illustrative: "Three billion years of evolution can't be wrong," [according to a genetic algorithm pioneer].... "It's the most powerful algorithm there is."

[7] The term *evolutionary* has undergone an interesting evolution of its own. By a novel turn, the term "mutation" in the GA terminology has become reinterpreted to refer to any form of change, including the purposeful change produced by a heuristic process. As a result, all methods that apply heuristics to multiple solutions, whether or not they incorporate strategies for combining solutions, are now considered kindred to genetic algorithms, and the enlarged collection is labeled "evolutionary methods." This terminology accordingly has acquired the distinction of embracing nearly every kind of method conceivable.

Scatter Search Extensions

- Tabu search memory is used to select current reference points from a historical pool (Glover (1989, 1994a)).
- Tabu search intensification and diversification strategies guide the generation of new points (Mulvey (1995), Zenios (1996), Fleurent et al. (1996)).
- Solutions generated as "vector combinations" are further improved by explicit tabu search guidance (Trafalis and Al-Harkan (1995), Kelly, Laguna and Glover (1996), Fleurent et al. (1996)).
- Directional rounding processes focus the search for feasible zero-one solutions allowing them to be mapped into convex subregions of hyperplanes produced by valid cutting plane inequalities (Glover (1995a)).
- Neural network learning is applied to filter out promising and unpromising points for further examination, and pattern analysis is used to predict the location of promising new solutions (Kelly, Laguna and Glover (1996)).
- Mixed integer programming models generate sets of diversified points, and yield refined procedures for mapping infeasible points into feasible points (Kelly, Laguna and Glover (1996)).
- Structured combinations of points take the role of linear combinations, to expand the range of alternatives generated (Glover (1994a)).

3.1 Modern Forms and Applications of Scatter Search

Recent implementations of scatter search (cited below) have taken advantage of the implicit learning capabilities provided by the tabu search framework, leading to refined methods for determining reference points and for generating new points. Current scatter search versions have also introduced more sophisticated mechanisms to map fractional values into integer values. This work has produced new theorems about searches over spaces of zero-one integer variables. Special models have also been developed to allow both heuristic and exact methods to transform infeasible trial points into feasible points. Finally, scatter search has been generalized to produce a broader of methods called *path relinking* methods, which offer a wide range of mechanisms for creating productive combinations of reference solutions. A brief summary of some of these developments appears in the table on the following page.

Implementation of various components of these extensions have provided advances for solving general nonlinear mixed discrete optimization problems with both linear and nonlinear constraints.

Path Relinking

From a spatial orientation, the process of generating linear combinations of a set of reference points may be characterized as generating paths between and beyond these points (where points on such paths also serve as sources for generating additional paths). This leads to a broader conception of the meaning of creating *combinations* of points: by natural extension, we may conceive such combinations to arise by generating paths between and beyond selected points in neighborhood space, rather than in Euclidean space (Glover (1989, 1994b)).

The character of such paths is easily specified by reference to attribute-based memory, as used in tabu search. In particular, it is only necessary to select moves in a neighborhood space that perform the following simple function: upon starting from an *initiating solution*, the moves must progressively introduce attributes contributed by a *guiding solution* (or reduce the distance between attributes of the initiating and guiding solutions). The process invites variation by interchanging the roles of the initiating and guiding solutions, inducing each to move simultaneously toward the other as a way of generating combinations. (When the goal for combining the solutions can be expressed as an optimization model, algorithmic processes may appropriately be incorporated to generate the moves, as in the case of vocabulary building approaches subsequently described. Related approaches are also described in Aggarwal, Orlin and Tai (1996) and Balas and Niehaus (1996).)

Multiparent path generation possibilities emerge by considering the combined attributes provided by a set of guiding solutions, where these attributes are weighted to determine which moves are given higher priority. The generation of such paths in neighborhood space characteristically "relinks" previous points in ways not achieved in the previous search history, hence giving the approach its name of *path relinking.*

Neighborhoods for this process may differ from those used in other phases of search. For example, they may be chosen to *tunnel through* infeasible regions that may be avoided by other neighborhoods. Such possibilities arise because feasible guiding points can be coordinated to assure that the process will reenter the feasible region, without danger of becoming "lost."

The scope of strategies made available by path relinking is significantly affected by the fact that the term *neighborhood* has a broader meaning in tabu search than it typically receives in the popular literature on search methods. Often, the neighborhood terminology refers solely to methods that progressively transform one solution into another. Such neighborhoods are called *transition neighborhoods* in tabu search, and are considered as merely one component of a collection of neighborhoods that also include *constructive* and *destructive* neighborhoods[8].

In addition, tabu search characteristically endows such a collection of neighborhoods with the ability to operate in regions beyond those visited by standard procedures for generating solutions. Strategic oscillation approaches in TS, for example, include variations that build solutions beyond the point of "complete construction," and more generally introduce complementary constructive and destructive processes that go beyond standard boundaries in both directions. By selecting neighborhoods that are relevant to a given problem setting, drawing on this expanded interpretation of a neighborhood, path relinking automatically provides solution combination procedures that are appropriate for specific contexts. This is in noteworthy contrast to the situation encountered with genetic algorithms, where each new class of problems initiates a search for "new crossovers," in order to overcome the limitations of classical models. (The possibility of creating combinations that exploit information from the problem

[8] This distinction seems either "superficial" or "perverse," according to different perspectives, yet disregarding it has led to a variety of misconceptions in the literature, and to designs for solution methods that are inappropriately limited.

setting is also sometimes still overlooked in GA approaches, due to the lingering preference for models that are "context free.")

Vocabulary Building

Recent discoveries point to a feature of the path relinking model that deserves special consideration, involving a connection to the tabu search strategy of *vocabulary building*. The basic idea of this approach is to identify meaningful fragments of solutions, rather than focusing solely on full vectors, as a basis for generating combinations. A pool of such fragments is progressively enriched and assembled to create larger fragments, until ultimately producing complete trial solutions. In some settings these fragments can be integrated into full solutions by means of optimization models (Glover (1992), Glover and Laguna (1993)). Procedures using this design have been developed with highly successful outcomes by Rochat and Taillard (1995) and Kelly and Xu (1995). Approaches that heuristically assemble fragments have also been successfully implemented by Taillard et al. (1995) and Lopez, Carter and Gendreau (1996).

Vocabulary building effectively may be conceived as an instance of path relinking. There are two key objectives: (1) to identify a good collection of reference points, in this case consisting of "partial solutions" (which include the fragments), and (2) to identify paths in neighborhood space that will unite components of these partial solutions, with suitable attendant modifications, to produce complete solutions. (Again it is important to keep in mind that neighborhood spaces include constructive and destructive spaces, as well as transition spaces. Attributes of different partial solutions may be imperfectly compatible, and hence the synthesis of such partial solutions can benefit from multiple or "compound" transformations to create effective linkages.) As a special instance, of course, solution fragments may be united by linear combinations, as in scatter search.

The crucial element of adaptive memory that permeates these alternative modes of combination, and binds them to other tabu search strategies, affords challenging opportunities for research into the nature and meaning of "intelligent combinations." To the extent that the evolutionary label is coming to be applied to increasingly broad domains, researchers who have previously adopted a GA perspective may find a shared interest in exploring the connections that emerge from complementary elements of the TS framework.

4 TS/AMP VIGNETTES

This final section provides a collection of "vignettes" that briefly summarize applications of tabu search in a variety of settings. These vignettes are edited versions of reports by researchers and practitioners who are responsible for the applications[9].

4.1 Protein Conformation Lattice Model Using Tabu Search

The determination of the three-dimensional structure of a protein from a given sequence of amino acids is one of the most challenging unsolved problems in the science of molecular biology. There have been many computer models designed to solve the protein folding problem. All computer models, though employing different types of energy minimization, can be expressed as the global optimization of a non-convex potential energy function. The basic difficult in solving these models is the existence of multiple local minimizers. Recently, there have been various approaches used to solve these models arising from protein folding.

Lattice models have been used by many researchers to describe the protein folding mechanism. This is motivated from two aspects of research interests. On one side, scientists (with practical insight) are hoping to use some lattice structures to obtain initial solutions of protein conformations, with the assumption that an optimal or near-optimal native state can be obtained by relaxing the monomers around the lattice structure. On the other side, acknowledging that we do not understand the mechanism of protein folding, scientists are seeking to understand protein conformations by using lattice models.

Pardalos, Liu and Xue (1995) design an algorithm for a class of lattice models using tabu search and test their approach with a chain of 27 monomers. The algorithm was developed in C and tested using the same data for a fundamental test set published by the American Mathematical Society. Among the protein sequences tested, only a few sequences fail to match the best results previously

[9] A debt of gratitude is owed to the individuals whose contributions have made this summary possible. (Deficiencies in describing their work are solely due to the current editing, and should not be interpreted to reflect shortcomings of the original reports. In a number of cases, the work cited presents only a small sampling of significant contributions by the authors referenced.)

reported. In all other cases (from a set of 200 examined) the tabu search method obtains results as good or better than the previous best.

4.2 Optimization of B-ISDN Telecommunication Networks

Costamagna, Fanni and Giacinto (1995) develop a Tabu Search algorithm for topological optimization of broad band communication networks whose structure is based on a single exchange and on a number of multiplexing centers.

The topology of the network is represented by an undirected graph G(N,A). The set of nodes N represents locations of both existing or possible multiplexers, and location of users. The set of arcs A represents the possible communication links that may be used to connect the users, the multiplexers and the exchanges among them (constituting a cable conduit graph). This graph contains all the information about the area in which the network must be built.

The design problem consists of choosing a spanning tree T of G that connects all the users to the multiplexers through the distribution network, and the multiplexers to the exchange through the transport network, allowing the overall cost of the plant be minimized.

An empirical study was performed comparing TS with three other methods: a Simulated Annealing (SA) method, a genetic algorithm (GA) and a heuristic "Add & Drop" procedure, in terms of computational time and cost. The TS approach reached better configurations in a time equal to or lower than that required by other algorithms. The work has been supported by Marconi, Genova, Italy. Thesis awards have been also granted by S.I.P. S.p.A. (now Telecom Italia), and the Italian public telephone company, which also provided cost data and evaluation of results.

4.3 Tabu Search for Location Analysis: The P-Median Problem

Many location analysis problems are prime targets for tabu search techniques. One such binary decision problem, the p-median problem, can be stated as follows: given a graph G = (V,E), the goal to find a set of nodes, S, of size p, such that the weighted sum of the distances from the remaining nodes (those of V-S) to the set S is minimized. Rolland, Schilling and Current (1995) provide a tabu search algorithm that utilizes single node transfers with a simple short-term memory structure, and a strategic oscillation scheme that allows the procedure to search through an infeasible solution space. Long-term memory structures are used to penalize moves that occur inordinately often. The resulting tabu search procedure outperformed all known heuristics both with respect to solution quality and computational effort.

4.4 Scheduling in Manufacturing Systems

Effective scheduling in manufacturing systems leads to the reduction of manufacturing costs (inventory costs, labor costs, etc.) and improves the operational efficiency of management. The most frequently used and extensively studied problems in the literature are (A) the job shop problem and (B) the flow shop problem. In addition, a basic model for a broad family of cases called flexible flow line scheduling problems is given by the problem known as (C) the flow shop problem with parallel machines. Industrial applications arise in computer systems, telecommunication networks, and the chemical and polymer industries.

Nowicki and Smutnicki (1993, 1994, 1995) have developed effective tabu search methods for problems A, B, and C to optimize the makespan criterion. These algorithms employ a classical insertion neighborhood which is significantly reduced by a candidate list strategy for removing useless moves, in order to concentrate on "the most promising part" of the neighborhood.

The proposed algorithms employ a short-term memory tabu list which stores attributes of visited solutions, represented by selected pairs of adjacent jobs on a machine. Linked intensification and diversification occurs by storing the best solutions collected during the search on a list of limited length. An extended sequence of unproductive steps triggers a 'back jump' on the search trajectory to the nearest elite solution, which is recovered together with its associated search history as a basis for re-initiating the search.

Implementations made on a PC are able to improve significantly the best known solution found by other algorithms. Computation times are only a few minutes for instances of A&B problems containing 10,000 operations, and for instances of C problems containing 3,000 operations. An extensive comparative study shows the significant superiority of TS over other approaches including iterative improvement, genetic search, simulated annealing, threshold accepting, constraint satisfaction, neural networks, and other local search methods (see Vaessens, Aarts and Lenstra (1995)).

4.5 Tabu Search for Designing Optical Telecommunication Networks

The increased complexity and globalization of today's world has been accompanied by the emergence of optical networks as flexible, fast, efficient, and reliable media for transferring information. In this domain, congestion minimization presents one of the main challenges of telecommunication network design. The problem often includes the goals of devising efficient routing and management techniques in case of network failures. An improved approach for minimizing congestion in optical networks, based on tabu search, has been developed by Skorin-Kapov and Labourdette (1995).

Further motivation for this work arises because changing traffic conditions create a need for fast algorithms to re-arrange logical connections. Algorithms that quickly obtain very high quality solutions are mandatory in order to optimize the use of network with respect to a given criterion. The goal is to find the logical connection diagram and routing of flow which minimizes the maximum congestion on a link. (This goal also effectively increases the relative capacity of the network.)

The tabu search approach for this problem has generated improved solutions for data sets established to provide comparative benchmarks. The approach not only improves on previous results, but has been calibrated by Skorin-Kapov and Labourdette to identify performance characteristics for different parameter values, and on different patterns of input data. The outcomes also yield guidelines for solving larger problems.

4.6 Automated Guided Vehicle Systems Flowpath Design Applications

Automated Guided Vehicle Systems (AGVSs) have been of great interest to industry for the last two decades. The number of applications of these systems has increased to a point where AGVSs are considered to be a basic concept in material handling. Although initial applications of AGVSs were generally limited to warehouses, in recent years an increasing number of applications in manufacturing systems have been reported.

Chiang and Kouvelis (1994a) address the flowpath design issue of AGVSs. The authors concentrate on the design of unidirectional flowpaths (where vehicles are restricted to travel only on one direction along a given segment of the flowpath), and develop different versions of simulated annealing and tabu search algorithms for the design of unidirectional AGVSs. Extensive computational results indicate that a tabu search implementation with the use of a frequency based memory structure dominates all tested heuristics in terms of solution quality, with an impressive average performance over 45 test problems of less than 0.85% deviation from optimality.

4.7 Graph Theory: Uniform Graph Partitioning

The uniform graph partitioning problem may be described as follows. Given a graph G = (V,E), where |V| = m = 2n, we seek a partition of V into two node sets V1 and V2 such that V = V1 + V2 and |V1| = |V2| = n. The goal is to identify such a partition that minimizes the sum of the cost of edges (i,j) where i ∈ V1 and j ∈ V2. Rolland, Pirkul and Glover (1995) have developed a tabu search algorithm for this problem that outperformed all other heuristics tested (including the reported best versions of simulated annealing and the Kernighan-Lin approach), both with respect to solution quality and computational requirements. A key element of the tabu search was a strategic oscillation that drove the search through infeasible solution configurations, causing the cardinality of the node sets to grow and shrink in coordinated waves. (Additional recent important tabu search developments for this problem are provided by Dell'Amico and Maffioli (1996).)

4.8 Tabu Search for Audit Scheduling

Scheduling problems often become increasingly difficult as they acquire greater realism. Audit scheduling adds complexities to traditional scheduling. Characteristically, in such problems the processing units (the auditors):

1. have unequal processing times

2. are not all fit (educated) for all jobs
3. are not always available (yet must work at least a minimum amount, and at most a maximum amount of time)
4. are movable (at a cost), and can transfer between various projects: hence introducing sequence dependent setup costs and times

Dodin, Elimam, and Rolland (1995) develop a tabu search procedure that utilizes traditional dispatching rules (such as forward loading) with the short- and long-term memory structures of tabu search. The tabu search intensifies the search using short-term memory, and diversifies the search using controlled dispatching, long-term memory and candidate lists. Computational tests show the tabu search has approach produces schedules superior to those obtained by heuristics traditionally applied to these problems.

4.9 Mapping Tasks to Processors to Minimize Communication Time in a Multiprocessor System

Connectionist machines are attracting widespread attention for their value as an embodiment of massively parallel computer architecture. This is particularly true for solving combinatorial optimization problems arising in a variety of engineering applications. At the same time, the goal of designing and implementing a connectionist machine as effectively as possible introduces challenging optimization problems.

An important problem is to minimize the communication time required by a connectionist machine. Communication time often is a substantial determinant of overall cost and efficiency. In a significant class of applications, such as finite element analysis, the communication pattern is static. The memory locations defining the source and destinations of messages do not change in these applications, but only the communicated data varies. Improved designs for allocating processors to chips according to the structure of their communication pattern offer considerable potential for savings in cost and time.

Chakrapani and Skorin-Kapov (1995) have developed an effective tabu search method for the problem of mapping tasks to processors to minimize communication time in a multiprocessor system. The method incorporates a parallel processing implementation which includes tabu search memory and guidance mechanisms for iteratively selecting pairs of tasks and swapping their processor assignments. The implementation employs two levels of parallelism. First, the candidate tasks to be swapped are identified in parallel. Second, more than one pair of tasks are swapped in a single iteration. This strategy is designed to operate with efficient approximations that allow inaccurate (incomplete) information for evaluating moves. The authors propose a diversification strategy which makes the search robust under these circumstances. Due to its robust parallel implementation, the algorithm can be used to develop heuristics for "quasi-dynamic" communication patterns, in which the task graph changes slowly with time.

4.10 Multiprocessor Task Scheduling in Parallel Programs

When parallel application programs are executed on MIMD machines, the parallel portion of the application can be speeded up according to the number of processors allocated to it. In a homogeneous architecture, where all processors are identical, the sequential portion of the application will have to be executed in one of the processors, considerably degrading the execution time. In a heterogenous structure, where a faster processor, responsible for executing the serial portion of the parallel application and is tightly coupled to other smaller processors, higher performance may be achieved. The procedure of assigning tasks to processors (task scheduling) is more complex in the heterogeneous case, where the processors have distinct processing speeds.

Porto and Ribeiro (1995a) have applied the tabu search metaheuristic to the task scheduling problem in a heterogeneous multiprocessor environment under precedence constraints. A series of different tabu search parameters and strategies were studied side-by-side with a variety of task precedence graphs (topology, number of tasks, serial fraction, service demand of each task) and system configurations (number of processors, architecture heterogeneity measured by the processor power ratio). The algorithm showed itself to be very robust and effective, systematically improving by approximately 25% the makespan of the solutions obtained by the best greedy algorithm used to provide an initial solution.

4.11 Probabilistic Diversification and Intensification in Local Search for Vehicle Routing

Rochat and Taillard (1995) develop a probabilistic TS technique to diversify, intensify and parallelize almost any local search for almost any VRP. This technique makes the local search more robust since it converges more often solutions whose quality is close to that of the best known solution. This technique has several advantages: First, it is relatively easy to design a local search that locally finds good tours, but it is hard to design a search that finds good tours for all customers simultaneously; the proposed technique makes it possible to overcome this difficulty and to design a fairly robust method more easily. Second, this technique may be applied to several types of VRPs, for example those including the following constraints:

- Time windows for the customer deliveries.
- Differentiated vehicles (cost of use per kilometer, volume capacity, carrying capacity).
- Constraints on the tours (maximum length, driver breaks, customers that cannot be reached by any vehicle).
- Backhauls.
- Multiple depots.

Third, this technique may easily be parallelized with an arbitrary number of processors (not depending on problem size).

The Rochat and Taillard approach exploits two primary perspectives, as follows.

The first comes from probabilistic tabu search, which is founded on the idea of translating information generated by the search history, coupled with current measures of attractiveness, into evaluations that are monotonically mapped into probabilities of selection. Operating in a neighborhood framework, the approach then successively selects among available alternatives according to a probability assignment that is strongly biased to favor the choice of higher evaluations.

The second main perspective that underlies the R&T approach derives from one of the most basic types of intensification strategies. The heart of this approach lies in generating solutions by reference to the notions of strongly determined and consistent variables.

It is shown that efficient first level tabu searches for vehicle routing problems may be significantly improved with this technique. Moreover, the solutions produced by this technique may often be improved by a post-optimization technique presented in this paper too, which embodies an effective means for applying a vocabulary building strategy in this context. The solutions of nearly 40 problem instances of the literature have been improved. This technique may also be applied to other local searches or other VRPs.

4.12 Optimization of Electromagnetic Structures With Tabu Search

Fanni, Giacinto and Marchesi (1996) develop a Tabu Search strategy to optimize the design of a magnet for Magnetic Resonance Imaging (MRI). This is an important biomedical device whose optimal design is sought by many companies, such as general Electric, Siemens and Oxford Instruments. Among different magnetic structures, MRI magnet systems are a tough benchmark for optimization procedures.

The goal of the problem considered was to design coils to yield a 'homogeneous' magnetic field in a fixed region, according to an appropriate function. For each coil the position and thickness have to be determined. To apply a TS based method, Fanni, Giacinto and Marchesi discretize the range of variation of each variable dividing it in sub-ranges, yielding a finite alphabet. The neighborhood of a solution consists of all the configurations obtained by considering all the possible symbols for each variable, keeping the others constant. Short term memory prevents repetitions of configurations of the coils (in terms of symbols of the finite alphabet). In non improving phases, frequency based memory also penalizes choices of moves that drive toward configurations often visited. Finally, local minimization using golden search is applied after the choice of the move. Computational tests show the TS approach performs more efficiently than an SA and a GA approach, and requires less than half as much solution time.

4.13 Multiprocessor Task Scheduling Using Parallel Tabu Search

Porto and Ribeiro (1995b) have designed and implemented parallelization tabu search strategies for the multiprocessor task scheduling problem. Parallelization relies exclusively on the decomposition of the solution space exploration. Four different parallel strategies were proposed and implemented on a 32-processor IBM SP1 parallel machine running PVM for varying problem sizes and number of processors: the master-slave model, with two different schemes for improved load balancing, and the single-program-multiple-data model (SPMD), with single-token and multiple-token message passing schemes. These two basic models mainly differ in the way information is exchanged between parallel tasks at the end of each iteration of the tabu search. The computational results confirmed the high adaptability of the TS algorithm to parallelization, showing that communication is not a burden to achieving almost ideal efficiency in the majority of the test problems. The task scheduling problem considered in this study is characterized by very large neighborhood structures that are costly to explore. However, the speedup achieved through simple parallelization techniques made possible a less restricted neighborhood search, which not only reduced computation time but produced better solutions for several test problems.

4.14 Tabu Search Applied to the Quadratic Assignment Problem (QAP), and Implementations for Connection Machines

The Quadratic Assignment Problem (QAP) is a classical NP-hard problem arising in many applications involving, for example, facility layout or VLSI design. Skorin-Kapov (1990, 1994) solves the quadratic assignment problem suboptimally using the so-called Tabu-Navigation procedure, obtaining improved outcomes over previous results. In the process of designing an efficient massively parallel algorithm for the QAP, Chakrapani and Skorin-Kapov (1992) first generalized the connectionist model proposed by Aarts and Korst (1989) for the Traveling Salesman Problem (TSP) to solve the QAP. This was the first study replacing simulated annealing with (deterministic) tabu search in a connectionist model. This was also the first study involving dynamically changing connection strengths for such problems. In a subsequent paper Chakrapani and Skorin-Kapov (1993a) developed a massively parallel tabu search algorithm and implemented it on the Connection Machine. The careful implementation on Connection Machine, a massively parallel computer architecture, proved to be extremely suitable: provided enough processors, the computational time grows with O(log n).

4.15 Facility Layout in Manufacturing

The design of the facility layout of a manufacturing system is critically important for its effective utilization: 20 to 50 percent of the total operating expenses in manufacturing are attributed to material handling and layout related costs. Use of effective methods for facilities planning can reduce these costs often by as much as 30 percent, and sometimes more. In general, the facility layout problem has been formulated as a quadratic assignment problem (QAP). The QAP is to find the optimal assignment of n candidate facilities (departments, machines, workstations) to n candidate sites, for the goal of minimizing the total layout costs (which includes the material handling cost, expressed as the product of workflow and travel distance, and a fixed cost associated with locating a facility at a specific location).

Chiang and Kouvelis (1994b) provide a new implementation of the tabu search metaheuristic to solve the QAP, with particular emphasis on facility layout problems, utilizing recency-based and long term memory structures, dynamic tabu size strategies, and intensification and diversification strategies. The tabu search algorithm quickly converges from an arbitrarily generated random solution. Computational experiments, including statistical analysis and library analysis, strongly support the superiority of the C & K tabu search heuristic compared to other procedures for this facility layout application.

4.16 Quadratic Semi-Assignment and Mass Transit Applications

The quadratic semi-assignment problem (QSAP) is related to the quadratic assignment problem by the requirement of assigning a set of n objects to any of m locations. The QSAP differs by allowing each location to be assigned none, one, or even more than one object, whereas the QAP requires a one-one mapping of objects to locations (m = n).

Voss (1992), Domschke et al. (1992) and Daduna and Voss (1995) develop dynamic tabu search approaches for the QSAP, in a series of applications for modeling and solving a schedule synchronization problem in a mass transit system, where the goal is to minimize the total transfer waiting times of passengers. Both from an economic and a social standpoint, reducing passenger waiting time is a major issue in the operation of mass transit systems.

The outcomes of the tabu search applications show that better schedules are produced than those obtained by previous approaches, which were based on simulated annealing. The Daduna and Voss study also reports the successful incorporation of the tabu search schedule synchronization procedures into an overall solution approach for changing a public mass transit system by introducing new bus lines. The resulting advances include the ability to perform sensitivity analysis more effectively, disclosing that small changes, appropriately determined, can create large improvements in both cost and quality of service.

4.17 Reactive Tabu Search in Combinatorial Optimization

Reactive Tabu Search (RTS) as developed by Battiti and Tecchioli (1992, 1994b), has been applied to a considerable range of optimization problems. Combinatorial problems studied with this approach include:

- Quadratic Assignment Problems
- N-K Models (derived from biological inspiration),
- 0-1 Knapsack and Multi-Knapsack Problems,
- Max-Clique Problems
- Biquadratic Assignment Problems

In many cases the results obtained with alternative competitive heuristics have been duplicated with low computational complexity, and without intensive parameter and algorithm tuning. In some cases (e.g., in the Max-Clique and Biquadratic assignment problems) significantly better results have been obtained. A comparison of RTS with alternative heuristics (Repeated Local Minima Search, Simulated Annealing, Genetic Algorithms and Mean Field Neural

Networks) is presented in B&T (1995c) and a comparison with Simulated Annealing on QAP tasks is contained in B&T (1994c), disclosing the effectiveness of the RTS approach relative to these alternative procedures.

4.18 Asynchronous Parallel Tabu Search for Integrated Circuit Design

The logical test of integrated circuits is one of the main phases of their design and fabrication. The pseudo-exhaustive approach for the logical test of integrated circuits consists in partitioning the original circuit to be tested into non-overlapping subcircuits with a small, bounded number of input gates, which are then exhaustively tested in parallel. Andreatta and Ribeiro (1994) developed an approximate algorithm for the problem of partitioning integrated combinational circuits, based on the tabu search metaheuristic. The circuits are modelled as directed acyclic graphs. The proposed algorithm contains several original features, including reduced neighborhoods; complex moves (similar to an ejection chain strategy); a multicriteria cost function and the use of a bin-packing heuristic as a post-optimization step. Computational results were compared with those obtained by the best algorithm previously published in the literature, with significant improvements. The average reduction rates have been on the order of 30% for the number of subcircuits in the partition, and of the order of 40% for the number of cuts required.

4.19 Asynchronous Multithread Tabu Search Variants

The use of alternative types of move attributes for the formation of the tabu lists, and multiple strategies for obtaining initial solutions, can very often enhance the quality of solutions obtained in TS approaches. As shown by Aiex, Martins, Ribeiro and Rodriguez (1996) the combination of different initial solution procedures with different types of move attributes may be usefully integrated in an asynchronous multithread procedure, in which each processor runs a tabu search algorithm with a different pair of initial solution procedure and move attributes.

Each time one of the search threads finds a new local optimum, it writes this solution in a pool of elite solutions, which is kept by a master processor in charge of search coordination. When a search thread is not able to improve its local best solution, it accesses the pool of solutions and randomly chooses one of them to restart the search. Global stopping criteria are used. Aiex et al. implement this asynchronous strategy for the circuit partitioning problem, using 10 processors, where one of them is the master in charge of the search coordination, and the other processors run nine different strategy combinations. The results obtained by this multithread version of the tabu search algorithm yield much better solutions than those obtained by the sequential tabu search algorithm, within very reasonable computational times. This multithread search procedure was implemented using two different parallel programming tools, PVM and Linda, also leading to comparative results concerning these tools. (Reviews of the parallel TS literature and evaluations of several parallel programming tools have been recently published by Martins, Ribeiro and Rodriguez (1996) and Toulouse, Crainic and Gendreau (1996). See also the Appendix for a discussion of strategies for exploiting multiple choice rules and neighborhoods.)

4.20 Tabu Search for Two Dimensional Irregular Cutting

The two-dimensional cutting problem is an optimization problem in which two-dimensional elements of arbitrary specified shapes are to be cut out of a rectangular material. The objective is to determine a cutting pattern that will minimize the amount of material used. The importance of the problem is growing due to its relation to packing, loading and partitioning, which have applications in multiple branches of industry.

Błażewicz and Walkowiak (1995) apply three variants of tabu search to this problem, extending earlier work of Błażewicz, Hawryluk and Walkowiak (1993). The first approach is a simple short term memory version of tabu search that incorporates a tabu list, an aspiration function and a single criterion for optimization. The next variant introduces a new type of evaluating function which combines several criteria to be optimized, together with an associated tabu condition. In the last version a probabilistic approach is used which translates the evaluation criteria into probabilities of selection. An exact algorithm for the subproblem of finding the placement of the polygon is incorporated to enhance the quality of solutions. The final version is also embedded in a parallel version of the algorithm by taking into account various tabu method parallelization schemes and geometric features of the algorithm and problem space.

Extensive computational comparisons disclose that all variants of the tabu search method tested, but most particularly the advanced ones, obtain significantly improved patterns for cutting layouts. In addition, the parallel implementation demonstrates the ability of the method to be exploited highly effectively in a multiprocessor environment.

4.21 Solving the Vehicle Routing Problem with Time Windows

In the face of today's global competition, the need for more efficient logistical planning has become a pressing issue for most manufacturing and distribution concerns. Barnes and Carlton (1995) present a reactive tabu search (RTS) approach to the vehicle routing problem with time windows (VRPTW). The VRPTW considered has available, at a single depot, m identical vehicles with a specified cargo capacity. Each of n nonidentical customers require a specified volume of cargo which must be delivered within a specified contiguous interval of time. Each customer must be visited exactly once and the objective is to find the feasible set of vehicle routes that minimize the total travel time.

The B & C study furnishes a brief review of the most recent literature associated with the VRPTW, presents their RTS algorithm and gives computational results for the algorithm when applied to a widely used benchmark test set of vehicle routing problems with time windows due to Solomon. The results were produced without any attempt at "tuning," and were obtained in a small fraction of the time required by current exact techniques. The proposed algorithm does not suffer from the computational limitations of exact approaches, which are unable to successfully attack larger problems. The algorithm experienced no difficulty in obtaining solutions to all 56 Solomon problems for the 100-customer sets.

4.22 The Damper Placement Problem on Space Truss Structures

NASA has conducted a set of laboratory experiments investigating the control of space structures. To facilitate these experiments, a large, flexible structure was assembled from truss elements and antenna support members and dubbed the Controls-structures-interactions Evolutionary Model (CEM). The CEM was designed to simulate characteristics of a large earth-observation platform and was dynamically tested in the NASA Langley Space Structures laboratory.

The overall structural motion of a flexible truss structure can be reduced by the use of structural dampers that both sense and dissipate vibrations. Kincaid and Berger (1993) develop a tabu search method for the problem of locating these dampers. The goal is to assure that vibrations arising from the control or operation of the structure and its payloads, or by cyclic thermal expansion and contraction of the space structure, can be damped as effectively as possible.

Given a strain energy matrix with rows indexed on the modes and the columns indexed on the truss members, the problem can be expressed as that of finding a set of p columns such that the smallest row sum, over the p columns, is maximized. The TS approach obtained high quality solutions, as verified by comparisons with designs previously proposed and also with upper bounds provided by the optimum value of an LP relaxation. Outcomes from the study led NASA engineers to reconsider their design assumptions, and in consequence to change the rigidity of support arms for the truss structure.

4.23 Mixed Integer, Multi-stage Stochastic Programming

A very large class of problems is characterized by a multi-stage decision process where the future is uncertain and some decisions are constrained to take on values of either zero or one (as in the decision of whether or not to open a facility at a particular location). Although some mathematical theory exists for such problems, no general purpose algorithms have been available to address them. Løkketangen and Woodruff (1996) introduce the notion of integer convergence for progressive hedging, and provide the first implementation of general purpose methods for finding good solutions to multi-stage, stochastic mixed integer (0,1) programming problems. The solution method makes use of Rockafellar and Wets' progressive hedging algorithm that averages solutions rather than data, and then applies a tabu search algorithm to obtain solutions to the induced quadratic, (0,1) mixed-integer sub-problems. Computational experiments verify the effectiveness of the new method across a range of problem instances. The software that the authors have developed reads standard (SMPS) data files.

4.24 Tabu Search for the Fixed Charge Transportation Problem

In a fixed charge transportation problem, a fixed ("all or none") cost is incurred whenever a route in the transportation network is used. Goods transported along that route are additionally subject to a unit variable cost. The underlying model is also applicable to plant and warehouse location problems, purchase/lease problems, and personnel hiring problems.

A tabu search approach was developed for this problem by Sun et al. (1995) using recency based and frequency based memories, two strategies for each of the intermediate and long term memory processes, and a network based implementation of the simplex method as the local search method. A computational comparison was performed to evaluate the performance of this approach on randomly generated problems of different sizes and of different ranges of magnitude of fixed costs relative to variable costs. Objective function values and CPU time were used as criteria to compare the performance of this procedure with that previously proposed methods consisting of an exact solution algorithm and a heuristic procedure.

The tabu search procedure obtained optimal and near-optimal solutions much faster than the exact solution algorithm for simple problems, and thoroughly dominated the exact algorithm for more complex problems. For example, in a set of 15 randomly generated problems in the class studied, restricting the problem size, the tabu search procedure found the optimal solutions for 12 problems, and the objective function value of the worst solution to the remaining 3 problems was less than 0.06% higher than that of the optimal solution. The exact solution procedure used an average of 5888 CPU seconds for these problems, while the tabu search procedure used an average of just 1.63 seconds. As problem size increased or as fixed costs became high relative to variable costs, the solution time for the exact algorithm became inordinate. Compared to the alternative heuristic approach, statistical results showed that the tabu search procedure found comparable solutions at least as good for very small and easy test problems, and found significantly better solutions for all other problems. For the small problems, the solution times used by both heuristics were similar, while for larger problems and for problems with higher fixed relative to variable costs, the tabu search procedure was 3 to 4 times faster than the competing heuristic.

4.25 Sub-symbolic Machine Learning (Neural Networks)

While derivative-based methods for training from examples have been used with success in many contexts (error backpropagation is an example in the field of neural networks), they are applicable only to differentiable performance functions and are not always appropriate in the presence of local minima. In addition, the calculation of derivatives is expensive and error-prone, especially if special-purpose VLSI hardware is used. Battiti and Tecchiolli (1995b) use a significantly different approach: the task is transformed into a combinatorial optimization problem (the points of the search space are binary strings), and solved with a reactive tabu search algorithm. To speed up the neighborhood evaluation phase a stochastic sampling of the neighborhood is adopted and a "smart" iterative scheme is used to compute the changes in the performance function caused by changing a single weight. The RTS approach escapes rapidly from local minima, it is applicable to non-differentiable and even discontinuous functions and it is very robust with respect to the choice of the initial configuration. In addition, by fine-tuning the number of bits for each parameter one can decrease the size of the search space, increase the expected generalization and realize cost-effective VLSI.

4.26 Vehicle Routing Problem With Time Windows Applications

The vehicle routing problem with time windows (VRPTW) can be used to model many real-world problems and has recently been the subject of intensive research. Applications of the VRPTW include bank deliveries, postal deliveries, industrial refuse collection, national franchise restaurant deliveries, school bus routing, and security patrol services.

Chiang and Russell (1995) have developed a reactive tabu search metaheuristic for the VRPTW that dynamically varies the size of the list of forbidden moves (in order to avoid cycles as well as an overly constrained search path). The method incorporates intensification and diversification strategies to achieve higher quality solutions. C & R also developed simulated annealing metaheuristics which achieve solutions that compare favorably with previously reported results. Computational tests of problems from the literature as well as of large-scale real-world problems show that several new best known solutions are achieved by both the tabu search and simulated annealing approaches.

However, tabu search outperforms simulated annealing in solution quality. The tabu search approach is especially effective in reducing fleet size requirements for routing problems constrained by time windows.

4.27 The Polymer Straightening Problem

Polymer chemists at NASA-Langley Research Center are interested in the crystallization of high-performance aromatic polyimides. The value of these polymers lies in their thermal stability, strength, and toughness. Aromatic polyimides are used to build high-performance carbon fiber composites for structural components in aircraft and spacecraft, which depends on their crystallization. A key problem is to determine a priori if there exists a conformation for which a given aromatic polyimide crystallizes.

The role of the optimizer in this application is to determine if a straight line conformation for a given polyimide exists among all possible combinations of allowable (minimum energy) torsion angles for the rotable bonds. The total number of combinations may be quite large, easily containing as many as 100 million possible conformations.

Kincaid, Martin and Hinkley (1995) develop a simple tabu search procedure to find a conformation that maximizes the cosine of the angle between the first bond and the projection of the last bond over all allowable conformations. The method was applied to the analysis of three polyimides of interest to NASA Langley Research Center, and succeeded in finding the optimal conformation in all three cases. (Normally, a research chemist could require as much as three years to perform such an analysis.) The ultimate goal of this research is to provide a technique that will serve as an aid to chemists in deciding what conformations are most likely to result in crystallizable structures when produced in a laboratory.

4.28 Tabu Search for Portfolio Management

An important issue in portfolio management is how to measure and handle risk. The challenge is to solve large problems that typically cannot be attacked with non-linear programming methods. Rolland (1996) develops a tabu search method that handles real valued decision variables by discretizing the problem space in 1% and 0.1% increments, and by further incorporating a greedy search to adjust the decision variables to real numbers "finer" than the 0.1% accuracy level. The moves alter their focus to comply with minimum variance and the target return. The approach identified optimal solutions to all the random and real-world problems it was tested on. Computation time was also very modest, even for large problems with 100 or more assets. This approach recently has been extended by Rolland and Johnson (1996) to be able to handle skewness computations and targets.

4.29 Modeling Generalized Capacity Requirements in Production Planning

Production planning problems that arise in real world applications are typically attended by capacity requirements. The incorporation of generalized capacity effects such as economies and diseconomies of scope and the learning-curve effect gives rise to a capacity-consumption function that is nonlinear in the tasks assigned to each facility. The resulting models for facility planning and loading decisions often involve nonlinear optimization problems in which some or all of the decision variables are integer-valued. The combined conditions of nonlinearity and discreetness makes these problems exceedingly difficult to solve.

Mazzola and Schantz (1995a) consider the resource allocation of a single facility under capacity-based economies and diseconomies of scope, and develop two models for this problem. In the first (more general) model the capacity of a single facility is considered to be a general function of the subset of tasks selected to be produced. In the second model the capacity is assumed to be consumed as a function of the number of tasks assigned to the facility. These can be viewed as single-facility production loading models that capture economies and diseconomies of scope within the production planning framework. The problems arising within each of these models generalize both the 0-1 knapsack problem and the 0-1 collapsing knapsack problem. Tabu search heuristics and branch-and-bound algorithms are defined for each model. Computational testing shows the tabu search heuristics to be extremely effective in obtaining high-quality solutions to these problems, including the more difficult problems that exhibit a high degree of nonlinear behavior.

This study is extended in Mazzola and Schantz (1995b) to the multiple-facility setting, focusing on under capacity-based economies (and diseconomies) of scope (MFLS), including applications of MFLS in hierarchical production planning, group technology, and professional services. MFLS is formulated as a nonlinear 0-1 mixed-integer programming problem which generalizes many well known and widely applicable optimization problems, such as the generalized assignment problem and the capacitated facility location problem. A tabu-search heuristic and a branch-and-bound algorithm are developed for MFLS, and computational testing of the solution procedures is discussed. Once again, tabu search proves to be an effective approach for heuristically solving MFLS, and is reported to be a powerful tool for capturing complex capacity requirements.

4.30 Continuous Optimization

A simple benchmark on a function with many suboptimal local minima is considered in Battiti and Tecchiolli (1994b), where a straightforward discretization of the domain is used. A novel algorithm for the global optimization of functions (C-RTS) is presented in Battiti and Tecchiolli (1995a), in which a combinatorial optimization method cooperates with a stochastic local minimizer. The combinatorial optimization component, based on reactive tabu search, locates the most promising boxes, where starting points for the local minimizer are generated. In order to cover a wide spectrum of possible applications with no user intervention, the method is designed with adaptive mechanisms: in addition to the reactive adaptation of the prohibition period, the box size is adapted to the local structure of the function to be optimized (boxes are larger in "flat" regions, smaller in regions with a "rough" structure).

4.31 Chunking in Tabu Search

Chunking — grouping basic units of information to create higher level units — is a critical component of human intelligence. The tendency for people to group information was described in the celebrated 1956 paper "The Magic Number Seven: Plus or Minus Two" by G.A. Mitchell. Human problem solvers, when faced with a hard problem, often proceed by linking and integrating features they perceive as related and germane to the solution process. However, people often prefer to organize problem data and problem solving methods in a hierarchical fashion. When possible, they decompose the problem into subproblems and solve those. When the problem cannot be decomposed, a common strategy is to form groupings of solution attributes so that the search space can be reduced and higher level relationships can be discovered or exploited.

Woodruff (1995, 1996) identifies special types of chunking to enhance tabu search memory structures, producing improved problem solving ability and giving useful supporting information for decision makers. Although the proposals are most natural in the context of the tabu search paradigm, they can also be employed in genetic algorithms and simulated annealing by omitting links to memory based constructions. This work outlines theory and proposals for learning about chunks and using them. Computational experience to date is briefly summarized to support the contention that chunking can be an important part of effective optimization algorithms.

4.32 Production Planning with Work-Force Learning

Mazzola, Neebe, and Rump (1995) consider a production planning problem in which the work-force productivity for each product depends on the amount of previous production of the product. This change in productivity is captured in the corresponding resource coefficients that occur in the work-force requirement constraint for each time period.

The model includes a learning effect in each period that depends on the level of production of each product in the preceding period. The corresponding work-force coefficient can increase, decrease, or remain the same, representing a forgetting, learning, or status-quo production effect. The resulting problem is modeled as a mixed-integer programming problem. In addition to establishing problem complexity and defining a branch-and-bound algorithm for solving the problem to optimality, this paper also examines heuristics for the problem. A forward pass, linear programming-based heuristic previously defined in the literature is examined and shown to produce arbitrarily bad solutions for this problem. The paper then proposes a new tabu search heuristic for the problem. Extensive computational experiments with the solution procedures establishes the effectiveness of the tabu-search approach in this problem setting.

The TS approaches of this study are concluded to provide an ability to handle new levels of modeling complexity, allowing for closer approximation of real-world phenomena. The findings indicate that this is particularly true for problems involving complex, nonlinear behavior involving discrete decision variables.

4.33 Tabu Search Applied to Hub-and-Spoke Communication Networks

The location of hub facilities is an important issue arising in the design of communication networks. Applications include: traffic networks (airline passengers flow and parcel delivery networks), as well as telecommunication networks (location of digital switching offices for Digital Data Service (DDS) networks, location of base stations for wireless networks).

In general, determining optimal locations of hub nodes and allocations of non-hub nodes to those hubs is an NP-hard combinatorial problem. For a widely used benchmark set of problems (the Civil Aeronautics Board (CAB) data set) efficient tabu search algorithms and lower bounds for a class of uncapacitated multiple and single allocation p-hub median problems have recently been developed which notably improve on results previously obtained. (Attempts to solve the problems of the CAB data set have been undertaken in more than 70 research papers.)

Skorin-Kapov and Skorin-Kapov (1994) have developed an efficient tabu search heuristic for the single allocation p-hub median problem, which models the situation when n nodes can interact only via a set of fully interconnected hubs. The hubs are uncapacitated, and their number is initially prescribed. Using the amount of flow and the cost per unit of flow between any two nodes in a network, one has to decide on the location of hubs, and on the allocation of each non-hub node to one of the hubs. The problem can be formulated as a quadratic integer program with a nonconvex objective function. The new tabu search approach, in addition to being efficient, obtains a number of new best solutions for the CAB data set.

Skorin-Kapov et al. (1995) provide a novel way to further take advantage of these improved heuristic outcomes by using high quality heuristic solutions to derive lower bounds. Accompanying this, they provide tight linear programming relaxations for the hub location and some other relevant uncapacitated p-hub median problems that allow the CAB data set to be solved to optimality for the first time. By this means, they verified that the solutions obtained by the tabu search approach were in fact optimal for all of the test problems.

By exploiting the LP solution, and the best known heuristic solutions derived from tabu search, integrality was quickly achieved by adding a partial set of integrality constraints. This result, in turn, proved the optimality of the best known tabu search heuristic solutions. Moreover, it provided a new approach of using the best known heuristic solution as a guidance in adding a partial set of integrality constraints to achieve integer solutions for those instances whose linear programming relaxations resulted with fractional solutions.

4.34 VLSI Systems with Learning Capabilities

In contrast to the exhaustive design of systems for pattern recognition, control, and vector quantization, an appealing possibility consists of specifying a general architecture, whose parameters are then tuned through Machine Learning (ML). ML becomes a combinatorial task if the parameters assume a discrete set of values: a reactive tabu search algorithm developed by Battiti et al. (1994a, 1994b) permits the training of these systems with low number of bits per weight, low computational accuracy, no local minima "trapping", and limited sensitivity to the initial conditions.

A board with the TOTEM chip used for Machine Learning applications — A project for IRST aims at developing special-purpose VLSI modules to be used as components of fully autonomous massively-parallel systems for real-time adaptive applications. Because of the intense use of parallelism at the chip and system level and the limited precision used, the obtained performance is competitive with that of state-of-the-art supercomputers (at a much lower cost), while a high degree of flexibility is maintained through the use of combinatorial algorithms. In particular, neural nets can be realized. In contrast to many "emulation" approaches, the developed VLSI completely reflects the combinatorial structure used in the learning algorithms. The first chip of the project (TOTEM, partially funded by INFN and EU (Esprit project MInOSS) and designed at IRST achieves a performance of more than one billion multiply-accumulate operations. Applications considered are in the area of pattern recognition (Optical Character Recognition), events "triggering" in High Energy Physics [A+], control of non-linear systems (Battiti and Tecchiolli (1995c), compression of EEG signals [B+]. Test boards for ISA, PCI or VME buses with software and technical documentation are available at IRST.

4.35 A Tabu Search Approach for Routing and Distribution

Rochat and Semet (1994) consider a real-life vehicle routing and distribution problem that occurs in a major Swiss company producing pet food and flour. In contrast with usual hypothetical problems, a large variety of restrictions must be considered. The main constraints involve the accessibility and the time windows at customers, the carrying capacities of vehicles, the total duration of routes and the drivers' breaks. The optimization problem for the transport plan consists in elaborating a set of routes that minimizes the total travel distance while satisfying the indicated constraints.

The general scheme used to solve this real-life VRP first applies a straightforward construction procedure to generate an initial solution which provides a starting point for the tabu search procedure. The key features of the TS approach consist of a constraint relaxation strategy for diversification and an intensification strategy. The relaxation of constraints makes it possible to expand the solution space, diversifying the search by examining infeasible solutions as well as feasible ones. The intensification strategy plays a complementary role, and leads the search to visit solutions close to the best solution found so far by rendering some routes tabu. Computational results show that the TS method yields solutions dominating those of the constructive heuristic even when the total number of iterations is small. Thus, good solutions are obtained in a reasonable amount of CPU time. Moreover, the study shows that embedding the TS algorithm in decision support software can be particularly useful. The fact that the TS approach generates multiple solutions that have approximately the same length of routes as the best makes it possible to propose several transport plans to the user. Comparisons of the solutions produced with the routes actually covered by the company disclose that the total distance traveled is reduced significantly by these solutions.

4.36 Tabu Search for Scheduling a Flow-Line Manufacturing Cell

Effective scheduling of flow-lines for manufacturing cells improves the operational efficiency of manufacturing processes, leading to reductions in setup costs, labor costs, tooling and inventory costs. This leads to further reductions in throughput times and a corresponding increase in the shipment of on-time deliveries. A tabu search method for this problem has been proposed and successfully implemented by Skorin-Kapov and Vakharia (1993).

A manufacturing cell consists of a group of similar machines located in close proximity to one another and dedicated to the manufacture of a specific number of part families. Part families consist of a set of jobs with similar processing requirements. In this context, a feasible schedule S consists of a sequence of part families and a sequence of jobs within each family in a manufacturing cell. The tabu search heuristic of Skorin-Kapov and Vakharia efficiently schedules a pure flow-line manufacturing cell under varying parameter conditions (given F families, M machines and N(f) jobs in family f).

A collection of alternative tabu search strategies (designed to test different aspects of tabu search) was compared against state-of-the-art simulated annealing heuristic that was tailored to solve this problem. Results from testing multiple data sets with alternative ratios of family set up times to job processing times showed the clear superiority of tabu search for these scheduling problems.

4.37 A Hybrid Scatter Genetic Tabu Approach for Continuous Global Optimization

A hybrid scatter genetic tabu search approach (HSGT) is proposed by Trafalis and Al-Harkan (1995) to solve an unconstrained continuous nonlinear global optimization problem. This approach combines the characteristics of the following metaheuristics: scatter search (SS), genetic algorithms (GAs), and tabu search (TS). The proposed approach has been tested against a simulated annealing (SA) algorithm and a modified version of a hybrid scatter genetic search approach (HSG) by optimizing twenty-one well known test functions. From the computational results, the HSGT approach proved to be quite effective in identifying the global optimum solution which makes the HSGT approach a promising approach to solve the general nonlinear optimization problem. In the hundred runs performed for each of the twenty-one functions, the HSGT approach performed better than the HSG and the SA approaches, except for one function. Also, the HSGT approach converged to a near global optimum in CPU times ranged between 1.3 seconds and 19.55 seconds. The algorithm was implemented in a GATEWAY 2000 (Pentium, P5-90) computer using the Microsoft FORTRAN PowerStation version 4.

4.38 Active Structural Acoustic Control

Active structural acoustic control is a method in which the control inputs used to reduce interior noise are applied directly to a vibrating structural acoustic system. The ultimate goal is to use active acoustic control to decrease the interior noise in propeller driven jet aircraft. Kincaid (1995) studies the instance of this problem in which the objective consists of damping noise generated by a single exterior source in the interior of a cylinder.

The model requires a determination of the force inputs and sites for piezoelectric actuators so that (1) the interior noise is effectively damped; (2) the level of vibration of the cylinder shell is not increased; and (3) the power requirements needed to drive the actuators are not excessive.

A tabu search approach was developed to determine the best set of actuator sites to meet the three specified objectives. Experiments confirmed that the TS procedure is able to uncover better solutions than those selected based solely upon engineering judgement. In addition, the high quality solutions generated by tabu search, when minimizing interior noise, do not further excite the cylinder shell. Thus, it was possible to meet objective (2) without imposing an additional constraint or forming a multi-objective performance measure. The TS solutions also led to identifying natural groupings that require fewer control channels and that permit a simpler control system.

4.39 Tabu Search for Automatic Graph Drawing

Graphs are commonly used as a basic modeling tool in areas such as project management, production scheduling, line balancing, business process reengineering, and software visualization. Drawings of graphs are called maps and their value for modeling and analysis is widely heralded in various fields of the economic, social and computational sciences.

The main quality desired for maps is readability: A map is readable if its meaning is easily captured by the way it is drawn. It is extremely difficult to make a readable map by hand of a graph that represents a real system, even when the graph size is relatively small. Therefore, an automatic procedure for drawing graphs by computer is indispensable for generating readable maps quickly.

An important problem in the area of graph drawing is to minimize arc crossings in a hierarchical digraph. It is customary to draw a hierarchical digraph by placing the vertices on a set of equally spaced horizontal or vertical lines called layers and then drawing the arcs as straight-line segments.

Valls, Marti and Lino (1995) provide a Tabu Thresholding (TT) approach for the problem of minimizing the number of arc crossings in a 2-layer hierarchical digraph (a bipartite digraph). The procedure combines elements of probabilistic tabu search, candidate list strategies and thresholds, yielding a simplified implementation of basic TS ideas that does not make explicit use of memory structures.

The computational study has been carried out on a set of 250 randomly generated problems of varying sizes and densities. The TT algorithm has been compared with the Greedy Switching (GS) algorithm and the Splitting (S) algorithm, which are reported to be the best methods in the literature previously available for graph drawing. Outcomes from the TT methods are also compared with the optimal solutions for the test problems, in those cases where the problems are small enough to permit them to be solved by state-of-the-art exact methods. Results show that in the 130 test problems where an optimum is available, the TT solution is optimal. Moreover in each of the 250 examples tested, the TT solution and generally is superior (and never inferior) to the best of those given by the GS and S heuristics.

4.40 Bipartite Graph Drawing with Tabu Search

Marti (1995) has developed a TS heuristic for the problem of minimizing the number of arc crossings in a bipartite graph. To perform an aggressive search for the global optimum, the author has considered intensification, diversification, influential moves and strategic oscillation elements of tabu search.

The procedure has three different search states: normal, influential and opposite, and oscillates among them according to the search history. In each state there are two alternately applied phases, an intensification phase and a diversification phase. The moves defined in the intensification are based on a positioning function while those defined in the diversification are based on permuting consecutive vertices. The use of different moves reinforces the non-monotonic search strategy. The criteria for differentiating between "improving" and "disimproving" moves within the oscillation strategy are not limited to the objective function evaluation, but consider factors of move influence, as determined by context and the search history.

Two variants of the general TS procedure are developed and compared with the GS, SP, BC and SM methods and with the Tabu Thresholding (TT) heuristic. The computational results show that both of these TS variants perform better than the other methods, closely followed by the TT procedure, which in turn also performs better than the methods remaining.

4.41 Layered Graph Drawing

Laguna, Marti and Valls (1995) propose a TS algorithm for the general k-layer graph drawing problem (k greater than 2). Existing solution methods for this problem are based on simple ordering rules for single layers that may lead to inferior drawings. The Tabu Search implementation consists of an intensification phase that seeks local optimal orderings of layers using an insertion move, and two levels of diversification. The first level of diversification is a strategy for selecting layers for intensification, while the second one escapes local optimality by means of switching moves. The authors utilize two different termination criteria (TABU1 and TABU2).

Computational testing was performed on a set of 200 randomly generated instances, including graphs with up to 571 vertices and 2,241 arcs. Comparisons were performed with procedures that have shown to be effective for arc crossing minimization, i.e., the barycentric and the semi media methods with switching (BC+SW, SM+SW).

The results of the experiments show that in terms of solution quality the procedures are ranked in the order TABU2, TABU1, BC+SW, and SM+SW. In terms of computational time, the tabu search version TABU1 is quite competitive with the procedures based on simple ordering rules plus switching, in spite of yielding significantly better outcomes. This allows TABU1 to be considered as a powerful procedure for real-time drawing (e.g., drawing on a computer screen). When still higher quality drawings are important, at the cost of additional computational time, TABU2 obtains solutions with fewer arc crossings with a maximum running time of 209 seconds.

4.42 Finding a Good Forest Harvest Schedule Satisfying Green-Up Constraints

A distinct change in public attitude toward the environment has led to demands for a forest management paradigm shift from one of a dominant timber use to one in which forests are managed for multiple values. Among the most important implications of such a shift in policy on forest management was the introduction of adjacency or green-up constraints. Green-up constraints are imposed to avoid large clear cut areas. For example, in British Columbia legislation requires that a block of forest which has been harvested must reach a mean tree height of three meters before any adjacent block can be harvested.

One result of these green-up constraints is that the forest harvesting scheduling problem, which previously was often formulated as a linear program, becomes combinatorial in nature. Current harvest scheduling codes, like FORPLAN, Timber RAM and others, are unable to generate harvest schedules satisfying the green-up constraints. Brumelle et al. (1996) formulate forest harvesting problems with green-up constraints arising in the Tangier watershed in British Columbia as multicriteria discrete optimization problems.

The study considers two harvest scheduling problems associated with the Tangier watershed. The small problem focuses on a 219 cut-block subset of the watershed located at the southern end. The northern boundary coincides with that of the proposed Serenity Peaks Wilderness Area. The larger problem included all 491 cut-blocks comprising the entire watershed, which will be appropriate should the proposed park not eventuate.

Tabu search is used to investigate the trade-offs between different criteria, which were chosen as the total volume of lumber cut, the period to period deviation from even-flow of lumber during a harvest rotation and adjacency violations. The tabu search methodology easily obtained good solutions to these problems, and was shown to be much superior to a biased random search method which is cited as one of the most effective methods to obtain good schedules satisfying green-up constraints. In fact, the tabu search method generates schedules which harvest more timber than the upper bound of the confidence interval suggested by previous empirical and algorithmic analysis.

APPENDIX A

THE MOST NEGLECTED TABU SEARCH STRATEGIES — PROMISING AVENUES FOR FUTURE RESEARCH

This appendix briefly reviews several key strategies in tabu search that are often neglected (especially in beginning studies), but which are important for producing the best results.

Our purpose is to call attention to the relevance of particular elements that may be missed in a lengthier or more formal exposition. In addition, observations about useful directions for future research are included.

A comment regarding implementation: first steps do not have to include the most sophisticated variants of the ideas described below, but the difference between "some inclusion" and "no inclusion" can be significant. Implementations that incorporate simple instances of these ideas will often disclose the manner in which refined implementations can lead to improved performance.

A.1 CANDIDATE LIST STRATEGIES

Efficiency and quality can be greatly affected by using intelligent procedures for isolating effective candidate moves, rather than trying to evaluate every possible move in a current neighborhood of alternatives. This is particularly true when such a neighborhood is large or expensive to examine. The gains to be achieved by using candidate lists have been widely documented, yet many TS studies overlook their relevance.

Careful organization in applying candidate lists, as by saving evaluations from previous iterations and updating them efficiently, can also be valuable for reducing overall effort. Time saved in these ways allows a chance to devote more time to higher level features of the search.

While the basic theme of candidate lists is straightforward, there are some subtleties in the ways candidate list strategies may be used. Considerable benefit can result by being aware of fundamental candidate list approaches, such as the *Subdivision Strategy*, the *Aspiration Plus Strategy*, the *Elite Candidate*

List Strategy, the *Bounded Change Strategy* and the *Sequential Fan Strategy* (see, for example, Glover (1995b)).

An effective integration of a candidate list strategy with the rest of a tabu search method will typically benefit by using TS memory designs to facilitate functions to be performed by the candidate lists. This applies especially to the use of frequency based memory. A major mistake of some TS implementations, whether or not they make use of candidate lists, is to consider only the use of recency based memory. Frequency based memory — which itself takes different forms in intensification phases and diversification phases — can not only have a dramatic impact on the performance of the search in general but also can often yield gains in the design of candidate list procedures.

A.2 PROBABILISTIC TABU SEARCH

Several studies have suggested the value of a probabilistic version of TS, where evaluations (including reference to tabu status) are translated into probabilities of selection, strongly skewed to favor higher evaluations. Findings from such studies support the notion that probabilities may partly substitute for certain functions of memory (hence reduce the amount of memory needed) but also suggest that probabilities may have a role in counteracting "noise" in the evaluations.

In well designed TS implementations, the gains of probabilistic TS over deterministic TS are chiefly in accelerating the rate at which good solutions are discovered in earlier stages of search. Overall, some settings appear more exploitable by probabilistic TS and others appear more exploitable by deterministic TS. The most effective forms of each type of approach depend strongly on identifying and implementing TS strategies of the type described below.

A.3 INTENSIFICATION APPROACHES

Intensification strategies, which are based on recording and exploiting elite solutions or, characteristically, specific features of these solutions, have proved very useful in a variety of applications. Some of the relevant forms of such strategies and considerations for implementing them are as follows.

(1) The simplest intensification approach is the strategy of recovering elite solutions in some order, each time the search progress slows, and then using these solutions as a basis for re-initiating the search. The list of solutions that are candidates to be recovered is generally limited in size, often in the range of 20 to 40 (although in parallel processing applications the number is characteristically somewhat larger). The size chosen for the list in serial TS applications also corresponds roughly to the number of solution recoveries anticipated to be done during the search, and so may be less or more depending on the setting. When an elite solution is recovered from the list, it is removed, and new elite solutions are allowed to replace less attractive previous solutions — usually dropping the worst of the current list members. However, if a new elite solution is highly similar to a solution presently recorded, instead of replacing the current worst solution, the new solution will compete directly with its similar counterpart to determine which solution is saved.

This approach has been applied very effectively in job shop and flow shop scheduling, in vehicle routing, and in telecommunication design problems. One of the best approaches for scheduling applications (see Section 3) keeps the old TS memory associated with the solution, but makes sure the first new move away from this solution goes to a different neighbor than the one visited after encountering this solution the first time. Another effective variant does not bother to save the old TS memory, but uses a probabilistic TS choice design.

The most common strategy is to go through the list from best to worst, but in some cases it has worked even better to go through the list in the other direction. In this approach, it appears effective to allow two passes of the list. On the first pass, when a new elite solution is found that falls below the quality of the solution currently recovered, but which is still better than the worst already examined on the list, the method still adds the new solution to the list and displaces the worst solution. Then a second pass, after reaching the top of the list, recovers any added solutions not previously recovered.

(2) The other primary intensification strategy is to examine elite solutions to determine the frequency in which particular solution attributes occur (where the frequency is typically weighted by the quality of the solutions in which the attributes are found).

This strategy was originally formulated in the context of identifying "consistent" and "strongly determined" variables — where, loosely speaking, consistent variables are those more frequently found in elite solutions, while strongly determined variables are those that would cause the greatest disruption by changing their values (as sometimes approximately measured by weighting the frequencies based on solution quality). The idea is to isolate the variables that qualify as more consistent and strongly determined (according to varying thresholds), and then to generate new solutions that give these variables their "preferred values." This can be done either by rebuilding new solutions in a multistart approach or by modifying the choice rules of an ongoing solution effort to favor the inclusion of these value assignments.

Keeping track of the frequency that elite solutions include particular attributes (such as edges of tours, assignments of elements to positions, narrow ranges of values taken on by variables, etc.) and then favoring the inclusion of the highest frequency elements, effectively allows the search to concentrate on finding the best supporting uses and values of other elements. A simple variant is to "lock in" a small subset of the most attractive attributes (value assignments) — allowing this subset to change over time or on different passes.

A Relevant Concern: In the approach that starts from a current (good) solution, and tries to bring in favored elements, it is important to introduce an element that yields a best outcome from among the current contenders. If an attractive alternative move shows up during this process, which does not involve bringing in one of these elements, aspiration criteria may determine whether such a move should be taken instead. Under circumstances where the outcome of such a move appears sufficiently promising, the approach may be discontinued and allowed to enter an improving phase (reflecting a decision that enough intensification has been applied, and it is time to return to searching by customary means).

Intensification of this form makes it possible to determine what percent of "good attributes" from prior solutions should be included in the solution currently generated. It also gives information about which subsets of these attributes should go together, since it is preferable not to choose attributes during this process that cause the solution to deteriorate compared to other choices. This type of intensification strategy has proved highly effective in the settings of vehicle routing and zero-one mixed integer optimization.

(3) Memory and Intensification: It is clearly somewhat more dangerous to hold elements "in" solution than to hold them "out" (considering that a solution normally is composed of a small fraction of available elements — as where a tree contains only a fraction of the edges of a graph). However, there is an important exception, previously intimated. As part of a longer term intensification strategy, elements may be selected very judiciously to be "locked in" on the basis of having occurred with high frequency in the best solutions found. In that case, choosing different mutually compatible (and mutually reinforc-

ing) sets to lock in can be quite helpful. This creates a *combinatorial implosion* effect (opposite to a combinatorial explosion effect) that shrinks the solution space to a point where best solutions over the reduced space are likely to be found more readily.

The key to this type of intensification strategy naturally is to select an appropriate set of elements to lock in, but the chances appear empirically to be quite high that some subset of those with high frequencies in earlier best solutions will be correct. Varying the subsets selected gives a significant likelihood of picking a good one. (More than one subset can be correct, because different subsets can still be part of the same complete set.) Aspiration criteria make it possible to drop elements that are supposedly locked in, to give this approach more flexibility.

(4) Relevance of Clustering for Intensification: A search process over a complex space is likely to produce clusters of elite solutions, where one group of solutions gives high frequencies for one set of attributes and another group gives high frequencies for a different set. It is important to recognize this situation when it arises. Otherwise there is a danger that an intensification strategy may try to compel a solution to include attributes that work against each other. This is particularly true in a strategy that seeks to generate a solution by incorporating a collection of attributes "all at once," rather than using a step by step evaluation process that is reapplied at each move through a neighborhood space. (Stepping through a neighborhood has the disadvantage of being slower, but may compensate by being more selective. Experimentation to determine the circumstances under which each of these alternative intensification approaches may be preferable would be quite valuable.)

A strategy that incorporates a block of attributes together may yield benefits by varying both the size and composition of the subsets of high frequency "attractive" attributes, even if these attributes are derived from solutions that lie in a common cluster, since the truly best solutions may not include them all. (An example of the relevance of clustering and related conditional analysis is provided in Topic VI of the Appendix.)

A.4 DIVERSIFICATION APPROACHES

Diversification processes in tabu search are sometimes applied in ways that limit their effectiveness, due to overlooking the fact that diversification is not just "random" or "impulsive," but depends on a purposeful blend of memory and strategy. As noted in Section 2, recency and frequency based memory are both relevant for diversification, and stem in part from proposals to use such ideas in surrogate constraint procedures. In this setting, the impetus is not simply to achieve diversification, but to derive appropriate weights in order to assure that evaluations will lead to solutions that satisfy required conditions. Accordingly, it is important to account for elements such as how often, to what extent, and how recently, particular constraints have been violated, in order to determine weights that produce more effective valuations.

The implicit *learning effects* that underlie such uses of recency, frequency and influence are analogous to those that motivate the procedures used for diversification (and intensification) in tabu search. Early strategic oscillation approaches exploited this principle by driving the search to various depths outside (and inside) feasibility boundaries, and then employing evaluations and directional search to move toward preferred regions.

In the same way that these early strategies bring diversification and intensification together as part of a continuously modulated process, it is important to stress that these two elements should be interwoven in general. A common mistake in many TS implementations is to apply diversification without regard for intensification. "Pure" diversification strategies are appropriate for truly long term strategies, but over the intermediate term, diversification is generally more effective if it is applied by heeding information that is also incorporated in intensification strategies. In fact, intensification by itself can sometimes cause a form of diversification, because intensifying over part of the space allows a broader search of the rest of the space. A few relevant concerns are as follows.

(1) Diversification and intensification links: A simple and natural diversification approach is to keep track of the frequency that attributes occur in non-elite solutions, and then to periodically discourage the incorporation of attributes that have modest to high frequencies (giving greater penalties to larger frequencies). The reference to non-elite solutions tends to avoid penalizing attributes that would be encouraged by an intensification strategy.

More generally, for a "first level" balance, an Intermediate Term Memory matrix may be used, where the high frequency items in elite solutions are not penalized by the long term values, but may even be encouraged. The tradeoffs involved in establishing the degree of encouragement, or the degree of reducing the penalties, represents an area where a small amount of preliminary testing can be valuable. This applies as well to picking thresholds to identify high frequency items. (Simple guesses about appropriate parameter values can often yield benefits, and tests of such initial guesses can build an understanding that leads to increasingly effective strategies.)

By extension, if an element has never or rarely been in a solution generated, then it should be given a higher evaluation for being incorporated in a diversification approach if it was "almost chosen" in the past but didn't make the grade. This observation has not been widely heeded, but is not difficult to implement, and is relevant to intensification strategies as well.

(2) Implicit conflict and the importance of interactions: Current evaluations also should not be disregarded while diversification influences are activated. Otherwise, a diversification process may bring elements together that conflict with each other, make it harder rather than easier to find improved solutions.

For example, a design that gives high penalties to a wide range of elements, without considering interactions, may drive the solution to avoid good combinations of elements. Consequently, diversification should be carried out for a limited number of steps, accompanied by watching for and sidestepping situations where indiscriminately applying penalties would create incompatibilities or severe deterioration of quality. To repeat the theme: even in diversification, attention to quality is important. And as in "medical remedies," sometimes small doses are better than large ones.

(3) An approach called "Reactive Tabu Search" (RTS) developed by Battiti and Tecchiolli (1992, 1994b) deserves consideration as a way to achieve a useful blend of intensification and diversification. RTS incorporates hashing in a highly effective manner to generate attributes that are very nearly able to differentiate among distinct solutions (that is, very few solutions contain the same hashed attribute). Accompanying this, B&T use an automated tabu tenure, which begins with the value of 1 (preventing a hashed attribute from being reinstated if this attribute gives the "signature" of the solution visited on the immediately preceding step). This tenure is then increased if examination shows the method is possibly cycling, as indicated by periodically generating solutions that produce the same hashed attribute.

The tabu tenure, which is the same for all attributes, is increased exponentially when repetitions are encountered, and decreased gradually when repetitions disappear. Under circumstances where the search nevertheless encounters an excessive number of repetitions within a given span (i.e., where a moving frequency measure exceeds a certain threshold), a diversification step is activated, which consists of making a number of random moves proportional to a moving average of the cycle length.

The reported successes of this approach invite further investigations of its underlying ideas and related variants. As a potential bases for generating such variants, attributes created by hashing may be viewed as fine grain attributes, which give them the ability to distinguish among different solutions. By contrast, "standard" solution attributes, which are the raw material for hashing, may be viewed as coarse grain attributes, since each may be contained in (and hence provide a signature for) many different solutions. Experience has shown that tabu restrictions based on coarse grain attributes are often advantageous for giving increased vigor to the search. (There can exist a variety of ways of defining and exploiting attributes, particularly at coarser levels, which complicates the issue somewhat.) This raises the question of when particular degrees of granularity are more effective than others.

It seems reasonable to suspect that fine grain attributes may yield greater benefits if they are activated in the vicinity of elite solutions, thereby allowing the search to scour "high quality terrain" more minutely. Of course, this effect may also be achieved by reducing tabu tenures for coarse grain attributes — or basing tabu restrictions on attribute conjunctions — and using more specialized aspiration criteria. Closer scouring of critical regions can also be brought about by using strongly focused candidate list strategies, such as a sequential fan candidate list strategy. (Empirical comparisons of such alternatives to hashing clearly would be of interest.)

However, another type of alternative to hashing also exists, which is to create new attributes by processes that are not so uniform as hashing. A potential drawback of hashing is its inability to distinguish the relative importance (and appropriate influence) of the attributes that it seeks to map into others that are fine grained. A potential way to overcome this drawback is to make use of vocabulary building (Section 2) and of conditional analysis (Topic VI of the Appendix). See Voss (1993), Carlton and Barnes (1995a, 1995b) and Woodruff (1995, 1996) for useful related observations.

(4) Ejection chain approaches: Ejection chain methods provide an implicit blending of diversification and intensification by generating compound moves out of simpler components. Such approaches have provided breakthroughs in handling certain types of tough problems, particularly those related to optimization over graphs (see for example, Dorndorf and Pesch (1994), Laguna et al. (1995), Pesch and Glover (1995), Rego and Roucairol (1996), Rego (1996a, 1996b)). TS memory structures can be used at two levels with ejection chains, both at a simple internal level which operates primarily as a bookkeeping function to avoid duplicate patterns (as complex moves are woven from simpler ones), and at an external level that guides the successively generated com-

pound moves to go beyond conditions of local optimality. So far ejection chain studies have chiefly focused on internal as opposed to external levels of control, and new discoveries may be expected by broadening this focus. In addition, opportunities exist in many settings for applications of ejection chains where such forms of compound neighborhoods have so far remained uninvestigated.

A.5 STRATEGIC OSCILLATION

A considerable amount has been written on strategic oscillation and its advantages. However, one of the uses of this approach that is frequently overlooked involves the idea of oscillating among alternative choice rules and neighborhoods. An important aspect of strategic oscillation is the fact that there naturally arise different types of moves and choice rules that are appropriate for negotiating different regions and different directions of search. Thus, for example, there are many constructive methods in graph and scheduling problems, but strategic oscillation further leads to the creation of complementary "destructive methods" which can operate together with their constructive counterparts. Different criteria emerge as relevant for selecting a move to take on a constructive step versus one to take on a destructive step. Similarly, different criteria apply according to whether moves are chosen within a feasible region or outside a feasible region (and whether the search is moving toward or away from a feasibility boundary).

The variation among moves and evaluations introduces an inherent vitality into the search that provides one of the sources underlying the success of strategic oscillation approaches. This reinforces the motivation to apply strategic oscillation to the choice of moves and evaluation criteria themselves, selecting moves from a pool of possibilities according to rules for transitioning from one choice to another. In general, instead of picking a single rule, a process of invoking multiple rules provides a range of alternatives that run all the way from "strong diversification" to "strong intensification."

This form of oscillation has much greater scope than may at first be apparent, because it invokes the possibility of simultaneously integrating decision rules and neighborhoods, rather than only visiting them in a strategically determined sequence. To understand the considerations involved, it is useful to trace a brief history of approaches that offer prototypes for such designs.

Among the first of such approaches, an early study by Crowston et al. (1963) examined ways to generate improved decision rules for job shop scheduling, using a strategy of combining various standard local decision rules such as least-waiting- time, first-come/first-serve, longest-time-remaining-to-completion, and so forth. The decision rules were combined in two ways: (1) by choosing probabilistically from the set of component rules at each decision point, and (2) by integrating the rules into a "parametric rule" rather than by alternating among them. The probabilistic combination approach began by giving equal probabilities to selecting the rules, and then revised the probabilities of selection according to the quality of the solution produced — increasing the probability of applying a particular rule at a particular juncture if the current resulting schedule turned out to be one of higher than usual quality. The parametric rule approach first re-expressed the component rules to give them a common metric, and then formulated a single composite rule as a weighted combination of the components. The weight parameters were then systematically varied to find a preferred combination, following the philosophy that it is more meaningful to heed the input of different rules simultaneously, rather than to alternately give priority to one or another input in isolation from the rest. (A bow and arrow analogy was postulated, to suggest that it is better to simultaneously account for factors such as wind direction, distance from the target, length of the draw, and so forth, than to simply alternate among heeding these factors.) Both of these two types of approaches improved over using any single decision rule, and the parametric rule approach proved especially effective.

The probabilistic and parametric approaches were later unified and extended (Glover and McMillan (1986)) to allow both rules and moves (hence neighborhoods) to be integrated by means of a "generalized voting" concept. In applying this concept, the probabilistic element governs aspects of timing while the parametric element allows different rules to vote on an outcome implicitly, by restructuring and joining the rules into a master rule. These forms of integration can naturally be extended by making use of strategic oscillation and path relinking approaches. Details of such extensions, based on generating structured combinations of solutions, appear in Glover (1994a).

Such concepts are beginning to find counterparts in investigations being launched by the computer science community. The "agent" terminology is being invoked in such applications to characterize different choice mechanisms and neighborhoods as representing different agents. Relying on this representation, different agents then are assigned to work on (or "attack") the problem serially or in parallel. The CS community has begun to look upon this as a significant innovation — unaware of the literature where such ideas were introduced a decade or more ago — and the potential richness and variation of these ideas still seems not to be fully recognized. (For example, there have not yet been any studies that consider the idea of "strategically sequencing" rules and neighborhoods, let alone those that envision the notion of parametric integration. The further incorporation of adaptive memory structures to enhance the application of such concepts also lies somewhat outside the purview of most current CS proposals.) At the same time, TS research has neglected to conduct empirical investigations of the broader possibilities. This is clearly an area that deserves fuller study.

A.6 CLUSTERING AND CONDITIONAL ANALYSIS

To reinforce the theme of identifying opportunities for future research, we provide an illustration to clarify the relevance of clustering and conditional analysis, particularly as a basis for intensification and diversification strategies in tabu research.

An Example: Suppose 40 elite solutions have been saved during the search, and each solution is characterized as a vector $\mathbf{x}$ of zero-one variables x_j, for $j \in N = \{1, \ldots, n\}$. Assume the variables that receive positive values in at least one of the elite solutions are indexed x_1 to x_{30}. (Commonly in such circumstances, n may be expected to be somewhat larger than the number of positive valued variables, e.g., in this case, reasonable values may be $n = 100$ or 1000.)

For simplicity, we restrict attention to a simple weighted measure of consistency which is given by the frequency that the variables x_1 to x_{30} receive the value 1 in these elite solutions. (We temporarily disregard weightings based on solution quality and other aspects of "strongly determined" assignments.) Specifically, assume the frequency measures are as shown in the following table:

Variables $x_j = 1$	Number of Solutions
x_1 to x_{10}	24
x_{16} to x_{20}	21
x_{21} to x_{25}	17
x_{26} to x_{30}	12

Since each of x_1 to x_{15} receives a value of 1 in 24 of the 40 solutions, these variables tie for giving "most frequent" assignments. An intensification strategy that favors the inclusion of some number of such assignments would give equal bias to introducing each of x_1 to x_{15} at the value 1. (Such a bias would typically be administrated either by creating modified evaluations or by incorporating probabilities based on such evaluations.)

To illustrate the relevance of clustering, suppose the collection of 40 elite solutions can be partitioned into two subsets of 20 solutions each, whose characteristics are summarized in the following table.

Subset 1 (20 solutions)		Subset 2 (20 solutions)	
Variables $x_j = 1$	No. of Solutions	Variables $x_j = 1$	No. of Solutions
x_{11} to x_{15}	20	x_{16} to x_{20}	20
x_{21} to x_{25}	16	x_6 to x_{10}	16
x_1 to x_5	12	x_1 to x_5	12
x_6 to x_{10}	8	x_{26} to x_{30}	8
x_{26} to x_{30}	4	x_{11} to x_{15}	4
x_{16} to x_{20}	1	x_{21} to x_{25}	1

A very different picture now emerges. The variables x_1 to x_{15} no longer appear to deserve equal status as "most favored" variables. Treating them with equal status may be a useful source of diversification, as opposed to intensification, but the clustered data provide more useful information for diversification concerns as well. In short, clustering gives a relevant contextual basis for determining the variables (and combinations of variables) that should be given special treatment.

Conditional Relationships

To go a step beyond the level of differentiation provided by cluster analysis, it is useful to sharpen the focus by referring explicitly to interactions among variables. Such interactions can often be identified in a very straightforward way, and indeed can form a basis for more effective clustering. In many types of problems, the number of value assignments (or the number of "critical attributes") needed to specify a solution is relatively small compared to the total number of problem variables. (For example, in routing, distribution and telecommunication applications, the number of links contained in feasible constructions is typically a small fraction of those contained in the underlying graph.) Using a 0-1 variable representation of possibilities, it is not unreasonable in such cases to create a *cross reference* matrix, which identifies variables (or coded attributes) that simultaneously receive a value of 1 in a specific collection of elite solutions.

To illustrate, suppose the index set $P = \{1, \ldots, p\}$ identifies the variables x_j that receive a value of 1 in at least r solutions from the collection of elite solutions under consideration. (Apart from other strategic considerations, the parameter r can also be used to control the size of p, since larger values of r result in smaller values of p.)

Then we create a $p \times p$ symmetric matrix M whose entries m_{ij} identify the number of solutions in which x_i and x_j are both 1. (Thus, row M_i of M represents the sum of the solution vectors in which $x_i = 1$, restricted to components x_j for $j \in P$.) The value m_{ii} identifies the total number of elite solutions in which $x_i = 1$, and the value m_{ij}/m_{ii} represents the "conditional probability" that $x_j = 1$ in this subset of solutions. Because p can be controlled to be of modest size, as by the choice of r and the number of solutions admitted to the elite set, the matrix M is not generally highly expensive to create or maintain.

By means of the conditional probability interpretation, the entries of M give a basis for a variety of analyses and choice rules for incorporating preferred attributes into new solutions. Once an assignment $x_j = 1$ is made in a solution currently under consideration (which may be either partly or completely constructed), an updated conditional matrix M can be created by restricting attention to elite solution vectors for which $x_j = 1$. (Restricted updates of this form can also be used for look-ahead purposes.) Weighted versions of M, whose entries additionally reflect the quality of solutions in which specific assignments occur, likewise can be used.

Critical event memory (Glover (1995c), Glover and Kochenberger (1996)), provides a convenient mechanism to maintain appropriate variation when conditional influences are taken into account. The "critical solutions" associated with such memory in the present case are simply those constituting a selected subset of elite solutions. Frequency measures for value assignments can be obtained by summing these solution vectors for problems with 0-1 representations and the critical event control mechanisms can then assure assignments are chosen to generate solutions that differ from those of previous elite solutions.

Conditional analysis, independent of such memory structures, can also be a useful foundation for generating solution fragments to be exploited by vocabulary building processes, as discussed in Section 2.

A.7 REFERENT-DOMAIN OPTIMIZATION

Referent-domain optimization is based on introducing one or more optimization models to strategically restructure the problem or neighborhood, accompanied by auxiliary heuristic or algorithmic process to map the solutions back to the original problem space. The optimization models are characteristically devised to embody selected heuristic goals (e.g., of intensification, diversification or both), within the context of particular classes of problems.

There are several ways to control the problem environment as a basis for applying referent-domain optimization. A natural control method is to limit the structure and range of parameters that define a neighborhood (or the rules used to navigate through a neighborhood), and to create an optimization model that operates under these restricted conditions.

Example 1: The use of specially constructed neighborhoods (and aggregations or partitions of integer variables) permits the application of mixed integer programming (MIP) models to identify the best options from all moves of depth at most k (or from associated collections of at most k variables). When k is sufficiently small, such MIP models can be quite tractable, and produce moves considerably more powerful than those provided by lower level heuristics.

Example 2: In problems with graph-related structures, the imposition of directionality or non-looping conditions gives a basis for devising generalized shortest path (or dynamic programming) models to generate moves that are optimal over a significant subclass of possibilities. This type of approach gives rise to a combinatorial leverage phenomenon, where a low order effort (e.g., linear or quadratic) can yield solutions that dominate exponential numbers of alternatives. (See, e.g., Glover (1992), Rego and Roucairol (1996).)

Example 3: A broadly applicable control strategy, similar to that of a relaxation procedure, but more flexible, is to create a proxy model that "resembles" the original problem of interest, and which is easier to solve. Such an approach must be accompanied with a method to transform the solution to the proxy model into a trial solution for the original problem. A version of such an approach, which also can be used to induce special structure into the proxy model, can be patterned after layered surrogate/lagrangian decomposition strategies for mixed integer optimization.

Referent-domain optimization can also be applied in conjunction with the *target analysis* learning approach, to provide a basis for creating more effective solution strategies (Glover and Laguna (1992), Glover (1995b), Løkketangen and Glover (1996)). In this case, a first stage learning model, based on controlled solution attempts, identifies a set of desired properties of good solutions, together with *target solutions* (or *target regions*) that embody these properties. Then a second stage model is devised to generate neighborhoods and choice rules to take advantage of the outcomes of the learning model. Useful strategic possibilities are created by basing these two models on a proxy model for referent-domain optimization, to structure the outcomes so that they may be treated by one of the control methods indicated in the foregoing examples.

REFERENCES

[1] Aarts, E.H.L. and J.H.M. Korst (1989). "Boltzmann Machines for Travelling Salesman Problems," *European Journal of Operational Research*, 39,79-95.

[2] Ackley, D. (1987). *A Connectionist Model for Genetic Hillclimbing*, Kluwer, Dordrecht. Academic Publishers.

[3] Aggarwal, C.C., J.B. Orlin, and R.P. Tai (1996). "Optimized Crossover for the Independent Set Problem," MIT Technical Report, to appear in *Operations Research*.

[4] Aiex, R.M., S.L. Martins, C.C. Ribeiro and N.R. Rodriguez (1996). "Asynchronous parallel strategies for tabu search applied to the partitioning of VLSI circuits," Research report, Department of Computer Science, Catholic University of Rio de Janeiro.

[5] Al-Mahmeed, A.S. (1996). "Tabu Search Combination and Integration," In *Meta- Heurisitcs: Theory and Applications*. pp. 319-330, I.H. Osman & J.P. Kelly (eds.), Kluwer Academic Publishers.

[6] Andreatta, A.A. and C.C. Ribeiro (1994). "A Graph Partitioning Heuristic for the Parallel Pseudo-exhaustive Logical Test of VLSI Combinational Circuits," *Annals of Operations Research* 50, 1-36.

[7] Bäck, T., F. Hoffmeister and H. Schwefel (1991). "A Survey of Evolution Strategies," In *Proceedings of the Fourth International Conference on Genetic Algorithms*, eds. R. Belew and L. Booker, pp. 2-9. Morgan Kaufmann, San Mateo, CA.

[8] Balas, E., and W. Niehaus (1996). "Optimized Crossover-Based Genetic Algorithms for the Maximum Cardinality and Maximum Weight Clique Problems," Management Science Research Report #MSRR-613, submitted to *Journal of Heuristics.*

[9] Barnes, J. Wesley (1993). "Solving the Multiple-Machine Weighted Flow Time Problem Using Tabu Search," *IIE Transactions*, Vol 25, Number 2.

[10] Barnes, J.W. and W.B. Carlton (1995). "Solving the Vehicle Routing Problem with Time Windows Using Reactive Tabu Search," presented at the Fall INFORMS Conference in New Orleans, Louisiana, Tuesday, October 31, 1995, in review by the *IIE Transactions.*

[11] Battiti, R., P. Lee, A. Sartori, and G. Tecchiolli (1994a). "Totem: A Digital Processor for Neural Networks and Reactive Tabu Search," In *Fourth International Conference on Microelectronics for Neural Networks and Fuzzy Systems*, MICRONEURO 94, pages 17-25.

[12] Battiti, R. P. Lee, A. Sartori, and G. Tecchiolli (1994b). "Combinatorial Optimization for Neural Nets: RTS Algorithm and Silicon," Technical Report no. 9406-04, IRST, Trento, IT.

[13] Battiti, R. and G.Tecchiolli (1992). "Parallel Based Search for Combinatorial Optimization: Genetic Algorithms and Tabu Search," *Microprocessor and Microsystems*, 16:351-367.

[14] Battiti, R. and G. Tecchiolli (1994b). "The Reactive Tabu Search," *ORSA Journal on Computing*, 6(2):126-140.

[15] Battiti, R. and G. Tecchiolli (1994c). "Simulated Annealing and Tabu Search in the Long Run: A Comparison on QAP Tasks," *Computer and Mathematics with Applications*, 28(6):1-8.

[16] Battiti, R. and G. Tecchiolli (1995a). "The Continuous Reactive Tabu Search: Blending Combinatorial Optimization and Stochastic Search for Global Optimization," *Annals of Operations Research* (in press).

[17] Battiti, R. and G. Tecchiolli (1995b). "Training Neural Nets with the Reactive Tabu Search," *IEEE Transactions on Neural Networks*, 6(5):1185-1200

[18] Battiti, R. and G. Tecchiolli (1995c). "Local Search with Memory: Benchmarking RTS," *Operations Research Spektrum*, 17(2/3):67-86.

[19] Beltran, H.F. and D. Skorin-Kapov (1994). "On Minimum Cost Isolated Failure Immune Networks," *Telecommunication Systems*, 3, 183-200.

[20] Błăzewicz, J., P. Hawryluk and R. Walkowiak (1993). "Using a Tabu Search Approach for Solving the Two-dimensional Irregular Cutting Problem," *Annals of Operations Research*, 41, 313-325.

[21] Błăzewicz, J. and R. Walkowiak (1995). "A Local Search Approach for Two-dimensional Irregular Cutting," *OR Spectrum*, 17, 93-98.

[22] Brumelle, S., D. Granot, M. Halme, and I. Vertinsky (1996). "A Tabu Search Algorithm for Finding a Good Forest Harvest Schedule Satisfying Green-up Constraints," Forest Economics and Policy Analysis Research Unit The University of British Columbia Vancouver, B.C., Canada

[23] Carlton, William B. and J. Wesley Barnes (1995). "A Note on Hashing Functions and Tabu Search Algorithms," To appear in the *European Journal of Operational Research*.

[24] Carlton, William B. and J. Wesley Barnes (1996). "Solving the Traveling Salesman Problem with Time Windows Using Tabu Search," to appear in *IIE Transactions*.

[25] Chakrapani, J. and J. Skorin-Kapov (1992). "Connectionist Approaches to the Quadratic Assignment Problem," *Computers and Operations Research*, 19(3/4), 287-295.

[26] Chakrapani, J. and J. Skorin-Kapov (1993). "Connection Machine Implementation of a Tabu Search Algorithm for the Traveling Salesman Problem," *Journal of Computing and Information Technology*, 1(1), 29-36.

[27] Chakrapani, J. and J. Skorin-Kapov (1993a). "Massively Parallel Tabu Search for the Quadratic Assignment Problem," *Annals of Operations Research*, 41, 327-341.

[28] Chakrapani, J. and J. Skorin-Kapov (1995). *Mapping Tasks to Processors to Minimize Communication Time in a Multiprocessor System, The Impact of Emerging Technologies of Computer Science and Operations Research*, Kluwer Academic Publishers, 45-64.

[29] Chiang, W.-C. and P. Kouvelis (1994a). "Simulated Annealing and Tabu Search Approaches for Unidirectional Flowpath Design for Automated Guided Vehicle Systems," *Annals of Operation Research*, Vol. 50.

[30] Chiang, W.-C. and P. Kouvelis (1994b). "An Improved Tabu Search Heuristic for Solving Facility Layout Design Problems," *International Journal of Production Research* (forthcoming).

[31] Chiang, W.-C. and R. Russell (1995). "A Reactive Tabu Search Metahuristic for the Vehicle Routing Problem with Time Windows," submitted to *ORSA Journal on Computing*, Working Paper, Department of Quantitative Methods and Management Information Systems.

[32] Costamagna, F., A. Fanni and G. Giacinto (1995). "Tabu Search for the Optimization of B-ISDN Telecommunication Networks," Tech. Report, Dept. of Electrical Eng., University of Cagliari, no. 60.

[33] Crainic, T.G., M. Toulouse and M. Gendreau (1993a). "A Study of Synchronous Parallelization Strategies for Tabu Search," Publication 934, Centre de recherche sur les transports, Universite de Montreal, 1993.

[34] Crainic, T.G., M. Toulouse, and M. Gendreau (1993b). "Appraisal of Asynchronous Parallelization Approaches for Tabu Search Algorithms," Publication 935, Center de recherche sur les transports, Universite de Montreal, 1993.

[35] Crowston, W.B., F. Glover, G.L. Thompson and J.D. Trawick (1963). "Probabilistic and Parametrick Learning Combinations of Local Job Shop Scheduling Rules," *ONR Research Memorandum No. 117*, GSIA, Carnegie-Mellon University, Pittsburg, PA.

[36] Daduna, J.R., and S. Voss (1995). "Practical Experiences in Schedule Synchronization," *Lecture Notes in Economics and Mathematical Systems*, 430, 39-55.

[37] Dammeyer, F., and S. Voss (1993). "Dynamic Tabu List Management Using the Reverse Elimination Method," *Annals of Operations Research*, 41, 31-46.

[38] Davis, L. (1989). "Adapting Operator Probabilities in Genetic Algorithms," In *Proceedings of the Third International Conference on Genetic Algorithms*, pp. 61- 69, Morgan Kaufmann, San Mateo, CA.

[39] Dell'Amico, M. and M. Trubian (1993). "Applying Tabu Search to the Job-Shop Scheduling Problem," *Annals of Operations Research*, Vol. 41, 231-252.

[40] Dell'Amico, M. and Francesco Maffioli (1996). "A New Tabu Search Approach to the 0-1 Equicut Problem," In *Meta-Heuristics: Theory & Applications*, pp. 361-378, I.H. Osman & J.P. Kelly (eds.), Kluwer Academic Publishers.

[41] Dodin, B., A.A. Elimam and E. Rolland (1996). "Tabu Search in Audit Scheduling," submitted to the special issue of *European Journal of Operation Research* on Tabu Search.

[42] Domschke, D., P. Forst and S. Voss (1992). "Tabu Search Techniques for the Quadratic Semi-assignment Problem," In: G. Fandel, T. Gulledge and A. Jones (eds.), *New Directions for Operations Research in Manufacturing*, Springer, Berlin, 389 - 405.

[43] Dorndorf, U. and E. Pesch (1994). "Fast Clustering Algorithms," *ORSA Journal on Computing*, 6, 141-153.

[44] Eiben, A.E., P.-E. Raue and Zs. Ruttkay (1994). "Genetic Algorithms with Multi-Parent Recombination," In Y. Davidor, H.-P Schwefel, and R. Manner (Eds.) *Proceedings of the Third International Conference on Parallel Problem Solving from Nature (PPSN)* (pp. 78-87). New York: Springer-Verlag.

[45] Eschelman, L.J. and J.D. Schaffer (1992). "Real-Coded Genetic Algorithms and Interval-Schemata," Technical Report, Phillips Laboratories.

[46] Fanni, A., G. Giacinto and M. Marchesi (1996). "Tabu Search for Continuous Optimization of Electromagnetic Structures," submitted to *Int. Workshop on Optimization and Inverse Problems in Electromagnetism*, Brno, Czech Republic, June.

[47] Fleurent, C., F. Glover, P. Michelon and Z. Valli (1996). "A Scatter Search Approach for Unconstrained Continuous Optimization," *Proceedings of the 1996 IEEE International Conference on Evolutionary Computation*, pp. 643-648.

[48] Fréville, A. and G. Plateau (1986). "Heuristics and Reduction Methods for Multiple Constraint 0-1 Linear Programming Problems," *European Journal of Operational Research*, 24, 206-215.

[49] Fréville, A. and G. Plateau (1993). "An Exact Search for the Solution of the Surrogate Dual of the 0-1 Bidimensional Knapsack Problem," *European Journal of Operational Research*, 68, 413-421.

[50] Glover, F. (1965). "A Multiphase-dual Algorithm for the Zero-one Integer Programming Problem," *Operations Research*, 13, 879-919.

[51] Glover, F. (1968). "Surrogate Constraints," *Operations Research*, 16, 741-749.

[52] Glover, F. (1975). "Surrogate Constraint Duality in Mathematical Programming," *Operations Research*, 23, 434-451.

[53] Glover, F. (1977). "Heuristics for Integer Programming Using Surrogate Constraints," *Decision Sciences*, 8, 156-166.

[54] Glover, F. (1989). "Tabu Search, Part I," *ORSA Journal on Computing*, 1, 190-206.

[55] Glover, F. (1992). "Ejection Chains, Reference Structures and Alternating Path Methods for Traveling Salesman Problems," University of Colorado. Shortened version published in *Discrete Applied Mathematics*, 1996, 65, 223-253.

[56] Glover, F. (1994a). "Tabu Search for Nonlinear and Parametric Optimization (with Links to Genetic Algorithms)," *Discrete Applied Mathematics*, 49, 231-255.

[57] Glover, F. (1994b). "Genetic Algorithms and Scatter Search: Unsuspected Potentials," *Statistics and Computing*, 4, 131-140.

[58] Glover, F. (1995a). "Scatter Search and Star-Paths: Beyond the Genetic Metaphor," *OR Spektrum*, 17:125-137.

[59] Glover, F. (1995b). "Tabu Search Fundamentals and Uses." Graduate School of Business, University of Colorado, condensed version published in *Mathematical Programming: State of the Art*, 1994, Birge & Murty, eds., 64-92.

[60] Glover, F. (1995c) "Tabu Thresholding: Improved Search by Nonmonotonic Trajectories," *ORSA Journal on Computing*, Vol. 7, No. 4, 426-442.

[61] Glover, F. and G. Kochenberger (1996). "Critical Event Tabu Search for Multidimensional Knapsack Problems," In *Meta-Heuristics: Theory & Applications*, I.H. Osman and J.P. Kelly (eds.), 407-427, Kluwer Academic Publishers, Norwell, MA.

[62] Glover, F. and M. Laguna (1993). "Tabu Search," In *Modern Heuristic Techniques for Combinatorial Problems*. ed. C. Reeves, pp. 70-141. Blackwell Scientific Publishing, Oxford.

[63] Glover, F. and C. McMillan (1986). "The General Employee Scheduling Problem: An Integration of Management Science and Artificial Intelligence," *Computers and Operations Research*, 15:5, 563-593.

[64] Greenberg, H.J. and W.P. Pierskalla (1970). "Surrogate Mathematical Programs," *Operations Research*, 18, 924-939.

[65] Greenberg, H.J. and W.P. Pierskalla (1973). "Quasi-conjugate Functions and Surrogate Duality," *Cahiers du Centre d'Etudes de Recherche Operationelle*, 15, 437-448.

[66] Hansen, P., B. Jaumard and Da Silva (1992). "Average Linkage Divisive Hierarchical Clustering," to appear in *Journal of Classification.*

[67] Hansen, P., B. Jaumard and M. Poggi di Aragao (1992). "Mixed Integer Column Generation Algorithms and the Probabilistic Maximum Satisfiability Problem," *Proceedings of the 2nd Integer Programming and Combinatorial Optimization Conference*, Carnegie Mellon.

[68] Hertz, A., B. Jaumard, and M. Poggi di Aragao (1991). "Topology of Local Optima for the K-Coloring Problem," to appear in *Discrete Applied Mathematics.*

[69] Jaumard, B., P. Hansen, and M. Poggi di Aragao (1991). "Column Generation Methods of Probabilistic Logic," *ORSA Journal on Computing*, 3, 135-148.

[70] Holland, J.H. (1975). *Adaptation in Natural and Artificial Systems.* University of Michigan Press, Ann Arbor, MI.

[71] Karwan, M.H. and R.L. Rardin (1976). "Surrogate Dual Multiplier Search Procedures in Integer Programming," School of Industrial Systems Engineering, *Report Series No. J-77-13*, Georgia Institute of Technology.

[72] Karwan, M.H. and R.L. Rardin (1979). "Some Relationships Between Lagrangean and Surrogate Duality in Integer Programming," *Mathematical Programming*, 17, 230-334.

[73] Kelly, J.P., M. Laguna and F. Glover (1996). "Tabu Search and Scatter Search for Optimizing Simulation," School of Business, University of Colorado, Presented at the national INFORMS Meeting, Washington, D.C.).

[74] Kelly, J.P. and J. Xu (1995). "Tabu Search and Vocabulary Building for Routing Problems," Graduate School of Business Administration, University of Colorado at Boulder.

[75] Kincaid, R.K., (1995). "Actuator Placement for Active Sound and Vibration Control of Cylinders," *NASA ASEE Report.*

[76] Kincaid, R.K. and R.T. Berger (1993). "The Damper Placement Problem on Space Truss Structures," *Location Science*, Vol. 1, No. 3, pp. 219-234.

[77] Kincaid, R.K., A.D. Martin, and J.A. Hinkley (1995). "Heuristic Search for the Polymer Straightening Problem," *Computational Polymer Science*, Vol. 5, pp. 1-5.

[78] Laguna, M., J.P. Kelly, J.L. Gonzalez-Velarde and F. Glover (1995). "Tabu Search for the Multilevel Generalized Assignment Problem," *European Journal of Operations Research*, 82, 176-189.

[79] Laguna M., R. Marti and V. Valls (1995). "Arc Crossing Minimization in Hierarchical Digraphs with Tabu Search," Tech Report, Graduate School of Business, University of Colorado, Campus Box 419, Boulder, CO 80309.

[80] Lee, P., A. Sartori, G. Tecchiolli and A. Zorat (1995). "A Parallel Processor for Neural Networks," 1995 Symposium on VLSI Circuits - Digest of Technical Papers, 8-10 June, 1995, Kyoto, *IEEE CAT. No. 95 CH 35780*, p. 81-82.

[81] Løkketangen, A. and F. Glover (1996). "Surrogate Constraint Methods with Simple Learning for Satisfiability Problems," *Proceedings of the DIMACS workshop on Satisfiability Problems: Theory and Applications*, D-Z. Du, J. Gu and P. Pardalos, editors.

[82] Løkketangen, A. and D. L. Woddruff, (1996). "Progressive Hedging and Tabu Search Applied to Mixed Integer (0,1) Multi-Stage Stochastic Programming," Forthcoming in *Journal of Heuristics*.

[83] Lopez, L., M.W. Carter and M. Gendreau (1996). "The Hot Strip Mill Production Scheduling Problem: A Tabu Search Approach," Centre de recherche sur les transports, Universite de Montreal.

[84] Marti, R. (1996). "An Aggressive Search Procedure for the Bipartite Drawing Problem," In *Meta-Heuristics: Theory & Applications*, 97-113, I.H. Osman and J.P. Kelly (eds.), Kluwer Academic Publishers.

[85] Martins, S.L., C.C. Ribeiro and N.R. Rodriguez (1996). "Parallel Programming Tools for Distributed Memory Environments," (in Portuguese), *Research report MCC-01/96*, Department of Computer Science, Catholic University of Rio de Janeiro.

[86] Mazzola, J.B. and R.H. Schantz (1995a). "Single-facility Resource Allocation Under Capacity-based Economies and Diseconomies of Scope," *Management Science*, 41, 669-689.

[87] Mazzola, J.B. and R.H. Schantz (1995b). "Multiple-Facility Loading Under Capacity-Based Economies of Scope," Working Paper, Fuqua School of Business, Duke University, August.

[88] Mazzola, J.B., A.W. Neebe, and C.M. Rump (1995). "Multiproduct Production Planning in the Presence of Work-Force Learning," Working Paper, Fuqua School of Business, Duke University, August.

[89] Michalewicz, Z. and C. Janikow (1991). "Genetic Algorithms for Numerical Optimization," *Statistics and Computing*, 1, 75-91.

[90] Mühlenbein, H., M. Gorges-Schleuter and O. Kremer (1988). "Evolution Algorithms in Combinatorial Optimization," *Parallel Computing*, 7, 65-88.

[91] Mühlenbein, H. and D. Schlierkamp-Voosen (1994). "The Science of Breeding and its Application to the Breeder Genetic Algorithm," *Evolutionary Computation* 1, 335-360.

[92] Mühlenbein, H. and H.-M. Voigt (1996). "Gene Pool Recombination in Genetic Algorithms," In *Meta-Heurisitics: Theory & Applications*, I.H. Osman and J.P. Kelly, eds., 53-62, Kluwer Academic Publishers.

[93] Mulvey, J. (1995). "Generating Scenarios for the Towers Perrin Investment Systems," Technical Report SOR Report, Princeton University, Department of Civil Engineering and Operations Research, Princeton University, Princeton, NJ, to appear *Interfaces*.

[94] Nowicki E. and Smutnicki C., (1993). "A Fast Taboo Search Algorithm for the Job Shop," Report ICT PRE 8/93, Technical University of Wroclaw. *Management Science*. (in print), 1996.

[95] Nowicki, E. and C. Smutnicki (1994). "A Fast Taboo Search Algorithm for the Flow Shop," Report ICT PRE 8/94, Technical University of Wroclaw. *European J. Oper. Res.* (in print), 1996.

[96] Nowicki, E. and C. Smutnicki (1995). "The Flow Shop with Parallel Machines: A Tabu Search Approach," Report ICT PRE 30/95, Technical University of Wroclaw.

[97] Pardalos, P.M., X. Liu and G. Xue (1995). "Protein Conformation of Lattice Using Tabu Search," Unpublished.

[98] Pesch, E. and F. Glover (1995). "TSP Ejection Chains," Graduate School of Business, University of Colorado, Boulder, to appear in *Discrete Applied Mathematics*.

[99] Porto, S.C.S. and C.C. Ribeiro (1995a). "A Tabu Search Approach to Task Scheduling on Heterogeneous Processors Under Precedence Constraints," *International Journal of High-Speed Computing* 7, 45-71.

[100] Porto, S.C.S. and C.C. Ribeiro (1995b). "Parallel Tabu Search Message-passing Synchronous Strategies for Task Scheduling Under Precedence Constraints," to appear in *Journal of Heuristics*.

[101] Potvin, J. Y., T. Kervahut, B. L. Garcia and J. M. Rousseau (1996). "The Vehicle Routing Problem with Time Windows - Part I: Tabu Search," *INFORMS Journal on Computing* Vol 8, No. 2, 158-164.

[102] Rego, C. (1996a). "Relaxed Tours and Path Ejections for the Traveling Salesman Problems," Universidade Portucalense, Porto, Portugal.

[103] Rego, C. (1996b). "A Subpath Ejection Method for the Vehicle Routing Problem," Universidade Portucalense, Porto, Portugal.

[104] Rego, C. and C. Roucairol (1996). "A Parallel Tabu Search Algorithm Using Ejection Chains for the Vehicle Routing Problem," In *Meta-Heuristics: Theory & Applications*, 661-675, I.H. Osman and J.P. Kelly, (eds.), Kluwer Academic Publishers.

[105] Rochat, Y. and F. Semet (1994). "A Tabu Search Approach for Delivering Pet Food and Flour in Switzerland," *J. Opl. Res. Soc.*, Vol 45, No. 11 1233-1246.

[106] Rochat, Y. and E. Taillard (1995). "Probabilistic Diversification and Intensification in Local Search for Vehicle Routing," *Journal of Heuristics*, Vol. 1, No. 1, 147-167.

[107] Rolland, E. (1995). "A Tabu Search Method for Constrained Real-Number Search: Applications to Portfolio Selection," Working paper, The A. Gary Anderson Graduate School of Management, University of California, Riverside.

[108] Rolland, E. and H. Johnson (1996). "Skewness and the Mean-Variance Frontier: A Tabu Search Approach," Working paper, The A. Gary Anderson Graduate School of Management, University of California, Riverside.

[109] Rolland, E., H. Pirkul and F. Glover (1995). "Tabu Search for Graph Partitioning," in *Annals of Operations Research*, Special issue on Metaheuristics in Optimization.

[110] Rolland, E., D.A. Schilling and J.R. Current (1996). "The P-Median Problem: Efficient Solutions using Tabu Search," to appear in *European Journal of Operational Research.*

[111] Skorin-Kapov, J. (1990). "Tabu Search Applied to the Quadratic Assignment Problem," *ORSA Journal on Computing*, Vol.2, No.1, 33-45.

[112] Skorin-Kapov, J. (1994). "Extensions of a Tabu Search Adaptation to the Quadratic Assignment Problem," *Computers and Operations Research*, 21(8), 855-865.

[113] Skorin-Kapov, J. and J.-F. Labourdette (1995). "On Minimum Congestion in Logically Rearrangeable Multihop Lightwave Networks," forthcoming in *Journal of Heuristics.*

[114] Skorin-Kapov, D., J. Skorin-Kapov and M. O'Kelly (1995). "Tight Linear Programming Relaxations of p-Hub Median problems," (forthcoming in) *European Journal of Operational Research.*

[115] Skorin-Kapov, D. and J. Skorin-Kapov (1994). "On Tabu Search for the Location of Interacting Hub Facilities," *European Journal on Operational Research*, 73(3), 501-508.

[116] Skorin-Kapov, J. and A. Vakharia (1993). "Scheduling a Flow-Line Manufacturing Cell: A Tabu Search Approach," *International Journal of Production Research*, 31(7), 1721-1734.

[117] Spears, W.M. and K.A. DeJong (1991). "On the Virtues of Uniform Crossover," In *4th International Conference on Genetic Algorithms*, La Jolla, CA.

[118] Sun, M., J.E. Aronson, P.G. McKeown and D. Drinka (1995). "A Tabu Search Heuristic Procedure for the Fixed Charge Transportation Problem," Working Paper No. 95-414, Terry College of Business, The University of Georgia, May.

[119] Taillard, E., P. Badeau, M. Gendreau, F. Guertin and J-Y. Potvin (1995). "A New Neighborhood Structure for the Vehicle Routing Problem with Time Windows," Centre de Recherche sur les Transports, Publication CRT-95-66.

[120] Taillard, E. (1993). "Parallel Iterative Search Methods for Vehicle Routing Problems," *Networks*, Vol. 23, 661-673.

[121] Trafalis, T., and I. Al-Harkan (1995). "A Continuous Scatter Search Approach for Global Optimization," Extended Abstract in: *Conference in Applied Mathematical Programming and Modeling (APMOD'95)*, London, UK, 1995.

[122] Toulouse, Michael, Teodor G. Crainic and Michel Gendreau (1996). "Communication Issues in Designing Cooperative Multi-Thread Parallel Searchers" In *Meta- Heuristics: Theory & Applications*, pp. 503-522, I.H. Osman and J.P. Kelly (eds.), Kluwer Academic Publishers.

[123] Ulder, N.L.J., E. Pech, P.J.M. van Laarhoven, H.J. Bandelt and E.H.L. Aarts (1991). "Genetic Local Search Algorithm for the Traveling Salesman Problem," In *Parallel Problem Solving from Nature*, eds. R. Maenner and H.P. Schwefel, pp. 109-116. Springer-Verlag, Berlin.

[124] Vaessens, R.J.M, E.H.L. Aarts and J.K. Lenstra (1995). "Job Shop Scheduling by Local Search," *Memorandum COSOR 94-05*, Eidhoven University of Technology.

[125] Valls V., R. Marti and P. Lino (1995). "A Tabu Thresholding Algorithm for Arc Crossing Minimization in Bipartite Graphs," *Annals of Operations Research*, Vol. 60, *Metaheuristics in Combinatorial Optimization*.

[126] Voss, S. (1992). "Network Design Formulations in Schedule Synchronization," *Lecture Notes in Economics and Mathematical Systems*, 386, 137-152.

[127] Voss, S. (1993). "Tabu Search: Applications and Prospects," In: D.-Z. Du und P.M. Pardalos (Hrsg.), *Network Optimization Problems: Algorithms, Applications and Complexity*, World Scientific, Singapore, 333 - 353.

[128] Voss, S. (1994). "Concepts for Parallel Tabu Search," *Technische Hochschule Dormstadt*, Germany.

[129] Whitley, D., V.S. Gordon and K. Mathias (1994). "Lamarckian Evolution, the Baldwin Effect and Function Optimization," in *Proceedings of the Parallel Problem Solving from Nature*, 3, pp. 6-15. New York: Springer-Verlag.

[130] Whitley, D. (1993). *Foundations of Genetic Algorithms 2*, Morgan Kaufmann.

[131] Woodruff, D. (1996). "Proposals for Chunking and Tabu Search," Submitted to *European Journal of Operational Research*.

[132] Wright, A.H. (1990). "Genetic Algorithms for Real Parameter Optimization," In *Foundations of Genetic Algorithms*, ed. G. Rawlins, pp. 205-218. Morgan Kaufmann, Los Altos, CA.

[133] Zenios, S. (1996). "Dynamic Financial Modeling and Optimizing the Design of Financial Products," Presented at the National INFORMS Meeting, Washington, D.C., May 1996.

2

NEURAL NETWORKS IN PRACTICE: SURVEY RESULTS

Bruce L. Golden
Edward A. Wasil*
Steven P. Coy
Cihan H. Dagli**

College of Business and Management
University of Maryland
College Park, Maryland 20742

**Kogod College of Business Administration*
American University
Washington, D.C. 20016

***Engineering Management Department*
University of Missouri-Rolla
Rolla, Missouri 65401

Neural networks will soon lose their "sex appeal."
Neural networks will rapidly "infiltrate" everyday life.
Alianna J. Maren (1990)

ABSTRACT

In early 1995, we prepared a questionnaire to determine how companies are using neural networks to solve their problems. We mailed this questionnaire to approximately 1050 individuals. Some of these were academics, but most worked in industry. The questions focus on these key issues: (1) neural network environment,

(2) neural network results, and (3) recent project success. We received 131 responses. In this paper, we report on our findings.

1 INTRODUCTION

An artificial neural network (or neural network for short) can be thought of as a computer program that is designed to recognize patterns and learn "like" the human brain. The name neural network is based on the (loose) analogy to neurons in the human brain. The field of neural networks (also known as neural computing) involves pattern recognition, learning, and computing. It has been around since the 1940's, but it languished in the 1970's, and then re-emerged as an exciting research area in the mid 1980's, attracting a wide variety of engineers and scientists (especially physicists). See Maren et al.[3] for a comprehensive review of the field.

Neural networks have been successfully used to perform prediction, classification, and optimization and have been embraced by numerous operations researchers in recent years (for example, see Sharda[5], Potvin[4], and Ignizio and Burke[1]).

Given the glamour associated with neural networks, we were interested in learning more about their practical impact. In early 1995, we prepared a two-page (three section) questionnaire to determine how companies are using neural networks to solve their problems. We mailed this questionnaire to approximately 1050 individuals. We tried to select individuals with direct responsibility for modeling real-world problems with neural networks.

In particular, our mailing list was compiled from the INNS (International Neural Network Society) membership list and the ANNIE (Artificial Neural Networks in Engineering) conference mailing list. It includes company employees, consultants, and academics, however, it is biased in favor of engineers and against academics; we wanted our list to consist of people who build neural network models with real-world (rather than artificial) data. Also, the list contains a mix of foreign and domestic addresses. We received 131 responses (124 of the 131 are actively engaged in neural networks practice and research).

We point out that an earlier questionnaire on neural networks was designed and analyzed by Lin[2], while at NASA. Lin sent out approximately 100 survey forms and received 28 responses, 11 of which came from NASA researchers. The purpose of his survey was to compare the modeling techniques used by researchers at NASA with the modeling techniques used by researchers who were not employed by NASA.

2. NEURAL NETWORK ENVIRONMENT

The first six questions in our survey pertain to the neural network environment found in each respondent's company. We designed these questions to gain a general understanding of how neural network modeling fits into a respondent's activities. In particular, we asked each respondent to determine the frequency of use, the number of modeling successes, the type of models used, and the type of software and hardware platforms used in constructing neural network models.

2.1 Number and General Purpose of Problems Modeled

We found that most respondents made frequent use of neural network models. On average, respondents modeled nearly five problems from 1994 to early 1995. The minimum number of problems modeled was one and the maximum number of problems modeled was 50. On average, respondents viewed about three of the five models as successful. See Table 1 for a summary of these results.

Respondents used neural networks for a wide variety of applications. We found that nearly two-thirds of respondents (63.7%) used neural networks for more than one type of application. Over 78% of respondents indicated that they used neural networks for classification. Of these, 20.1% used neural networks exclusively for classification. More respondents used neural networks for optimization than for prediction, which we found surprising. In terms of exclusive use, however, more respondents (11.3%) used neural networks exclusively for prediction than used neural networks for optimization (2.4%). Table 2 summarizes these results. Control and function approximation applications were not listed on the questionnaire; however, we added these two applications to Table 2 because they received multiple responses.

	Average	Standard deviation
Number of problems modeled	4.91	7.06
Number of successful models	3.42	3.38
Percentage of successful models	69.7	

Table 1. Number of problems respondents modeled using neural networks in 1994-1995.

Neural network applications	Percentage response	Percentage of respondents who modeled this type of application exclusively
Classification	78.2	20.1
Optimization	30.6	2.4
Prediction	21.4	11.3
Other*	15.3	0.0
Control	5.6	0.0
Function approximation	1.6	1.6

*Other includes segmentation, duplication of simulation functionality, missing data replacement, indexing, spatio-temporal pulse transform, signal linearization, model inversion, symbolic and statistical data integration, decision analysis, tracking, and image processing.

Table 2. Purpose of neural network modeling effort (multiple responses recorded).

2.2 Type of Neural Network

Next, we asked each respondent to indicate the type of neural network model used. Given the results summarized in Table 2, we expected that the majority used the multilayer perceptron (historically, this type of network has been used for classification and prediction). In fact, almost 85% of respondents reported that they

Neural networks	*Percentage response*
Multilayer perceptron	84.7
Other*	23.4
Kohonen	22.6
Hopfield	11.3
Adaptive resonance network	4.8
Counterbackpropagation	4.8
Bidirectional associative memory	3.2
Learning vector quantization	2.4
Fuzzy	1.6
Probabilistic	1.6
Radial basis function	1.6
Cascade correlation	1.6

*Other includes quantum neurodynamics, general regression, generalized factorial, cellular recurrent, genetically engineered, stochastic filter, parametric avalanche, nested classifier, Eckhorn linking field, Potts glass self-organizing system, clustering via integer subtraction, elastic, functional link, reinforcement learning, and adaptive critic.

Table 3. Type of neural network used (multiple responses recorded).

use this network architecture (see Table 3). The Other category was the second most frequent response (23.4%). There was a wide variety of responses in this category. These responses included applications such as quantum neurodynamics, cellular recurrent, genetically engineered, and nested classifier networks. The next most popular network architecture (22.6%) was the Kohonen self-organizing network. Learning vector quantization, probabilistic, radial basis function, and cascade correlation networks were not included on the questionnaire, but were written in by respondents. We included them in Table 3 because these four networks received multiple responses.

2.3 Hardware and Software Choices

Computer platforms used	*Percentage response*
Personal computer	66.9
Workstation	63.7
Mainframe	6.5
Neural net specific	6.5
Mini computer	5.7

Table 4. Computer hardware used (multiple responses recorded).

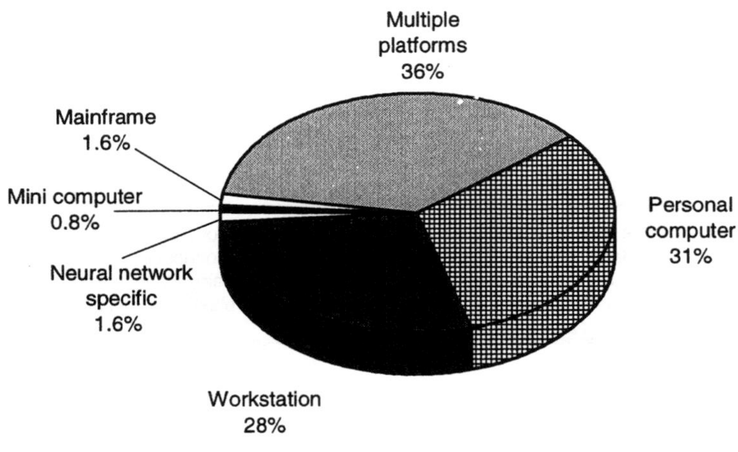

Figure 1. Computer hardware choices

In Table 4, we summarize results for the type of computer hardware that was used by respondents. Two-thirds of respondents (66.9%) used a personal computer in their modeling effort. Workstations were also very popular — 63.7% of respondents indicated that they used workstations. Few respondents (12.2%) used mini computers or mainframe computers in their work. A small number (6.5%) of respondents indicated that they used hardware dedicated to neural network computation. This hardware included special chips, boards, and dedicated parallel processing machines

Figure 1 shows the choice of computer hardware from the perspective of exclusive use. In particular, more respondents used multiple platforms (more than one choice of computer hardware) than used any single platform exclusively. We can see that the majority of those respondents who used only one platform performed their work on personal computers or workstations. Only 2.4% used a mainframe or mini computer exclusively.

We found that a respondent was more likely to develop neural network software in-house than to use a commercial neural network package (this includes public domain software). More than half (57.3%) of respondents indicated that they used software developed in-house exclusively, 23.4% used one or more commercial products, and 19.3% used both commercial products and software developed in house. Table 5 summarizes these results.

Of the 53 respondents who used commercially available software, 51.1% used NeuralWorks (see Table 6). The next most popular package (20.4%) was the MATLAB Fuzzy Logic Toolbox. The remaining 15 packages were far less popular (each was used by fewer than 7% of respondents).

	Percentage response
Respondent exclusively used software developed in-house	57.3
Respondent exclusively used commercial software	23.4
Respondent used both commercial software and software developed in-house	19.3

Table 5. Software used in the modeling effort

Software packages	Percentage response*
NeuralWorks Professional 2+	51.1
MATLAB Fuzzy Logic Toolbox	20.4
BrainMaker	6.1
SNNS (Stuttgart Neural Network Simulator)	6.1
NeuroShell	6.1
Braincel	4.1
Neuralyst	4.1
ANSLM	2.0
DynaMind	2.0
Meta Neural	2.0
PDP (Parallel Distributed Processing)	2.0
Process Insights	2.0
Aspirin/MIGRAINES	2.0
ModelWare	2.0
Michizane	2.0
CNAPS Development Kit	2.0
Win-NN	2.0

*53 respondents used commercial or public domain software. Percentage response is a proportion of these 53 respondents.

Table 6. Software packages used (multiple responses recorded).

Of the 53 respondents who used commercially available software, 51.1% used NeuralWorks (see Table 6). The next most popular package (20.4%) was the MATLAB Fuzzy Logic Toolbox. The remaining 15 packages were far less popular (each was used by fewer than 7% of respondents).

2.4 Total Quality Management

Two of the authors (Golden and Wasil) have worked with Westinghouse Electric Corporation over the past four years on a variety of process manufacturing activities

Questions	*Yes*	*No*	*Percent Yes*
Did your company have a TQM program in place at the time of the application?	41	80	33.8
If yes, did the neural network modeling effort contribute to that program?	11	30	26.8

Table 7. TQM and the modeling effort.

(for details, see Sun et al.[6]). Essentially, they tried to apply neural network technology to monitor, control, and continuously improve the quality of wire bonds in microcircuits (microcircuits are the building blocks of virtually all electronic systems). Their efforts fit well within Westinghouse's variability reduction and total quality management (TQM) programs. Based on their experience, we conjectured that other neural network modeling efforts, especially in engineering, might also contribute to a company-wide TQM program.

We found that one-third of the respondents' companies had a TQM program in place and, for those that did, roughly one out of four (26.8%) neural network projects contributed to that program. See Table 7 for these results. It may be that larger companies are more likely to have TQM programs in operation; however, we were unable to confirm this from the survey results.

3. NEURAL NETWORK RESULTS

Respondents were asked to circle one of five responses (from strongly agree to strongly disagree) to seven statements about various aspects of the modeling effort, neural network performance, and supporting environment. The results are summarized in Table 8.

The first two statements consider the data used in the modeling effort. In the first statement, we wanted to determine whether respondents believed that input variable instability (a given input varies as a function of time) was a significant problem in their work. Most respondents (64.6%) believed that the values of the variables were reasonably stable over time, while few (15.6%) believed that variable instability was a problem. The second statement considered the data collection phase. Over half (56.1%) of the respondents agreed or strongly agreed that data collection was the most expensive part of the modeling effort. Fewer than one-quarter (22.8%) of the respondents disagreed or strongly disagreed.

Statements	Percentage response				
	Strongly agree	*Agree*	*Neither agree nor disagree*	*Disagree*	*Strongly disagree*
1. The values of the variables were reasonably stable over time.	19.8	44.8	19.8	13.8	1.8
2. Data collection was the most expensive part of the modeling effort.	28.5	27.6	21.1	21.1	1.7
3. The neural network modeling effort had strong corporate support.	13.9	31.1	35.2	14.8	5.0
4. The results produced by the neural network model were better than the status quo.	43.0	40.5	13.2	2.5	0.8
5. The neural network modeling effort was viewed as a success by management.	28.5	47.2	22.8	1.5	0.0
6. The future looks bright for using neural networks to model company problems.	30.9	39.8	20.3	4.9	4.1
7. The results of the neural network model compared favorably to the results of a statistical method or an optimization technique.	39.8	43.7	9.7	3.9	2.9

Table 8. Percentage breakdown of responses to neural network results statements.

The third statement considered the level of corporate support for the modeling effort. Less than half (45%) of the respondents believed that the neural network modeling effort had strong corporate support. Over one-third (35.2%) of the respondents

neither agreed nor disagreed with the statement, while about one-fifth (19.8%) of the respondents disagreed or strongly disagreed.

The fourth statement considered neural network performance when compared to the performance of more traditional methods at the respondent's company. Most respondents (83.5%) believed that the results produced by the neural network model were better than the status quo. Only 3.3% of respondents believed that the neural network results were worse.

Most respondents (75.7%) believed that management viewed the modeling effort as a success, while only 1.5% of respondents felt that management did not consider the modeling effort a success. Furthermore, the majority (70.7%) of respondents believed that the future looks bright for using neural networks to model company problems. Few (9%) disagreed or strongly disagreed.

Most respondents (83.5%) agreed that the results of the neural network model compared favorably to the results of a statistical method or an optimization technique. Only 6.8% believed that the neural network results were not favorable.

4. NEURAL NETWORK SUCCESS STORY

We asked each respondent to briefly describe the most recent company problem modeled with a neural network that was viewed as a success. We supplied a broad definition of success. To be considered a success, either the model performed well during training or testing or the result of the modeling effort was a great benefit to the company.

4.1 Type of Application

Table 9 summarizes the types of applications respondents viewed as successes, the average time to develop a successful neural network model, and the percentage of successful models still in use. To aid in our analysis, we grouped similar successful applications into categories. For example, materials science includes applications such as modeling the behavior of wet cement as it turns in a drum. Image and pattern recognition contains all applications in which the network must interpret or classify an image. Examples in this category include optical character recognition, fingerprint classification, and lip reading. Process and quality control refers to applications where the neural network detects out-of-control conditions in the manufacturing process or in the quality of the products produced. Remote sensing and analysis and remote sensing and control differ in the manner with which the network uses the remote data. In the former case, the remote data are analyzed or classified. The interpretation and classification of satellite data is a good example. Remote sensing and control refers to applications where data are analyzed in real time and commands are issued based on the analysis. Robotic control is a good example from this category.

In Table 9, the two categories labeled Classification and Prediction contain applications that clearly have some classifying or predicting element. The applications that we placed into these two categories do not fit well into any of the other categories, such as Biomedical. For example, we placed the forecasting of football spreads into the Prediction category. The Other category is a catch-all for 10 applications that do not lend themselves to direct categorization in our list.

The average development time for applications ranged from three months for optimization applications to slightly more than 26 months for military applications. The average development time for all successful applications was over 11 months. Since many of the application categories have small counts (for example, the Optimization category only had two responses), we urge caution when interpreting the average development time for an application.

Successful applications	*Frequency*	*Percentage response*	*Average development time (months)*	*Percentage still in use*
Acoustic signature classification	4	3.9	13.50	50.0
Biomedical	10	9.6	10.10	80.0
Chaos modeling	3	2.9	8.33	33.3
Classification	8	7.7	7.22	75.0
Geological	4	3.8	13.25	100.0
Materials science	7	6.7	8.00	85.7
Military	4	3.8	26.25	75.0
Noise filtering	2	1.9	12.00	50.0
Optimization	2	1.9	3.00	50.0
Other*	10	9.6	10.70	90.0
Image and pattern recognition	17	16.4	13.74	70.6
Prediction	10	9.6	6.90	70.0
Process and quality control	10	9.6	13.70	100.0
Remote sensing and analysis	5	4.9	14.44	100.0
Remote sensing and control	6	5.8	11.83	33.3
Time series	2	1.9	4.50	100.0
Average			11.35	76.0

*Other includes transformation of an image to a time signal, bathymetric surface data inversion, strategy engine for game development, verification of the accuracy of flight instrumentation test data, visual speech training aid, cascade correlation of potable water demand, modeling vehicle comfort, reinforcement learning for nuclear power ascent, real estate price estimation, and population modeling.

Table 9. Respondents' most successful application.

Most (an average of 76%) of the successful neural network models are still in use. In four categories (see the right-most column of Table 9) all of the models are still in use. This represents 20.2% (21 of 104) of the successful applications. In five categories, no more than half of the models are still in use. This represents 16.3% (17 of 104) of the successful applications.

4.2 Obstacles Overcome and Benefits Received from Successful Applications

In the final section of the questionnaire, we asked each respondent to comment on the major obstacles encountered during the development of a recent successful application and on the benefits realized from this application.

4.2.1 Greatest Obstacles

Respondents listed several obstacles encountered during model development. Difficulties in the data collection phase were cited most frequently. Obstacles related to data collection included gaining access to classified (military) data, data entry, and sparse data. Once the data were collected, new problems such as feature selection (choice of appropriate input variables) and model development were reported.

Respondents listed several problems related to training. These problems included too much data, noisy data, training takes too long, and early convergence to local minima while training. Many respondents cited management resistance, including difficulties in acquiring adequate funding for the project, as their greatest obstacle. Educating skeptical customers and operators (operators are individuals who use the model; for example, an operator in a quality control application might be the quality control inspector or machine operator) was difficult for many respondents. Implementation and maintenance of the final model were also problems for some respondents.

4.2.2 Major Benefits

Respondents reported a wide variety of benefits from their successful applications. Among the benefits, improved accuracy over past methods was cited most often. Respondents also reported that their successful neural network applications improved productivity, found a good solution faster than other methods, saved lives (this application was a radar tracking system), aided in planning time reduction (this application optimized the placement of electronic components on a circuit board), allowed for real-time classification of data, and increased profits. Many respondents gave examples of improvements over traditional modeling methods. One respondent reported that the neural network model was a better on-line predictor of quality than other methods. Another reported that the neural network required less development time than traditional methods, while another respondent reported that the neural network was more robust than competing statistical methods.

5. NEURAL NETWORKS IN PRACTICE: SIX EXAMPLES

In this section, we present six interesting examples of neural networks in practice. The first two examples deal with corporate attitudes toward modeling with neural networks. The next two examples suggest that modeling with neural networks is more of an art than a science. The final two examples deal with divergent views about the usefulness of neural network models in the same application area. In each example, we briefly sketch the neural network environment, describe the nature of the successful application, and summarize the comments offered by the respondent.

5.1 Corporate Attitude towards Neural Networks

The two examples in this section discuss the corporate culture surrounding neural network applications and management's acceptance of the final results of a neural network modeling effort. Respondents claimed that company colleagues viewed neural network models as mysterious black boxes. Upper-level managers did not fully believe that the final models worked as intended and, because of this, they were reluctant to accept the final results. Several respondents suggested that a name change (from neural network modeling to empirical modeling or nonlinear statistical modeling) might be helpful since the name neural networks turned off company executives and colleagues.

5.1.1 Example 1

Neural Network Environment. From 1994 to early 1995, a computer scientist developed nine neural network models for prediction, classification, and optimization problems. He used multilayer perceptron and Kohonen networks and developed the networks with in-house software. Of the nine models, the computer scientist viewed eight models as successes.

Successful Application. The most recent successful application involved an image and pattern recognition problem. The computer scientist along with a colleague developed a neural network model of this problem in two weeks. Their model outperformed a statistical model that was developed over the course of two years by another company team (we refer to it as the competing team). The neural network model was 10 times faster than the competing team's model and it was more accurate on test data.

Comments. Company managers were pleased with the neural network results but were unwilling to redirect money that they had already committed to the competing team and its model. Managers killed the neural network modeling effort because they were skeptical of the results (How could one obtain such good results in such a short development time?) and they hated to acknowledge the waste of two years of effort expended by the competing team.

5.1.2 Example 2

Neural Network Environment. From 1994 to early 1995, a geoscientist developed seven neural network models for prediction and classification problems. He used multilayer perceptron, Kohonen, cascade correlation, and genetically engineered networks. He generated the neural network models with in-house developed software and the NeuralWorks software package. Of the seven models, the geoscientist viewed five models as successes.

Successful Application. The most recent successful application was in reservoir management and involved the prediction of the permeability of core samples. This prediction problem is extremely difficult to solve using conventional methods. The neural network models outperformed a wide variety of statistical models including linear, polynomial, and multiple regression models.

Comments. The geoscientist reported that the neural network models in this application area were quite successful and resulted in four patents being awarded for this work. Despite this success, the geoscientist noted that he awaits a company-wide commitment to neural network technology in order to move forward.

5.2 The Art of Neural Network Modeling

The two examples in this section demonstrate that modeling problems with neural networks is still an art form. The two respondents modeled similar problems but their results and their overall reactions to neural networks were quite dissimilar.

5.2.1 Example 3

Neural Network Environment. From 1994 to 1995, a group of analysts developed over 50 neural network models for prediction, classification, and optimization problems. The analysts used recurrent neural networks, fuzzy neural networks, and generalized functional networks and developed the networks using commercial software and in-house software. Of the 50 models, the respondent viewed two models as successes.

Successful Application. The most recent successful application involved an acoustic signature classification problem. The respondent reported that this neural network model was carefully constructed. However, three other methods outperformed this neural network model along three dimensions — time, cost, and accuracy.

Comments. After developing 50 neural network models, the respondent reported that learning the limitations of these models was the greatest benefit. Consequently, the respondent would use advanced statistical techniques to model new problems, rather than develop neural network models. In fact, the respondent would "look elsewhere" if a company employee or consultant relied on a neural network as the primary modeling approach.

5.2.2 Example 4

Neural Network Environment. The chief executive officer (CEO) of a company that relies exclusively on neural network models reported that the company developed four models in the past year for prediction, classification, and optimization problems. The company used two types of network architectures — multilayer perceptron and parametric stochastic filter — and developed the computer software in-house. Of the four models, the respondent viewed three models as successes.

Successful Application. The most recent successful application used a parametric stochastic filter to model an acoustic signature classification problem. The CEO reported that the trained network achieved a 96% success rate.

Comments. The CEO reported three major obstacles to success: (1) timely data collection, (2) preprocessing of data, and (3) external data analysis. The CEO pointed out that the company has done neural network modeling since 1986 and stated that: "Neural networks is what we do."

5.3 Usefulness of Neural Network Models

Respondents had widely different opinions about the appropriate use of neural network models. Several respondents pointed out that they rarely used neural network models, since traditional statistical models were deemed more appropriate for the application areas under study. Many respondents reported that neural networks were the most useful modeling approaches among all of the statistical and mathematical programming approaches at their disposal. The two examples in this section describe widely divergent views about the usefulness of neural networks in modeling problems found in the insurance industry.

5.3.1 Example 5

Neural Network Environment. A systems analyst for an insurance company reported that the company had used neural networks to model prediction and classification problems over three years ago. The company used the multilayer perceptron architecture and developed the networks with in-house software. The analyst reported that he had not used neural networks in the past three years.

Successful Application. There were no recent applications of neural networks.

Comments. The respondent reported that the last application of a neural network model (an application that involved a credit problem) met with resistance from the home office. The home office did not cooperate in supplying data sets and eventually declared that a rule-based system (not the neural network model) successfully modeled the credit problem under study. Other than the credit problem, the respondent did not think that there were problems at the company (for example, actuarial problems) that lent themselves to being modeled by neural networks.

5.3.2 Example 6

Neural Network Environment. In the past year, a researcher for an insurance company developed seven neural network models for prediction, classification, and decision analysis problems. He used multilayer perceptron, recurrent, Hopfield, and Kohonen networks and developed the networks with in-house software and commercial software packages including NeuralWorks, NeuroShell, BrainMaker, and MATLAB. The respondent viewed all seven neural network models as successes.

Successful Application. The most recent successful application involved classifying underwriting decisions. The neural network results were compared to internal underwriting criteria as part of an audit process. The respondent reported that data acquisition was the greatest obstacle and that the neural network model helped to streamline and enhance the underwriting audit process.

Comments. The respondent pointed out that his insurance company has a strong commitment to modeling problems with cutting-edge approaches such as neural networks.

6. CONCLUSIONS

In this paper, we have reported the results of a survey of neural network practitioners and researchers. The survey was sent to approximately 1050 individuals drawn from a wide variety of fields. We received 131 responses and 124 of these responses were usable (seven questionnaires were received from individuals who were not active in neural network practice or research).

In compiling the survey results, we saw a wide variety of comments that ranged from high praise

> "[Neural networks] do a great job when compared [to other statistical and mathematical] approaches"

to complete disdain

> "Neural networks are a much over-blown technique of statistical analysis and modeling."

However, most comments, fell into the middle ground (somewhat closer to the sentiment of the first quotation) as the respondents enjoyed a moderate degree of success when they applied neural network models to their problems.

The questionnaire asked each respondent to answer questions in three areas: (1) neural network environment, (2) neural network results, and (3) most recent neural network success. Based on the responses to these questions, we observe the following:

Neural network environment

- Survey respondents used neural networks fairly frequently with a high rate of success.
- Classification was the most frequent problem type modeled by respondents.
- Personal computers and workstations and in-house-developed software were the most popular hardware and software choices among respondents.

Neural network results

- For the most part, neural network models outperformed traditional methods.
- While many managers were unsupportive during model development, most responded favorably to the results of the modeling effort.

Most recent neural network success

- Data collection was the major obstacle in developing a successful model.
- Improved accuracy was the major benefit of a successful model.

References

[1] J. Ignizio, and L. Burke, eds., "Neural Networks and Operations Research," *Computers & Operations Research,* 19 (3/4), 175-311 (1992).

[2] C. S. Lin, "An Overview on Development of Neural Network Technology," Space Technology Initiatives Office, NASA Langley Research Center, Hampton, VA (1993).

[3] A. Maren, C. Harston, and R. Pap, *Handbook of Neural Computing Applications*, Academic Press, San Diego, CA (1990).

[4] J-Y. Potvin, "The Traveling Salesman Problem: A Neural Network Perspective," *ORSA Journal on Computing,* 5 (4), 328-348 (1993).

[5] R. Sharda, "Neural Networks for the MS/OR Analyst: An Application Bibiliography," *Interfaces*, 24 (2), 116-130 (1994).

[6] X. Sun, B. Golden, O. Kwon, and E. Wasil, "Enhancing the Performance of Neural Network Models for the Wire Bonding Process," in *Intelligent Engineering Systems Through Artificial Neural Networks: Volume 4*, (C. Dagli, B. Fernandez, J. Ghosh, and R. Kumara, eds.), ASME Press, 1041-1047 (1994).

3

TRACTABLE THEORIES FOR THE SYNTHESIS OF NEURAL NETWORKS

V Chandru, M Vidyasagar*, V Vinay**

CS & Automation,
Indian Institute of Science
Bangalore–560 012, INDIA
email:chandru@csa.iisc.ernet.in

**Centre for AI and Robotics,*
Raj Bhavan Circle
Bangalore–560 001, INDIA
email:sagar@cair.ernet.in

***CS & Automation,*
Indian Institute of Science
Bangalore–560 012, INDIA
email:vinay@csa.iisc.ernet.in

ABSTRACT

The Radius of Direct attraction of a discrete neural network is a measure of stability of the network. It is known that Hopfield networks designed using Hebb's Rule have a radius of direct attraction of $\Omega(\frac{n}{p})$ where n is the size of the input patterns and p is the number of them. This lower bound is tight if p is no larger than 4. We construct a family of such networks with radius of direct attraction $\Omega(\frac{n}{\sqrt{p \log p}})$, for any $p \geq 5$. The techniques used to prove the result led us to *the first polynomial-time algorithm for designing a neural network with maximum radius of direct attraction* around arbitrary input patterns. The optimal synaptic matrix is computed using the ellipsoid method of linear programming in conjunction with an efficient separation oracle. Restrictions of symmetry and non-negative diagonal entries in the synaptic matrix can be accommodated within this scheme.

We also present a *stability theory for generalized analog neural networks* with energy functions that are multilinear polynomials. The main conclusions are that (i) the network is totally stable, and (ii) "almost all" trajectories of the network converge to a local minimum of the energy function. This is the largest class of functions for which sharp convergence properties are known.

1 INTRODUCTION

A recursive neural network is a discrete time, discrete valued dynamical system with a bipolar ($\{+1, -1\}$) state vector. The evolution of these networks is governed by a matrix of weights W called the synaptic matrix. These networks are of interest because of their use as associative memory in classification and pattern recognition. Given a specified set of binary vectors $\{y^1, y^2, \cdots, y^p\}$ the network behaves as an associative memory if it eventually reproduces as its output one of the inputs, say y^i, when triggered at the input by a vector z which is sufficiently close to y^i. Here "sufficiently close" means that the Hamming distance between z and the fixed point y^i is no more than the radius of the basin of attraction around y^i. The dynamical behaviour of recursive neural networks makes it difficult to analyse the radius of attraction around a given pattern. A more manageable measure of stability of the network is the *radius of direct attraction,* i.e., the radius of the sphere centered at each of the y^i such that any vector in the sphere converges to y^i in one step. In addition, the radius of direct attraction is invariant with the operation mode (synchronous, asynchronous or block-sequential) of the dynamical system. It is also desirable from a practical perspective to have a large radius of direct attraction since this would imply that the network has fewer candidates as spurious attractors.

A fundamental problem in recursive neural networks is the choice of the synaptic matrix W so as to impose the prescribed $\{y^1, y^2, \cdots, y^p\}$ as fixed points of the network with suitably large basins of attraction about them. The Hebb's rule ($W = \sum_{k=1}^{p}(y^k)(y^k)^t$) is a traditional technique, inspired by neurophysiological considerations, for choosing the weight matrix when the input patterns are mutually orthogonal (uncorrelated). It is well known that any such network has a radius of direct attraction of at least $\frac{n}{2p}$. It is not to difficult to argue that this lower bound is realized in all instances provided p is small (2,3 and 4). We show that this is not the case for higher values of p. We construct a family of networks under the Hebb's rule that has a radius of attraction of $\Omega(\frac{n}{\sqrt{p \log p}})$.

As we have noted above, Hebb's rule leads to network designs with substantive radius of direct attraction when the input patterns are uncorrelated. When the patterns are correlated, they may not even be fixed points of the network under

Hebb's rule. So there is a necessity for an algorithm that produces a weight matrix with large radius of direct attraction for arbitrary input patterns. In this paper, we present the first polynomial-time algorithm to synthesize a recursive neural network with maximum radius of direct attraction.

(MRDA) *Given vectors $y^1, \ldots, y^p$, where each y^i is an n - dimensional vector whose components are $\{+1, -1\}$, construct a weight matrix, W, such that the resulting network has maximum radius of direct attraction about the y^i.*

It is well known that a neural network whose weight matrix is symmetric and has a non-negative diagonal is free of cycles. Our algorithm can be easily extended to solve, in polynomial time, (MRDA) with these added restrictions. These results give us the necessary tools to design large neural networks that are stable.

Our approach to the design problem is based on linear programming and in particular on the ellipsoid method for linear programming. That network design is amenable to linear programming methodology has been noted before in the literature [3, 9, 5]. However, previous attacks have suffered from two shortcomings. The first is that in the linear programming model the number of linear constraints is explosive; explicit representations are unmanageable. Therefore, even testing the feasibility of a weight matrix W requires an oracle for checking the implicit inequalities. Efficient oracles for this purpose have hitherto not been realized.

In addition to dealing with the implicit specification of the linear constraints there is also the issue of speed of convergence of iterative schemes for computing the weights in the synaptic matrix W. Simple rules such as Hebb's Rule and the Projection Rule attempt to arrive at W with closed form computations and are woefully incomplete methods for normative design in the case of general input patterns. Other approaches based on so-called Perceptron Methods suffer from problems of oscillation. First order relaxation methods for linear inequalities such as the Ho-Kashyap [13] algorithm have been tried on network design but these methods have linear rates of convergence and require exponential time to guarantee convergence to an adequate W matrix.

We overcome both difficulties detailed above by designing a very efficient oracle for the linear programming representation of (MRDA). In addition, we use the ellipsoid method which by virtue of being a higher order relaxation method converges in polynomial time.

The input patterns of a discrete neural network may in fact be mapped to local minima of a suitably defined energy function. Bruck and Goodman [2] suggest that discrete neural networks may be used to find locally optimally solutions to problems such as MAX CUT. Schaffer and Yannakakis [11] prove that finding an attractor of a given network is $\mathcal{PLS}$-complete. Their results show the futility of modelling $\mathcal{PLS}$-complete problems in the discrete neural network framework. Schaffer and Yannakakis [11] also devised reductions that show that standard local search algorithms take an exponential time on problems such as MAX CUT. Local search corresponds to updating a single neuron in a neural network (asynchronous updating mode).

In an attempt to speed up the operation of the network, Hopfield [7] considers *analog* neural networks, where the constituent neurons have *graded* responses. Such networks are described by

$$C_i \dot{u}_i = -\frac{1}{R_i} u_i + \sum_{j=1}^{n} w_{ij} v_j + \theta_i, \ v_i = g_i(\lambda u_i), \ i = 1, \ldots, n, \tag{1.1}$$

where n is the number of neurons; v_i is the neural current and u_i is the neural voltage; θ_i is the external current input to the i-th neuron; C_i is the membrane capacitance and R_i is the neural resistance; λ is a scaling parameter, and w_{ij} is the weight of the interconnection from neuron j to neuron i. The function $g_i : \Re \rightarrow (0, 1)$ is continuously differentiable, strictly increasing, $g_i(x) \rightarrow 1$ as $x \rightarrow \infty$, and $g_i(x) \rightarrow 0$ as $x \rightarrow -\infty$. Note that, as the scaling parameter $\lambda \rightarrow \infty$, the function $x \mapsto g_i(\lambda x)$ approaches the "saturation" function $\mathrm{sgn}(x)$, which equals 0 if $x < 0$ and equals 1 if $x > 0$. In [7], Hopfield assumes that $w_{ij} = w_{ji}$ for all i, j, and that $w_{ii} = 0$ for all i. He then proposes the energy function

$$E_c(\mathbf{v}) = \sum_{i=1}^{n} \left[\frac{1}{\lambda R_i} \int_0^{v_i} g_i^{-1}(v)\, dv - \theta_i v_i - \frac{1}{2} \sum_{j=1}^{n} w_{ij} v_i v_j \right], \tag{1.2}$$

and shows that

$$\frac{d}{dt} E_c[\mathbf{v}(t)] \leq 0 \tag{1.3}$$

along trajectories of the system (1.1). Hirsch [8] shows, using the relationship (1.3), that the system (1.1) is *totally stable*; that is, every solution trajectory approaches an equilibrium. Salam *et al.* [12] fix a small technical flaw in the reasoning of [8] by establishing that no solution trajectory of (1.1) escapes to infinity (in the $\mathbf{u}$-space).

Our objective is to generalize the system description (1.1) to the case of an energy function that is not necessarily quadratic. It is shown that one can assume, without loss of generality, that the objective function is a multilinear polynomial. The generalized neural network dynamics are analyzed in this case. We show that (i) the network is totally stable, and (ii) "almost all" trajectories of the network converge to a local minimum of the energy function. Though the generalization of (1.1) to a higher order energy function is rather obvious, the analysis of the resulting network is not so.

As a consequence of this new stability theory, many combinatorial optimization problems, particularly in the class $\mathcal{PLS}$, may be modelled as analog neural networks in a natural way. We illustrate these ideas on the WEIGHTED CNF SAT problem which is known to be $\mathcal{PLS}$-complete [10].

2 PRELIMINARIES

In this section, we provide the necessary background in recursive neural networks. More details may be found in [9]. By a *vector* we usually mean an n − dimensional bipolar vector.

A recursive neural network is a network composed of artificial neurons with interconnections. The interconnections have weights associated with them. We let $W = (w^i_j)$ represent the weight matrix, where w^i_j is the weight associated with the interconnection from neuron j to neuron i. It is convenient to think of W as a set of row vectors w^i.

At any instant, a neuron is in one of two bipolar states, {1, -1}. Consider the state of the ith neuron $(1 \leq i \leq n)$ at time t, s_i. Then the state of the neuron at the next time instant is governed by the equation

$$s_i^{new} = sgn(\sum_{1 \leq j \leq n} w^i_j s_j).$$

Let $\mathbf{s} = (s_1, \ldots, s_n)^t$ denote the state vector of the neural network at some time instant. Then the updating rule may be defined as

$$\mathbf{s}^{new} = sgn(W\mathbf{s}).$$

We will use $y^1, ..., y^p$ to denote the n-dimensional input vectors which require to be stored as *fixed points* of the neural network. z is a fixed point of a neural network defined by weight matrix W if $z = sgn(Wz)$.

We will also need the notion of a *potential of a vector* z, $\phi(z)$. It is a column vector defined by the equation,

$$\phi(z) = Wz.$$

3 A FAMILY WITH LARGE RADIUS OF DIRECT ATTRACTION

In this section we will prove our result by explicitly constructing a set of Hopfield networks that have large radius of attraction. The family of networks have a very simple structure. Our bounds on the constructed family of networks are sharp. We however establish only a lower bound on the radius of direct attraction as it suffices to prove our result. We need the following lemma, which will also be used critically in our main algorithm.

Lemma 1 *Let* z *be any bipolar vector and* y *be one among the input vectors,* $\{y^1, \cdots, y^p\}$. *Further, let* $A = \{j : z_j \neq y_j\}$. *Then the* i^{th} *component of the vector* $\phi(z)$ *is,*

$$\phi_i(z) = w^i.z = \phi_i(y) - 2\sum_{j \in A} w^i_j * y_j.$$

Proof: On the components indexed by set A, y and z have opposite signs. Therefore one may write

$$\phi_i(z) = w^i.z = \sum_{j \notin A} w^i * y_j - \sum_{j \in A} w^i_j * y_j$$

and the lemma follows. ∎

Notice that $\phi_i(z)$ and $\phi_i(y)$ have the same sign provided $2\sum_{j \in A} w^i_j * y_j$ is "small". It easy to see that the cardinality of the set A is precisely the Hamming distance between the vectors z and y. Also, when W is computed using the Hebb's rule, $|w^i_j|$ is bounded by p. Hence,

Corollary 2 *Let* y *be an input vector. If* z *is within a distance of* $min_i(\frac{\phi_i(y)}{2*p})$ *from* y, z *is in the radius of direct attraction of* y. *In particular, if the input vectors are mutually orthogonal, then any vector within Hamming distance* $\frac{n}{2p}$ *is in the radius of direct attraction of* y.

Proof: The result follows from the previous lemma and the discussion above. ∎

We will show that the $\frac{n}{2p}$ bound is not tight in the sense that there is a family of Networks $\{N_p\}$ (under the Hebbian rule) that have large radius of direct attraction.

Consider the 2^{p-1} possible combinations of strings of length $p-1$ on $\{+1,-1\}$. Regard them as $p-1$ bipolar vectors with dimension 2^{p-1} by considering the ith entry (or bit) of each of the string as a vector. To this set, add the vector (with 2^{p-1} components) of all 1's. This completes the description of the set N_p.

For example, for $p = 3$, we have the three patterns $(1,1,1,1)^T, (1,1,-1,-1)^T$ and $(1,-1,1,-1)^T$.

It is easy to check that the set of vectors in N_p are mutually orthogonal. Let W_p be the associated weight matrix. We will agree to use W to simplify notation. We need the following simple lemma.

Lemma 3 *Any two rows of W are permutations of each other.*

It is therefore adequate to analyse, say, the first row of W. Without loss of generality fix the vector of all 1's to be y in Lemma 1. In the light of Lemma 1, we need to prove that the minimum number of components in w^1 that need to be chosen such that their sum exceeds $n/2$, is large. Fixing $n = 2^{p-1}$, a simple combinatorial argument (on the number of elements with a particular weight) suffices to show that we need

$$\min_q \{\sum_{k=0}^{q} \binom{p-1}{k}\} \text{ subject to } \sum_{k=0}^{q} \binom{p-1}{k} (p-2k) \geq 2^{p-2}.$$

We prove

Theorem 4

$$\min_q \{\sum_{k=0}^{q} \binom{p-1}{k}\} \geq \frac{2^{p-2}}{6\sqrt{p \log p}} \text{for } \sum_{k=0}^{q} \binom{p-1}{k} (p-2k) \geq 2^{p-2}.$$

To prove this result, we need the following lemma.

Lemma 5 *Let* $2l = p - 1$. *Fix* $r = \sqrt{(4l\log p)} + 1$.
Then $\sum_{0 \le k \le \frac{p-1}{2} - r} \binom{p-1}{k}(p-2k) \le 2^{p-3}$.

Proof: For $0 < x < 1$, it is easy to see $\log(\frac{1+x}{1-x}) \ge 2x$. Assume $x = \lambda/l$. Then $\log(\frac{l+\lambda}{l-\lambda}) \ge 2\frac{\lambda}{l}$. Now,

$$\sum_{\lambda=1}^{r} \log(\frac{l+\lambda}{l-\lambda}) \ge \frac{2}{l}\sum_{\lambda=1}^{r} \lambda \ge 4\log p.$$

Taking exponentials on both sides of the above equation, and by definition of $\binom{n}{k}$, one can infer

$$\frac{\binom{2l}{l}}{\binom{2l}{l-r}} \ge p^4.$$

Using the trivial upper bound $\binom{2l}{l} \le 2^{p-1}$, we get

$$\binom{2l}{l-r} \le \frac{2^{p-2}}{p^4}.$$

Therefore,

$$\sum_{0 \le k \le \frac{p-1}{2} - r} \binom{p-1}{k}(p-2k)$$

$$\le p * \frac{p-1}{2}\binom{2l}{l-r} \le 2^{p-3} \text{for} p \ge 2$$

We proved the lemma assuming that p is odd. It is easy to extend it for even p. ∎

Proof of Theorem 4: Consider

$$\sum_{l-r \le k \le q} \binom{p-1}{k}(p-2k).$$

This may be written as
$\sum_{k\le q}\binom{p-1}{k}(p-2k) - \sum_{k\le l-r}\binom{p-1}{k}(p-2k) \ge 2^{p-2} - 2^{p-3} \ge 2^{p-3}.$
Hence,

$$2^{p-3} \le \sum_{l-r\le k\le q}\binom{p-1}{k}(p-2k) \le (p-2l+2r)\sum_{k\le q}\binom{p-1}{k}$$

$$\le 3\sqrt{p\log p}\sum_{k\le q}\binom{p-1}{k}.$$

It follows that

$$\sum_{k\le q}\binom{p-1}{k} \ge \frac{2^{p-2}}{6\sqrt{p\log p}}.$$

∎

4 THE LINEAR PROGRAMMING FORMULATION

In this section, we describe a linear programming formulation for the maximum radius of direct attraction problem. The formulation is verbose as it can have an exponential number of inequalities in the number of variables.

We will concentrate on solving the following decision problem:
Given a set of input patterns, $\{y^1, \ldots, y^p\}$, is there a weight matrix W for which the radius of direct attraction is at least k? That is,

$$d(y^i, z) \le k \Rightarrow y^i = sgn(Wz) \;\; i = 1, \ldots, p.$$

where $d(y^i, z)$ denotes the Hamming distance between y^i and z.

It is a simple matter to check that if a vector z converges to a pattern y in one step ($y = sgn(Wz)$), then so does every vector on a shortest path from z to y. Another useful observation is that the condition $y = sgn(Wz)$ may equivalently be expressed as $\sum_{1\le j\le n} y_i w_j^i z_j > 0, \;\; 1 \le i \le n$. The latter observation follows from the fact that the product of two numbers with identical sign is always positive.

The above observations suggests that for each pattern y^i, we enumerate all vectors at Hamming distance k from y^i. For every such vector z, we may write n inequalities of the form

$$\sum_{1 \leq j \leq n} y_i w_j^i z_j \geq 1, \quad 1 \leq i \leq n.$$

As the weight matrix is scalable by any positive quantity, we replace the strict inequality as originally described by an inequality whose right-hand side is unity. If the above set of inequalities is feasible, then there exists a weight matrix. Otherwise the radius of direct attraction is strictly smaller than k. The total number of inequalities in the above format is $pn \binom{n}{k}$. Unfortunately, this quantity is exponential in the number of variables (e.g.: let k be $n/4$). The ellipsoid algorithm allows us to get around the somewhat naive formulation given above.

5 SEPARATION AND THE ELLIPSOID METHOD

Let us consider the problem of testing if a polyhedron $Q \in \Re^d$, defined by linear inequalities, is non-empty. For technical reasons let us assume that Q is rational, i.e. all extreme points and rays of Q are rational vectors or equivalently that all inequalities in some description of Q involve only rational coefficients. The ellipsoid method (in contrast with the simplex method and Karmarkar's algorithm) does not require the linear inequalities describing Q to be explicitly specified. It suffices to have an oracle representation of Q. Several different types of oracles can be used in conjunction with the ellipsoid method[4]. We will use the *strong separation oracle* described below.

Oracle: **Strong Separation**(Q,y)

Given a vector $y \in \Re^d$**, decide whether** $y \in Q$**, and if not find a hyperplane that separates** y **from** Q**; more precisely, find a vector** $c \in \Re^d$ **such that** $c^T y < \min\{c^T x \mid x \in Q\}$.

Theorem 6 ([4]) *There exists a polynomial $g(d, \phi)$ such that procedure* `Ellipsoid`*(Q) runs in time bounded by $T\, g(d, \phi)$ where ϕ is an upper bound on* *the size of linear inequalities in some description of Q and T is the maximum time required by the oracle* **Strong Separation**(Q, y) *on inputs y of size at most $g(d, \phi)$.* ∎

As a consequence, if we can construct a polynomial time separation oracle for the linear program formulated in the previous section, we would have a polynomial time algorithm to solve the decision version of (MRDA). We are now ready to describe the separation oracle. The construction of the oracle is based on Lemma 1.

Notes on Lemma 1:

(i) The cardinality of the set A is precisely the Hamming distance between the vectors z and y.

(ii) We require $\phi_i(z)$ to have the same sign as y_i. For this to happen, $\phi_i(z) * y_i = \phi_i(y_i) - 2\sum_{j\in A} w_j^i * y_j * y_i \geq 0$.
We introduce new variables $s_{ij} = 2w_j^i * y_j * y_i$. Clearly, z does not converge to y on component i if $\phi_i(y) * y_i < \sum_{j\in A} s_{ij}$.

These observations can be used to calculate the radius of direct attraction for a given weight matrix W. The obvious strategy is to arrange $s_{ij}, 1 \leq j \leq n$ in descending order and use greedy packing to determine the smallest cardinality set A that causes $\sum_{j\in A} s_{ij} > \phi_i(y) * y_i$.

The strong separation oracle for the decision version of (MRDA) uses the above procedure. Recall that the separation oracle needs to determine whether a given weight matrix W results in a network with radius of direct attraction at least k. If the answer is negative, we need to identify a separating inequality that strictly separates W from the polyhedron described by the linear inequalities of the linear programming formulation.

This completes the description of the oracle. It is easy to see that the procedure **RDA**(W, r) can in fact be implemented in $O(pn^2 \log n)$ time by using a sorting routine at the greedy step. Therefore, the oracle runs in polynomial time. In fact, as sorting and prefix sum are in $\mathcal{NC}^1$, it follows that the separation oracle is an $\mathcal{NC}^1$ oracle.

Procedure: **RDA**(*W*, *r*)
for all patterns $y = y^1, \ldots, y^p$ and a component i, $1 \leq i \leq n$
do
form a set S_i whose elements are $s_{ij}(1 \leq j \leq n)$ with weights $2 * w_j^i * y_j * y_i$.
calculate $t := \phi_i(y) * y_i$.
if $t < 0$ **then** y is itself not a fixed point.
  set $A(y,i) = \emptyset$.
else
  do choose greedily the elements of S_i with decreasing weight
  until their cumulative sum exceeds t.
Let $A(y,i)$ be the subset of elements chosen and $r(y,i)$ be the cardinality of the set.
Return $r - 1$, the radius of direct attraction of W, where $r = \min_{y,i} r(y,i)$.
End

Procedure: **Separation**(*W*, *k*)
call Procedure RDA(W,r);
if $(r \geq k)$ then output ''Feasible''
else $(r < k)$ and there is some $r(y,i) < k$. Then the inequality

$$\left(\sum_{s_{ij} \notin A(y,i)} w_j^i y_j - \sum_{s_{ij} \in A(y,i)} w_j^i y_j\right) * y_i \geq 1$$

is a separating inequality.

We now present the main algorithm. The ellipsoid algorithm, Ellipsoid($y^1, \ldots, y^p, k$) takes the input patterns and a value k as an input parameter. These inputs fix the linear programming formulation whose feasibility is to tested. The procedure either finds a weight matrix W which has a radius of direct attraction of k or declares the problem as infeasible (i.e. radius k is unattainable). Noting that the radius of direct attraction is at most $\frac{n}{2}$, the algorithm does a binary search on the interval $[0, \ldots, \frac{n}{2}]$. The binary search terminates with a weight matrix with the largest radius of direct attraction. Thence we have the following result.

Theorem 7 *For an arbitrary set of input bipolar patterns, the problem of designing a recursive neural network with maximum radius of direct attraction, about the input patterns, is soluble in polynomial time.*

The added restrictions that the weight matrix be symmetric with a non-negative diagonal can be encoded as additional (polynomially many) linear inequalities in our linear programming formulation. The separation oracle can explicitly check these inequalities.

6 STABILITY OF GENERALIZED ANALOG NEURAL NETWORKS

The results in this section are presented over the unit hypercube, $\{0,1\}^n$.

¿From here on, the problem under study is that of minimizing a function $E(\mathbf{x})$ over the *discrete* set $\{0,1\}^n$. Thus the exact expression for $E(\mathbf{x})$ is unimportant, and only the set of values of E at the 2^n points in $\{0,1\}^n$ is relevant. In other words, we are free to replace $E(\mathbf{x})$ by another function $F(\mathbf{x})$, so long as

$$E(\mathbf{x}) = F(\mathbf{x}) \text{ for all } \mathbf{x} \in \{0,1\}^n. \tag{6.4}$$

In Lemma 8 below, the freedom afforded by (6.4) is exploited to show that one can always assume, without loss of generality, that the objective function E is a *multilinear polynomial* in the components $x_1, \ldots, x_n$.

Lemma 8 *Suppose $E : \Re^n \to \Re$ is an arbitrary function. Then there exists a unique multilinear polynomial $F : \Re^n \to \Re$ such that (6.4) holds.*

Proof: Every multilinear polynomial in the variables $x_1, \ldots, x_n$ can be written in the form

$$F(\mathbf{x}) = \sum_{\mathbf{i} \in \{0,1\}^n} c_{\mathbf{i}}\, x_1^{i_1} \ldots x_n^{i_n}, \tag{6.5}$$

where 0^0 is taken as 1. Given the function E, determine the family of 2^n (not necessarily distinct) values $(E(\mathbf{x}), \mathbf{x} \in \{0,1\}^n)$. It is first shown that it is possible to select the coefficients $c_{\mathbf{i}}, \mathbf{i} \in \{0,1\}^n$, such that (6.4) holds. First, let $\mathbf{x} = \mathbf{0}$. Then, from (6.5),

$$F(\mathbf{0}) = c_{\mathbf{0}}. \tag{6.6}$$

Hence, set $c_{\mathbf{0}} = E(\mathbf{0})$. Next, let $\mathbf{x} = \{0 \ldots 0\ 1\ 0 \ldots 0\}$, where the 1 is in the j-th position. Then

$$F(\mathbf{x}) = c_{\mathbf{0}} + c_{\{0\ldots0\ 1\ 0\ldots0\}}. \tag{6.7}$$

Thus set

$$c_{\{0\ldots0\ 1\ 0\ldots0\}} = E(\{0\ldots0\ 1\ 0\ldots0\}) - c_{\mathbf{0}}. \tag{6.8}$$

In this way, the n coefficients $c_{\{1\ 0\ldots0\}}$ through $c_{\{0\ldots0\ 1\}}$ can be determined. Next, choose $\mathbf{x}$ to have exactly two nonzero components, and so on. At the end of the process, we will have chosen a set of coefficients $c_{\mathbf{i}}, \mathbf{i} \in \{0,1\}^n$ such that the function F defined by (6.5) satisfies (6.4). To show that this choice is unique, suppose $\bar{F}$ is another multilinear polynomial tha also satisfies (6.4) (with F replaced by $\bar{F}$ of course). Then $\Delta F(\mathbf{x}) = F(\mathbf{x}) - \bar{F}(\mathbf{x})$ is also a multilinear polynomial, and $\Delta F(\mathbf{x}) = 0$ for all $\mathbf{x} \in \{0,1\}^n$. Now a repetition of the above argument shows that all coefficients of ΔF are zero; that is, $\Delta F(\mathbf{x}) = 0$ for all $\mathbf{x} \in \Re^n$. ∎

For each $\mathbf{x} \in \Re^n$ and each index $i \in \{1, \ldots, n\}$, let $\bar{\mathbf{x}}_i$ denote the $(n-1)$-dimensional vector obtained by *omitting* the ith component of $\mathbf{x}$; that is,

$$\bar{\mathbf{x}}_i = [x_1 \ldots x_{i-1}\ ,\ x_{i+1} \ldots x_n]^t. \tag{6.9}$$

Suppose $E(\mathbf{x})$ is a multilinear polynomial. Then it is easy to see that there exist functions $\bar{E}_i : \Re^{n-1} \to \Re$ and $c_i : \Re^{n-1} \to \Re$ such that

$$E(\mathbf{x}) = x_i\ \bar{E}_i(\bar{\mathbf{x}}_i) + c_i(\bar{\mathbf{x}}_i). \tag{6.10}$$

This property proves useful below.

Next, a useful characterization of the local minima of multilinear polynomials is given.

Definition 2.1 A vector $\mathbf{x} \in \{0,1\}^n$ is said to be a *local minimum* of the objective function E if

$$E(\mathbf{x}) \leq E(\mathbf{y}), \text{ for all } \mathbf{y} \in N(\mathbf{x}), \tag{6.11}$$

where $N(\mathbf{x})$ denotes the set of all vectors in $\{0,1\}^n$ lying at a Hamming distance of one from $\mathbf{x}$. The vector $\mathbf{x}$ is said to be a *strict local minimum* of E if

$$E(\mathbf{x}) < E(\mathbf{y}), \text{ for all } \mathbf{y} \in \{0,1\}^n. \tag{6.12}$$

In what follows, we will discuss convergence of analog neural networks to strict local minima of E. The discussion in section 6.1 indicates how the theory extends to solving for local minima that are not necessarily strict.

Lemma 9 *Suppose E is a multilinear polynomial on $\Re^n$, and that $\mathbf{x} \in \{0,1\}^n$. Then the following statements are equivalent:*

1. *$\mathbf{x}$ is a strict local minimum of E.*
2. *$\mathbf{x}$ satisfies the "parity condition," that is, no component of $\nabla E(\mathbf{x})$ is zero, and*

$$[\nabla E(\mathbf{x})]_i < 0 \text{ if } x_i = 1, \; [\nabla E(\mathbf{x})]_i > 0 \text{ if } x_i = 0. \tag{6.13}$$

Proof: "(2) ⇒ (1)" Suppose Statement 2 is true. Select an arbitrary index $i \in \{1, \ldots, n\}$, and define $\mathbf{y} \in \{0,1\}^n$ by

$$y_i = 1 - x_i, \; y_j = x_j \text{ for all } j \neq i. \tag{6.14}$$

Then $\mathbf{y} \in N(\mathbf{x})$, and $\bar{\mathbf{y}}_i = \bar{\mathbf{x}}_i$, where $\bar{\mathbf{x}}_i$ is defined in (6.9). Hence, from (6.10), it follows that

$$E(\mathbf{y}) = y_i \, \bar{E}(\bar{\mathbf{x}}_i) + c_i(\bar{\mathbf{x}}_i) = (1 - x_i) \, \bar{E}(\bar{\mathbf{x}}_i) + c_i(\bar{\mathbf{x}}_i). \tag{6.15}$$

Therefore

$$E(\mathbf{y}) - E(\mathbf{x}) = (1 - 2x_i) \, \bar{E}(\bar{\mathbf{x}}_i). \tag{6.16}$$

Now observe from (6.10) that

$$\bar{E}(\bar{\mathbf{x}}_i) = \frac{\partial E}{\partial x_i} = [\nabla E(\mathbf{x})]_i. \tag{6.17}$$

Hence, if $\mathbf{x}$ satisfies the parity condition (6.13), it follows from (6.16) that $E(\mathbf{y}) > E(\mathbf{x})$. Since the index i is arbitrary, the conclusion is that $\mathbf{x}$ is a strict local minimum.

"(1) ⇒ (2)" We show instead that if Statement 2 is false, then so is Statement 1. Suppose accordingly that the parity condition is violated. This means that either $[\nabla E(\mathbf{x})]_i = 0$ for some i, or else (6.13) is violated for some i. Suppose the former, and define $\mathbf{y} \in N(\mathbf{x})$ as in (6.14). Then (6.16) shows that $E(\mathbf{y}) = E(\mathbf{x})$, so that $\mathbf{x}$ cannot be a *strict* local minimum of E. In the latter case, define $\mathbf{y}$ as in (6.14); then (6.16) shows that $E(\mathbf{y}) < E(\mathbf{x})$, so that $\mathbf{x}$ is not even a minimum. ∎

Suppose E is a given function on $\Re^n$. We propose an analog neural network described by

$$\dot{u}_i = -\frac{u_i}{\alpha_i} - [\nabla E(\mathbf{x})]_i, \; x_i = g_i(\lambda u_i), \; i = 1, \ldots, n. \tag{6.18}$$

Comparing (6.18) with (1.1), we find that the two "physical" constants C_i and R_i have been combined into a single time constant α_i. Notice that the above formulation is a proper generalization of the quadratic energy function [6] provided $w_{ii} = 0$ for all i, that is, if there are no self-interactions.

Theorem 10 *The neural network (6.18) is totally stable.*

Proof: Define the Lyapunov function

$$E_c(\mathbf{x}) = E(\mathbf{x}) + \sum_{i=1}^{n} \frac{1}{\lambda \alpha_i} \int_0^{x_i} g_i^{-1}(x)\, dx. \tag{6.19}$$

The function $\dot{E}_c(\mathbf{x})$, denoting the derivative of $E_c(\mathbf{x})$ along the solution trajectories of (6.18), is defined in the familiar way, namely (cf. [15] page 151)

$$\dot{E}_c(\mathbf{x}) = [\nabla E_c(\mathbf{x})]^t\, \dot{\mathbf{x}}, \tag{6.20}$$

where $\dot{\mathbf{x}}$ is obtained from (6.18). Differentiating (6.18) gives

$$\dot{x}_i = \frac{\partial g_i(\lambda u_i)}{\partial u_i}\, \lambda \dot{u}_i, \; i = 1, \ldots, n. \tag{6.21}$$

This can be expressed more compactly. Define $J : \Re^n \rightarrow \Re^{n\times n}$ to be a diagonal matrix with

$$J_{ii}(\mathbf{u}) = \frac{\partial g_i(u_i)}{\partial u_i}, \; i = 1, \ldots, n. \tag{6.22}$$

Note that J is just the Jacobian matrix of the map $\mathbf{u} \mapsto G(\mathbf{u}) = [g_1(u_1) \ldots g_n(u_n)]^t$. With this notation, we get

$$\dot{\mathbf{x}} = J(\lambda \mathbf{u})\, \lambda \dot{\mathbf{u}} = -\lambda\, J(\lambda \mathbf{u})\, [A^{-1}\mathbf{u} + \nabla E(\mathbf{x})], \tag{6.23}$$

where $A = \text{Diag}\{\alpha_1, \ldots, \alpha_n\}$. Now, from (6.19),

$$\nabla E_c(\mathbf{x}) = \nabla E(\mathbf{x}) + \lambda^{-1} A^{-1} G^{-1}(\mathbf{x})$$
$$= \nabla E(\mathbf{x}) + A^{-1}\mathbf{u} = -\lambda^{-1}\, J^{-1}(\lambda\mathbf{u})\, \dot{\mathbf{x}}, \tag{6.24}$$

where the last step follows from (6.23). Thus from (6.20),

$$\dot{E}_c(\mathbf{x}) = -\lambda^{-1}\dot{\mathbf{x}}^t\, J^{-1}(\lambda\mathbf{u})\dot{\mathbf{x}}. \tag{6.25}$$

Now observe that $J(\lambda\mathbf{u})$ is diagonal and positive definite. Hence

$$\dot{E}_c(\mathbf{x}) \leq 0 \text{ for all } \mathbf{x},\ \dot{E}_c(\mathbf{x}) < 0 \text{ if } \dot{\mathbf{x}} \neq 0. \tag{6.26}$$

The total stability of the system now follows as in [8, 12]. ∎

Thus far there is nothing surprising about the development, and Theorem 10 is valid for arbitrary objective functions. Next, we adapt the ideas of [14] to establish some useful properties of the system (6.18) in the case where the objective function $E(\mathbf{x})$ is a multilinear polynomial. Of particular interest are the location and the stability status of the equilibria of the network, that is, solutions of the equation $\dot{\mathbf{u}} = \mathbf{0}$. ¿From (6.18), it follows that the equilibria in the $\mathbf{x}$-space are solutions of the equation

$$\nabla E(\mathbf{x}) = \lambda^{-1} A^{-1} G^{-1}(\mathbf{x}), \tag{6.27}$$

and in the $\mathbf{u}$-space they are solutions of the equation

$$\mathbf{u} = A\, \nabla E[G(\lambda\mathbf{u})]. \tag{6.28}$$

As shown in [14], three types of equilibria of the network (in the $\mathbf{x}$-space) can arise in the high gain limit, namely: (i) equilibria that remain in the interior of the unit hypercube $[0,1]^n$, (ii) equilibria that approach a face of $[0,1]^n$, and (iii) equilibria that approach a corner of $[0,1]^n$. A solution of (6.28) with the property that $\mathbf{u}\to\mathbf{0}$ as $\lambda\to\infty$, but $\lambda\mathbf{u}$ approaches a finite vector, is of type (i). In this case, each component of $\mathbf{x}$ remains bounded away from 0 and 1. A solution such that $\mathbf{u}$ approaches a vector whose components are *all* nonzero, and hence each component of $\mathbf{x}$ approaches either 0 or 1, is of type (iii). A solution such that $\mathbf{u}$ approaches a vector, *some* of whose components are nonzero, is of type (ii). In this case, *some* components of $\mathbf{x}$ approach 0 or 1, while the rest remain bounded away from 0 and 1.

To prove the next theorem, we make some *slightly* stronger assumptions about the sigmoidal nonlinearities $g_i(\cdot)$. First, it is assumed that each g_i is *twice* continuously differentiable, not just once. This assumption is needed to apply the converse theorem on exponential stability. An equilibrium x_0 is exponentially stable if there exists constants α, β such that for every $x(0)$ sufficiently close to x_0,

$$||x(t) - x_0|| \leq \alpha \, ||x(0) - x_0|| \, e^{-\beta t} \text{ forall } t \geq 0.$$

(See [15] for definitions and characterizations.) Second, it is assumed that

$$\lim_{x \to \pm\infty} x \, g_i'(x) = 0, \text{ for all } i. \tag{6.29}$$

This is also a very mild assumption. The fact that $g_i(x)$ has a definite limit as $|x| \to \infty$ implies that

$$\liminf_{x \to \pm\infty} x \, g_i'(x) = 0, \tag{6.30}$$

because the function $1/x$ is not integrable over any infinite interval. Hence the only extra feature of (6.29) is that "lim inf" has been replaced by "lim." Both of these assumptions are quite mild, and are satisfied by the commonly used sigmoidal function $g_i(x) = 1/(1 + e^{-x})$.

Now we can state the main convergence results.

Theorem 11 *Consider the system (6.18), and suppose E is a multilinear polynomial. Under these conditions, the following statements are true about the equilibria (in the* **x***-space) of the network as $\lambda \to \infty$:*

1. *Any equilibrium that remains within the interior of $[0, 1]^n$ cannot be exponentially stable.*
2. *Any equilibrium that approaches a face of $[0, 1]^n$ cannot be exponentially stable.*
3. *Any equilibrium that approaches a corner of $[0, 1]^n$ is exponentially stable, and the corresponding corner of $[0, 1]^n$ is a strict local minimum of E. Conversely, every strict local minimum of E over $\{0, 1\}^n$ is a limit of an equilibrium of 6.18 as $\lambda \to \infty$ and this equilbrium is exponentially stable.*

Proof: The proof borrows heavily from the ideas of [14]. Each of the three types of equilibria is treated separately.

To prove Statement 1, suppose that as $\lambda \to \infty$, (6.28) has a solution such that $\mathbf{u} \to \mathbf{0}$ but $\lambda \mathbf{u}$ approaches a finite vector, call it $\mathbf{d}_0$. Then $\mathbf{x}_0 = G(\mathbf{d}_0)$ is a solution of (6.27). To determine the stability status of this equilibrium, we make two transformations of variables. First, instead of working in the $\mathbf{u}$-space, we work in the $\lambda\mathbf{u}$-space. Second, we change the time variable from t to $\tau = t/\lambda$. This reflects the fact that, as $\lambda \rightarrow \infty$, the time scale gets correspondingly foreshortened. Now (6.18) leads to

$$\frac{d}{dt}(\lambda\mathbf{u}) = -A^{-1}(\lambda\mathbf{u}) + \lambda\nabla E(\mathbf{x}),\ \mathbf{x} = G(\lambda\mathbf{u}), \tag{6.31}$$

$$\frac{d}{d\tau}(\lambda\mathbf{u}) = -\lambda^{-1}A^{-1}(\lambda\mathbf{u}) + \nabla E(\mathbf{x}),\ \mathbf{x} = G(\lambda\mathbf{u}). \tag{6.32}$$

By assumption, (6.32) has an equilibrium at $\lambda\mathbf{u} = \mathbf{d}_0$ as $\lambda \rightarrow \infty$. In other words, the vector $\mathbf{d}_0$ satisfies

$$\nabla E[G(\mathbf{d}_0)] = \mathbf{0}. \tag{6.33}$$

To determine the stability status of the equilibrium, we apply the linearization method (cf. [15], Sec. 5.5). Differentiating the right side of (6.32) with respect to $\lambda\mathbf{u}$ and setting $\lambda = \infty$ and $\lambda\mathbf{u} = \mathbf{d}_0$ gives

$$\begin{aligned} M_0 &= \frac{d}{d(\lambda\mathbf{u})}\{-\lambda^{-1}A^{-1}(\lambda\mathbf{u}) + \nabla E[G(\lambda\mathbf{u})]\}_{\lambda=\infty,\ \lambda\mathbf{u}=\mathbf{d}_0} \\ &= \{-\lambda^{-1}A^{-1} + \nabla^2 E[G(\lambda\mathbf{u})]\ J(\lambda\mathbf{u})\}_{\lambda=\infty,\ \lambda\mathbf{u}=\mathbf{d}_0} \\ &= \nabla^2 E[G(\mathbf{d}_0)]\ J(\mathbf{d}_0), \end{aligned} \tag{6.34}$$

where $\nabla^2 E$ is the Hessian matrix of E; that is,

$$[\nabla^2 E(\mathbf{x})]_{ij} = \frac{\partial^2 E}{\partial x_i \partial x_j}. \tag{6.35}$$

Now the multilinearity of E implies that

$$\frac{\partial^2 E}{\partial x_i^2} = 0, \text{ for all } i. \tag{6.36}$$

Thus the diagonal elements of $\nabla^2 E[G(\mathbf{d}_0)]$ are all zero. Since $J(\mathbf{d}_0)$ is a diagonal matrix, it follows that the diagonal elements of M_0 are also all zero. Hence the sum of the eigenvalues of M_0, equal to the sum of the diagonal elements of M_0, is also equal to zero. Hence it is *not* the case that all eigenvalues of M_0 have negative real parts. By [15] (page 246, Theorem (5.8.1)), the equilibrium at $\lambda\mathbf{u} = \mathbf{d}_0$ is exponentially stable if and only if all eigenvalues of M_0 have negative real parts. Hence the equilibrium at $\mathbf{d}_0$ is *not* exponentially stable.

Now consider Statement 2. Suppose there is an equilibrium of (6.18) in a face of $[0,1]^n$. Thus, as $\lambda\to\infty$, some components of $\mathbf{u}$ approach zero, while the rest approach a nonzero value. For notational convenience, suppose the first k components of $\mathbf{u}$ approach zero while the rest approach a nonzero value. This assumption can always be satisfied by renumbering the neurons. Partition $\mathbf{u}\in\Re^n$ as

$$\mathbf{u}=\begin{bmatrix}\mathbf{u}_a\\ \mathbf{u}_b\end{bmatrix},\ \text{where } \mathbf{u}_a\in\Re^k,\ \mathbf{u}_b\in\Re^{n-k}. \tag{6.37}$$

Then, as $\lambda\to\infty$, the assumption is that $\mathbf{u}_a\to\mathbf{0}_k$ but $\lambda\mathbf{u}_a$ approaches a nonzero vector, call it $\mathbf{d}_a$. Now $\mathbf{u}_b$ approaches a vector whose components are *all* nonzero; call it $\mathbf{d}_b$. Define the diagonal matrix

$$\Lambda=\begin{bmatrix}\lambda I_k & \mathbf{0}\\ \mathbf{0} & I_{n-k}\end{bmatrix}, \tag{6.38}$$

and define $\bar{\mathbf{u}}=\Lambda\mathbf{u}$. The dynamics of $\bar{\mathbf{u}}$ are obtained from (6.18), which leads to

$$\frac{d\bar{\mathbf{u}}}{dt}=\Lambda\dot{\mathbf{u}}=-A^{-1}\bar{\mathbf{u}}+\Lambda\,\nabla^2E[G(\lambda\Lambda^{-1}\bar{\mathbf{u}})], \tag{6.39}$$

where we take advantage of the fact that A^{-1} and Λ commute, because they are both diagonal matrices. Now, by assumption, as $\lambda\to\infty$, the system (6.39) has an equilibrium (in the $\bar{\mathbf{u}}$-space) at

$$\mathbf{d}:=\begin{bmatrix}\mathbf{d}_a\\ \mathbf{d}_b\end{bmatrix}, \tag{6.40}$$

which is a finite vector. To determine the stability status of this equilibrium, we linearize the system (6.39) around $\mathbf{d}_0$. Taking the Jacobian of the right side of (6.39) with respect to $\bar{\mathbf{u}}$ gives

$$M_0=-A^{-1}+\Lambda\,\nabla^2E[G(\lambda\Lambda^{-1}\bar{\mathbf{u}})]\;J(\lambda\Lambda^{-1}\bar{\mathbf{u}})\;\lambda\Lambda^{-1}. \tag{6.41}$$

Note that

$$\lambda\Lambda^{-1}=\begin{bmatrix}I_k & \mathbf{0}\\ \mathbf{0} & \lambda I_{n-k}\end{bmatrix}. \tag{6.42}$$

Hence, as $\lambda\to\infty$,

$$\lambda\Lambda^{-1}\bar{\mathbf{u}}\to\begin{bmatrix}\mathbf{d}_a\\ \pm\infty\end{bmatrix}, \tag{6.43}$$

where $\pm\infty$ is a shorthand for a vector such that *all* of its components become unbounded. Now, by (6.29) it follows that

$$g_i'(\lambda u_i)\cdot\lambda\to 0 \text{ as } \lambda\to\infty. \tag{6.44}$$

Hence

$$\lim_{\lambda \to \infty} J(\lambda \Lambda^{-1} \bar{\mathbf{u}}) \cdot \lambda \Lambda^{-1} = \begin{bmatrix} J_a(\mathbf{d}_a) & \mathbf{0} \\ \mathbf{0} & \mathbf{0} \end{bmatrix}, \tag{6.45}$$

where $J_a(\mathbf{d}_a) \in \Re^{k \times k}$ is a diagonal matrix whose ii-th component is $g_i'(d_i)$. Next, define

$$\lim_{\lambda \to \infty} \nabla^2 E[G(\lambda \Lambda^{-1} \bar{\mathbf{u}})] =: F. \tag{6.46}$$

Note that, as $\lambda \to \infty$, some components of the vector $\lambda \Lambda^{-1} \bar{\mathbf{u}}$ become unbounded, as shown in (6.43). However, the limit

$$\lim_{\lambda \to \infty} G(\lambda \Lambda^{-1} \bar{\mathbf{u}}) \tag{6.47}$$

is well-defined, and belongs to $[0,1]^n$. Hence F is also well-defined and finite. Also, by the multilinearity of E, it follows that all diagonal elements of F are zero. Now, partition A^{-1} and F commensurately with (6.45), and substitute into (6.41). This gives

$$\begin{aligned} M_0 \to & - \begin{bmatrix} A_a^{-1} & \mathbf{0} \\ \mathbf{0} & A_b^{-1} \end{bmatrix} + \begin{bmatrix} \lambda I_k & \mathbf{0} \\ \mathbf{0} & I_{n-k} \end{bmatrix} \begin{bmatrix} F_{aa} & F_{ab} \\ F_{ba} & F_{bb} \end{bmatrix} \begin{bmatrix} J_a & \mathbf{0} \\ \mathbf{0} & \mathbf{0} \end{bmatrix} \\ & = \begin{bmatrix} \lambda F_{aa} J_a - A_a^{-1} & \mathbf{0} \\ \lambda F_{ba} J_a & -A_b^{-1} \end{bmatrix}. \end{aligned} \tag{6.48}$$

As $\lambda \to \infty$, the matrix M_0 becomes unbounded. One way to deal with the situation is to change the time variable from t to $\tau = t/\lambda$. This has the effect (as in the proof of Statement 1) of replacing M_0 by

$$M_0/\lambda = \begin{bmatrix} F_{aa} J_a - \lambda^{-1} A_a^{-1} & \mathbf{0} \\ F_{ba} J_a & -\lambda^{-1} A_b^{-1} \end{bmatrix}. \tag{6.49}$$

This matrix is block-triangular. Hence its spectrum (set of eigenvalues) is the union of the spectra of the two diagonal blocks. Now the spectrum of $\lambda^{-1} A_b^{-1}$ is just $\{-1/\lambda \alpha_i, \; i = k+1, \ldots, n\}$; these are all negative numbers. On the other hand,

$$\lim_{\lambda \to \infty} F_{aa} J_a - \lambda^{-1} A_a^{-1} = F_{aa} J_a. \tag{6.50}$$

As noted previously, the diagonal elements of $F_{aa} J_a$ are all zero. Hence it is *not* possible for all the eigenvalues of this matrix to have negative real parts. By [15] (page 246, Theorem (5.8.1)), this implies that the equilibrium at $\mathbf{d}_0$ is *not* exponentially stable.

The proof of Statement 3 is quite easy. It is shown in [14] (Proposition 5.2) that any equilibrium of (6.18) that approaches a corner of $[0,1]^n$ is exponentially stable. Also, it is shown in [14] (Proposition 5.1) that an equilibrium of (6.18) approaches a vector $\mathbf{x} \in \{0,1\}^n$ if and only if $\mathbf{x}$ satisfies the so-called "parity condition." In the case of the system description (6.18), the parity condition of [14] (Proposition 5.1) becomes: no component of $\nabla E(\mathbf{x})$ is zero, and

$$[\nabla E(\mathbf{x})]_i < 0 \text{ if } x_i = 1, \; [\nabla E(\mathbf{x})]_i > 0 \text{ if } x_i = 0. \tag{6.51}$$

By Lemma 9, this is equivalent to $\mathbf{x}$ being a strict local minimum of E. ■

6.1 Local Minima and the Bias Term

Actually, one can see from the proof of Theorem 11 that any equilibria that remain within the interior of $[0,1]^n$ or approach a face of $[0,1]^n$ are likely to be *unstable*. Hence, except in some contrived situations, solution trajectories starting from almost all initial conditions will converge to a corner of $[0,1]^n$, i.e., to a *strict local minimum* of E. In order to use the network effectively for optimization we need strict local minima. A small bias term of the form $b^t x$ added to the energy function $E(\mathbf{x})$ succeeds in creating strict local minima at corners of $[0,1]^n$ without affecting the optimality structure globally. The stability analysis presented above goes through with slight modifications. A formal statement of these stronger convergence results follows. The proofs are quite involved and the interested reader is directed to [16].

Theorem 12 *Suppose E is a multilinear polynomial on $\Re^n$, and that Assumption (A2) is satisfied. Let $\{\lambda_j\}$ be any sequence of positive numbers approaching infinity. Under these conditions, there exists an $\epsilon > 0$ such that, for all $\mathbf{b}$ with $||\mathbf{b}|| < \epsilon$ except for those belonging to a set of measure zero, for all sufficiently large indices j, for all initial conditions $\mathbf{u}(0)$ except for those belonging to a set of measure zero, the solution trajectory $\mathbf{x}(t)$ converges to a vector $\mathbf{e} + \mathbf{o}(1/\lambda_j)$, where $\mathbf{e} \in \{0,1\}^n$ and $\mathbf{e}$ is a local minimum of E over $\{0,1\}^n$.*

In practice, one would use Theorem 13 as follows: In order to minimize E over $\{0,1\}^n$, one would choose a "small" bias vector $\mathbf{b}$, a "large" sigmoidal gain λ, a "random" initial condition $\mathbf{u}(0)$, and set the network (6.10) in motion. When the vector $\mathbf{x}(t)$ appears to be approaching some corner of the hypercube $[0,1]^n$, this vector is rounded off to the nearest binary vector $\mathbf{e}$. Then, one would check whether $\mathbf{e}$ satisfies the parity condition (6.9). If so, a local minimum of E will have been found. If the parity condition is *not* satisfied, then the process is repeated by increasing λ, and only as a last resort, changing the bias vector $\mathbf{b}$.

Thus, in practice, the neural network (6.10) generates *candidate* local minima, which must then be tested to determine whether they really are local minima. It may be pointed out that this is a common feature of almost all "interior point" methods of discrete optimization.

6.2 An Analog Net for Local MAX-SAT

In this section, we study an application of the theory developed thus far, namely, solving the local MAX-SAT problem.

The MAX-SAT problem is a variant of the SAT (Satisfiability) problem. The objective of the MAX-SAT problem is to determine a set of truth assignments to the variables in X such that the number of true clauses in the set $C_1, \ldots, C_k$ is maximized.

The local MAX-SAT problem is a local version of the MAX-SAT problem. The objective here is to determine a *local optimum*, that is, a set of truth assignments on X such that changing the truth value of a *single* variable among $X_1, \ldots, X_n$ cannot increase the number of true clauses. This problem is complete within a class known as PLS [10], which stands for *Polynomial Local Search*.

To formulate the MAX-SAT problem as an optimization problem, let us define binary variables $x_1, \ldots, x_n$ such that $x_i = 0$ if X_i is false, and $x_i = 1$ if X_i is true. With this definition, it is easy to see that the variable $\neg X_i$ corresponds to $(1 - x_i)$. Next, for each clause C_i, we define a function $E_i(\mathbf{x})$ such that $E_i(\mathbf{x}) = 1$ if C_i is false, and $E_i(\mathbf{x}) = 0$ if C_i is true. The overall objective function is defined as

$$E(\mathbf{x}) = \sum_{i=1}^{k} E_i(\mathbf{x}).$$

Then $E(\mathbf{x})$ equals the number of false clauses. It is easy to see that each $E_i(\mathbf{x})$ multilinear, whence so is $E(\mathbf{x})$.

To solve the local MAX-SAT problem using a discrete (two-state) neural network, one would proceed as follows: Let $\mathbf{x}(t) \in \{0,1\}^n$ denote the state of the network at time t. Thus $\mathbf{x}(t)$ represents the current guess (of truth assignments). Choose an index $i \in \{1, \ldots, n\}$ at random. Set

$$x_i(t+1) = \text{sat}\{-\nabla E[\mathbf{x}(t)]_i\},\ x_j(t+1) = x_j(t) \text{ for } j \neq i, \tag{6.52}$$

where

$$\mathrm{sat}(x) = \begin{cases} 0 & \text{if } x < 0 \\ 1 & \text{if } x > 0 \end{cases} . \tag{6.53}$$

The above updating rule corresponds to asynchronous operation of the network with the objective function E. Since E is multilinear, the above updating rule can be given another, equivalent, interpretation. Given $\mathbf{x}(t)$, choose an index i at random, and define $\mathbf{y} \in N[\mathbf{x}(t)]$ by

$$y_i = 1 - x_i(t), \; y_j = x_j(t) \text{ for } j \neq i. \tag{6.54}$$

The above updating rule guarantees that

$$E[\mathbf{x}(t+1)] \leq E[\mathbf{x}(t)]. \tag{6.55}$$

Hence, the state vector $\mathbf{x}(t)$ will "eventually" converge to a fixed point, i.e., a local minimum of E. However, Schäffer and Yannakakis [11] show that, in the worst case, such a search procedure can take an *exponential* number of steps to reach a local minimum. Indeed, they show that the problem of finding a fixed point of a discrete neural network is PLS-complete.

On the other hand, if we use an *analog* neural network of the form 6.18 to minimize E, then Theorem 12 guarantees that for almost all bias vectors of sufficiently small norm, for all sufficiently large sigmoidal gains, and for almost all initial conditions, the network state will converge to a local minimum of E.

REFERENCES

[1] Bland, R.G., Goldfarb, D., and Todd, M. J., "The Ellipsoid Method: A Survey", *Operations Research*, **29** (1981), 1039-1081.

[2] Bruck, J., and Goodman, J. W., "A Generalized Convergence Theorem for Neural Networks", **34** (1988), 1089-1092.

[3] Delsatre, P., and Kamp., Y., "Low Rank Matrices with given Sign Patterns", *SIAM J. Disc. Math.*, **2** (1989), 51-63.

[4] Grötschel, M., Lovász, L., and Schrijver, A., *Geometric Algorithms and Combinatorial Optimization*, Springer-Verlag, 1988.

[5] Hao, J., Tan, S., and Vanderwalle, J., "A New Approach to the Design of Hopfield Associative Memory", *International Joint Conference on Neural Networks*,(1991), 1705-1710.

[6] Hopfield, J. J., "Neural Networks and Physical Systems with Emergent Collective Computational Abilities", *Proc. Nat Acad. Sci (U.S.A)*,**79** (1982), 2554-2558.

[7] Hopfield, J. J., "Neurons with Graded Response have Collective Capabilities like those of two-state Neurons", *Proc. Nat Acad. Sci (U.S.A)*,**81** (1984), 3088-3092.

[8] Hirsh, M. W., "Convergence of Neural Network", *International Joint Conference on Neural Networks*,(1987), 115-125.

[9] Kamp, Y., Hasler, M., *Recursive Neural Networks as Associative Memory*, John Wiley and Sons, 1991.

[10] Krental, M. W., "On Finding Locally Optimal Solutions", *SIAM J. of Computing*,**19** (1990), 742-749.

[11] Schäffer, A. A., and Yannakakis, M., "Simple Local Search Problems that are Hard to Solve", *SIAM J. of Computing*,**20** (1991), 56-87.

[12] Salam, F. M. A., Wang, Y-W., and Choi, M-R., "On the Analysis of Dynamic Feedback Neural Networks", *IEEE Trans. on Circuits and Systems*,**38** (1991), 196-201.

[13] Tou, J. and Gonzalez, R., *Pattern Recognition Principles*, Addison-Wesley, 1974.

[14] Vidyasagar, M., "Location and Stability of High-Gain Equilibria of Nonlinear Neural Networks", *IEEE Trans. on Neural Networks*, bf 4 (1993), 660-672.

[15] Vidyasagar, M., *Nonlinear Systems Analysis*, 2nd ed., Prentice-Hall, Englewood Cliffs, NJ, 1992.

[16] Vidyasagar, M., "Minimum-Seeking Properties of Analog Neural Networks with Multilinear Objective Functions", *IEEE Transactions on Automatic Control*, **40** (1995), 1359-1375.

4

NEURAL NETWORK TRAINING VIA QUADRATIC PROGRAMMING

Theodore B. Trafalis
Nicolas P. Couellan

ABSTRACT

We develop two new training algorithms for feedforward supervised neural networks based on quadratic optimization methods. Specifically, we approximate the error function by a quadratic convex function. In the first algorithm, the new error function is optimized by an affine scaling method, which replaces the steepest descent method in the backpropagation algorithm. In the second algorithm, the steepest descent method is replaced by a trust region technique. Comparative numerical simulations for medical diagnosis problems show significant reductions in learning time with respect to the backpropagation algorithm.

1. INTRODUCTION

The training phase of a supervised feedforward Artificial Neural Network (ANN) can be defined as an optimization problem. In a strictly mathematical sense this is equivalent to a function approximation problem. Therefore we require the estimate of certain parameters which are called weights. There are various non-linear optimization methods that can be used for training of ANN (Hertz, Krogh and Palmer 1991). One of those is the gradient descent method on which the backpropagation method is based (Werbos 1974, Rumelhart, Hinton and Williams 1986). A major drawback of the current learning algorithms is the long training (learning time) for large scale problems (Shepanski 1988). Recent breakthroughs of interior point methods (IPM) for solving large scale constrained optimization problems (Lustig et al. 1994, Gonzaga 1991, 1992, Anstreicher et al. 1990) suggest that IPM can be used to develop new fast learning laws in ANN which can enlarge the set of problems which can be solved successfully. Some new ideas along these lines have already been studied such as training of ANN by a logarithmic barrier function (Trafalis and Sieger 1993, Trafalis and Tutunji 1994). Affine Scaling methods have been successful for solving large scale convex programming problems (Fang and Puthenpura, 1993). The purpose of this paper is to propose two new methods of training for ANN using an affine scaling method and a trust region technique respectively.

During the last few years since the introduction of backpropagation, an enormous amount of research has been reported to further understand, extend and apply this learning method. The general methodology for supervised learning in feedforward ANN is based on using local models of the error function which are usually either linear or quadratic (Battiti 1992, Barnard 1992). In the first case, the function is approximated by the first (constant) and second (linear) terms of a Taylor expansion; in the second case, the third (quadratic) term is also taken into account. When the approximation of the function by its second order Taylor expansion is exact, the global minimum can be found after exactly N single variable searches, where N is the number of the weights of the system (Bazaraa et al. 1993).

Recently linear programming (LP) formulations have been also suggested for training of supervised feedforward ANNs (Mangasarian 1993, Bennett et al. 1992, Trafalis and Couellan 1994, Roy et al. 1995). Historically LP models have been used for discriminant analysis (Mangasarian 1965, Smith 1968, Glover et al. 1988, Mangasarian 1990 and others). Our approach is fundamentally different from those of Mangasarian (1993) and Roy et al. (1995) based on convex quadratic optimization.

A natural question related to the complexity of learning is the following. Is there any algorithm which solves the learning problem in polynomial time? Recently Judd (1987,1990) showed that in the case of multilayer feedforward supervised learning architectures the optimization problem is NP complete. Therefore it seems highly improbable that there exists an algorithm which solves the learning problem of the above architectures in polynomial time. Hence, we have to use approximate techniques. A natural approach is to approximate the objective function of the learning optimization problem by a function which is quadratic with respect to the neuron's weights.

In this paper two new training methods based on quadratic optimization are presented to find the optimal weights of a multi-layer feedforward supervised neural network architecture. Specifically, we approximate the error function by a convex quadratic approximation. The quadratic approximation of the error function is

backpropagated at each layer of the ANN using a classical backpropagation scheme. The solution of the minimization problem is found at each layer using either an affine scaling method for quadratic optimization problems or a trust region technique.
This paper is organized as follows: Section 2 discusses the quadratic approximation problem, Section 3 discusses the Affine Scaling Algorithm for quadratic programming. Section 4 presents the trust region technique for quadratic programming. Section 5 describes the quadratic algorithm for the training of an ANN, Section 6 gives some computational results and Section 7 is the conclusion.

2. QUADRATIC APPROXIMATION PROBLEM

Consider a single layer neural network (See Fig. 1) with m inputs and n output neurons where the non-linearity of an output is described as follows :

$$y_i = f(u' w_i) \qquad i=1, ..., n$$

f is the activation function (non-linearity) of a neuron (Hyperbolic tangent)

$u:=[u_1,...,u_m]'$ is the input vector

$w_i:=[w_{1i},...,w_{mi}]'$ is the weight vector corresponding to the ith output neuron

For p different patterns, we can define the following matrices: $U:=[u(1),...,u(p)]\in \mathbf{R}^{m\times p}$ is defined as the matrix of input vectors and $D:=[d(1),...,d(p)]'$ as the matrix of desired outputs.

A supervised feedforward neural network can be modelled as an unconstrained least square optimization problem where the optimality criterion is a quadratic functional that describes the difference between the desired and the actual behavior of the ANN. The objective is to minimize this criterion with respect to the interconnection weights.

In our development the single layer neural network training optimization problem is defined as follows :

$$\min E_o(W) = \text{trace}((D - \Phi(U'W))' (D - \Phi(U'W))) \qquad (P_o)$$

where $\Phi(z) = [f(z_1),...,f(z_p)] \in \mathbf{R}^{p \times n}$ and $z \in \mathbf{R}^{p \times n}$. The ji^{th} component of the matrix $\Phi(z)$ is defined as $f(z_{ji})$ where f is the neuron's activation function.

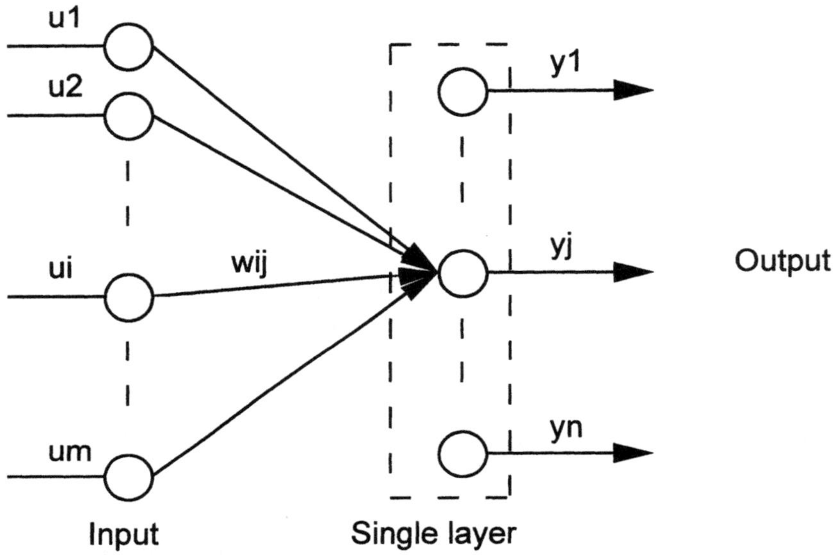

Figure 1

Instead of solving the previous problem we will solve the following quadratic approximation optimization problem :

$$\min E(W) = \text{trace}\,((V - U'W)'(V - U'W)) \qquad (P)$$
where V is such that $\Phi(V) = D$

It has been proven in Sartori and Antsaklis (1992) that the solution of the quadratic problem solves (with a small error) also the original problem.

We can also approximate problem (P_0) by a quadratic approximation using the Taylor's series expansion of E_o up to the second order. Unfortunately the Hessian of E_o is not necessarily positive semi-definite. We have selected the approximation (P) because of the convexity of the approximate error function E.

The quadratic problem (P) can be rewritten into the following form :

$$\min \sum_{i=1}^{n} w_i Q_o w_i + c_i' w_i + g_i \qquad \text{(QP)}$$

where $Q_o := UU'$, $c_i = -2v_i' U'$, and $g_i = v_i' v_i$.

It is easy to prove the following proposition :

Proposition : *The matrix Q_O is symmetric positive semi-definite and problem (P) is a convex quadratic optimization problem. If rank(U')=m then Q_O is positive definite and E(W) has a unique minimum point which is the global minimum.*

Proof : See for example (Bazaraa and Shetty 1993). By minimizing independently each term of the sum, (QP) is equivalent to the following n subproblems :

$$\min \frac{1}{2} w_i' Q w_i + c' w_i \qquad i=1,\dots,n \qquad \text{(QP)}_i$$

where $Q=2Q_O$.

3. THE AFFINE SCALING ALGORITHM

Next we describe an affine scaling algorithm for solving the unconstrained $(QP)_i$ problem. The algorithm solves the following problem:

$$\text{Minimize } f(x) = \frac{1}{2} x' Qx + c' x$$

We assume Q is positive definite. In our case, we can perturb the diagonal elements of the matrix Q to make it strictly positive definite (i.e. $Q \leftarrow Q + \mu I$).

The algorithm is based on the scaling of the variables at each iteration. At each step, we find the descent direction in the scaled space and come back to the original space to compute the move. Next, we explain the main computations of the algorithm.

We first consider the transformation of variables $\tilde{x} = Dx$. Any direction vector transforms then to $\tilde{d} = Dd$. In the transformed space, the steepest-descent direction is :

$$\tilde{d} = -\tilde{g} := -D^{-1} g \text{ where } g = \nabla f(x) = Qx + c .$$

In the original variables, the descent direction is given by:

$$Dd = -D^{-1}g,\ d = -D^{-2}g.$$

The step length on the direction d is obtained by solving the following problem: $\min_{\alpha \geq 0} f(x + \alpha d)$. Clearly, the optimal α is the solution of $\frac{\partial f(x + \alpha d)}{\partial \alpha} = 0$, and is given by :

$$\alpha = -\frac{d'(Qx + c)}{d'Qd}$$

One can see the importance of the positive definiteness of the matrix Q in the computation of the optimal step length.

The structure of the algorithm for solving $(QP)_i$ is given next.

Algorithm:

Step 1. Compute the starting point $x=(x_1,...,x_n)$. Initialize k=1.

Step 2. Form the matrix $D=\text{diag}\{x_1,...,x_n\}$.

Step 3. Calculate $d=-D^{-2}g$ with $g=Qx+c$.

Step 4. Compute the step length $\alpha = -\frac{d'(Qx + c)}{d'Qd}$

Step 5. Move to the new point $x=x+\alpha d$.

Step 6. If k large enough STOP. Otherwise, k=k+1, go to step 2.

4. THE TRUST REGION TECHNIQUE

So far, the weights were not constrained. However, there are problems where it is beneficial to consider symmetric box constraints. There are a lot of advantages in bounding the weights. During the training of the network the weights can become very large producing excessively large outputs. Then small propagation error corrections can result with small improvements to the weights. Therefore we can have *network paralysis* (Wasserman 1989). Bounding the weights we can also avoid floating-point overflow during the update of the weights. A priori information about weight limits for a particular problem can be easily incorporated into the problem. The main disadvantage with using box constraints is the large amount of constraints that may result.

We used an affine scaling method to solve the unconstrained approximate quadratic program (section 3). We will now use a trust region technique to solve the constrained approximate program (Flippo and Jansen 1992, Ye 1992). The variables are constrained in the unit sphere. The problem to be solved can be written as follows :

$$\min_x\{\psi(x)=\frac{1}{2}x^tQx+c^tx \mid x^tx\le 1\} \qquad \text{(QP)'}$$

Let $\lambda_1=\ldots=\lambda_k<\lambda_{k+1}\le\ldots\le\lambda_n$ be the eigenvalues of $Q=(q_{ij})$ $1\le i,j\le n$. The corresponding orthonormal eigenvectors are given by $\alpha_i\in\Re^n$. Let $\Lambda=diag\{\lambda_1,\ldots,\lambda_n\}$ and $P=(\alpha_1,\ldots,\alpha_n)$. The orthonormal decomposition of Q is given by $Q=P\Lambda P^t$. The first and second order optimality conditions are given by the following theorem:

Theorem: *x^* is an optimal solution of (QP)' if and only if for some $\mu^*\ge 0$, x^* satisfies the conditions:*

(1) $(Q+\mu^*I)x^*=-c$

(2) $(x^*)^tx^*\le 1$

(3) $\mu^*((x^*)^tx^*-1)=0$

(4) $(Q+\mu^*I)$ *is positive semidefinite*

Proof: (Flippo and Jansen (1992)). From the theorem, we have the following results: $\mu^*\ge 0$ and (4) implies $\mu^*\ge\max(0,-\lambda_1)$, (3) and (4) implies that if $\lambda_1>0$ then the solution can be in the interior of the sphere.

If $\lambda_1>0$ then the matrix Q is positive definite and has an inverse. The solution of the problem is then given by inverting the system (1) where $\mu^*=0$. In the following we will then focus on the cases where Q is not positive definite.

If $\lambda_1\le 0$ and $\mu>-\lambda_1$:

It is clear that $Q+\mu I$ is positive definite and (1) has a unique solution $x(\mu)$ which is a continuous function of μ. We define $\Phi(\mu)=x(\mu)^tx(\mu)-1$. Φ can be seen as the distance from the point to the boundary of the sphere. Using the orthonormal decomposition of Q, (1) can be written as $P(\Lambda+\mu I)P^tx=-c$ or $(\Lambda+\mu I)y=-\bar{c}$ where $\bar{c}=P^t c$ and $y=P^tx$.

For $\mu>-\lambda_1$, $\Phi(\mu)=\|y\|^2-1=\sum_{i=1}^{n}\frac{\bar{c}_i^2}{(\lambda_i+\mu)^2}-1$

If the optimum $x(\mu^*)$ is attained for $\mu^* \in (-\lambda_1,\infty)$, then computing a solution to the optimality conditions is equivalent to finding a root μ^* of $\Phi(\mu)$ in the region $(-\lambda_1,\infty)$. An upperbound for μ^* can be calculated by setting $\Phi(\mu^*)=0$, which gives: $\mu^* \leq \|c\| + n.\max_{i,j}|q_{ij}|$.

if $\lambda_1$²0 and $\mu=-\lambda_1$:

The system (1) is equivalent to $(\Lambda-\lambda_1 I)y=-\bar{c}$. We have the two following cases:

Case 1:

$$\text{if } \bar{c}_j \neq 0 \text{ for some } j \in \{1,\ldots,k\}$$
$$\Rightarrow \text{ No solution } (\lambda_j - \lambda_1 = 0 \text{ while } \bar{c}_j \neq 0)$$
$$\Rightarrow \lim_{\mu \to -\lambda_1} \Phi(\mu) = \infty$$

Case 2:

$$\text{if } \bar{c}_j = 0 \text{ for all } j = 1,\ldots,k$$
$$\Rightarrow (\lambda_i - \lambda_1)y_i = -\bar{c}_i \quad i = k+1,\ldots,n$$
$$\Rightarrow y_i = \frac{-\bar{c}_i}{\lambda_i - \lambda_1} \Rightarrow \text{(1) has many solutions}$$

solution with minimal norm:

$$y_i^{\min} = 0 \qquad i = 1,\ldots,k$$
$$y_i^{\min} = \frac{-\bar{c}_i}{\lambda_i - \lambda_1} \quad i = k+1,\ldots,n$$

From these various cases, one can develop the main steps of the algorithm. Since in our case $\lambda_1=0$, the algorithm can be simplified. The detailed algorithm is provided in the appendix.

5. QUADRATIC ALGORITHM FOR THE TRAINING OF ANN

Next we consider an ANN with one hidden layer. The idea of the method is based on the backpropagation algorithm. Specifically, we backpropagate the output error to form a desired output for each layer. We consider a quadratic convex error function at each layer and formulate the quadratic optimization problem of section 2. Then we solve it by using the affine scaling algorithm developed in section 3 or the trust region technique developed in section 4. We generalize the results obtained in section 2 for the case of multi-layer ANN. Specifically, each layer is taken as a single layer ANN and a link output/input is established between the different layers. Since we know the output that we would like to achieve at the output layer and the activation function f is 1-1, it is easy to define the matrix V such that $\Phi(V) = D$ for this layer (See section 2). Since at the hidden layer there is no desired output, we will approximate a desired output using the backpropagation scheme and calculate the corresponding matrix V(Krogh et al. 1991, Sartori and Antsaklis 1992).

Specifically, the error at the output layer for the ith neuron is defined as follows :

$E_i^o(j) = d_i(j) - y_i^o(j)$ where $Y^o = (\, y_i^o(j)\,)_{ij}$ is the output matrix.

Similarly we can write the error at the hidden layer for the ith neuron :

$$E_i^1(j) = d_i^1(j) - y_i^1(j)$$

where Y^1 is the output matrix at the hidden layer. Using the backpropagation of the output error, $E_i^1(j)$ can also be taken as :

$$E_i^1(j) = \sum_{h=1}^{n} \delta_h^o(j) w_{ih}^o \quad \text{where}$$

W^o is the output weight matrix

n is the number of output neurons

and $\delta_h^o(j) = [d_i(j) - y_i^o(j)][1 - y_i^o(j)]^2$

Then the desired output at the hidden layer is as follows:

$$d_i^1(j) = E_i^1(j) + y_i^1(j) \qquad (*)$$

Using (*) V^1 (V at the hidden layer) can be computed such that $\Phi(V^1) = D^1$. Therefore V^1 must satisfy the following equation:

$$f(v_i^1(j)) = y_i^1(j) + \sum_{h=1}^{n} \delta_h^o(j) w_{ih}^o$$

Since f is 1-1 we can invert the function f to calculate the components of V^1. Thus a convex quadratic optimization problem is defined for each layer.

General algorithm

Step 1 : Given D, compute V^2 (V at the output layer).

Step 2 : Initialize W^1 and W^2.

Step 3 : Test $E_0(W^1,W^2)$. If E_0 small enough then STOP, otherwise go to step 4.

Step 4 : Compute V^1 using the backpropagation scheme.

Step 5 : Compute W^1 by solving the quadratics (See section 2) using an affine scaling method (section 3) or the trust region technique (section 4) for quadratic optimization problem.

Step 6 : Compute U^2 (input of output layer) for the new weights W^1

Step 7 : Compute W^2 by solving the quadratics (See section 2) using an affine scaling method (section 3) or the trust region technique (section 4) for quadratic optimization problem.

Step 8 : Go to step 3.

A more detailed description of the algorithm is provided in the appendix.

6. COMPUTATIONAL RESULTS

The algorithms have been implemented in C++ on a DEC 5000/25 workstation. At each iteration, we compute only four steps of the affine scaling procedure instead of finding the global minimizer of the quadratics. This modification saves a lot of computing time and gives satisfying results. We use a perturbation μ=0.01 to ensure positive definiteness of the matrix Q+μI. This value of μ has been chosen empirically and gives very good results for all of the training problems. We trained a one hidden layer feedforward ANN using the Wisconsin Breast Cancer Database (Mangasarian et al. 1990). The problem is to train the neural network with pairs (observation, diagnosis) where the observation is represented by a vector of nine cellular attributes and the diagnosis is either 0 if benign or 1 if malignant. The database is organized in groups. Table 1 describes the data and the parameters of the algorithms. Table 2 gives the correctness achieved by the affine scaling algorithm and Table 3 reports the correctness of the trust region algorithm. Table 4 gives comparisons of CPU times and the training errors of our two algorithms with the backpropagation algorithm. We perform multiple runs (150) of the algorithms and compare their mean, min, max and standard deviation values.

From Table 4, we can observe that the trust region algorithm (TRA) achieves lower training error values than the affine scaling (ASA) or backpropagation algorithm (BP) . With the information on the CPU time and the final training error, we can conclude that for the cancer database, TRA converges much faster than BP. In the worst cases where TRA takes as much time as BP, TRA reaches a lower error, therefore the rate of convergence of TRA is higher. ASA does not perform as well as TRA, but we can also conclude that ASA converges faster than BP. The training error achieved is always lower than for BP, while the CPU times are of the same order for both algorithms.

Table 1. Parameters for Wisconsin cancer database.

Database group number	*Numbers of patterns*	*No. of hidden neurons*	*Number of runs*	*Stopping criterion(*)*
1	70	8	150	E<5, N>50
2	31	8	150	E<1, N>50
3	17	8	150	E<0.15, N>50
4	48	8	150	E<2, N>50
5	49	8	150	E<2, N>50

(*) E is the training error after one run and N is the number of iterations for the run.

Table 2. Affine scaling algorithm - Correctness.

Database group number	*Number of incorrect trainings over 150 runs*	*Resulting correctness %*
1	0	100
2	12	92
3	15	90
4	7	95.3333
5	2	98.6666

Table 3. Trust region algorithm - Correctness.

Database group number	*Number of incorrect trainings over 150 runs*	*Resulting correctness %*
1	0	100
2	0	100
3	1	99.3333
4	0	100
5	0	100

Table 4. Comparison of algorithms for Wisconsin cancer database.

Database group number	*Mean*	*CPU*	*time (s)*	*Mean*	*training*	*error*
	Affine-scaling	*Trust region*	*Back - prop*	*Affine-scaling*	*Trust region*	*Back - prop*
1	1.59955	0.49333	1.2	3.80157	1.59955	4.72646
2	0.45367	0.82	0.3	0.94913	0.45367	0.909332
3	0.02281	0.79333	0.2	0.13377	0.02281	0.148569
4	0.82023	0.46	1.1	1.72433	0.82023	3.17057
5	0.49682	0.46	0.2	1.64858	0.49682	1.79402

Table 5. IRIS and WINE training CPU time with Affine scaling algorithm.

Database Name	*Number of hidden neurons*	*Number of weights*	*Number of patterns*	*Min CPU time (s)*	*Max CPU time (s)*	*Mean CPU time (s)*	*Standard deviation*
IRIS	20	120	150	<1	10	2	2.84623
WINE	25	375	178	1	26	7.56	7.27292

The training error achieved by both affine scaling and trust region algorithm always seems to correspond to a local minimum of the original error function. In the Appendix, Figure 2 shows the output of the ANN for two of the groups of the cancer database. Clearly, the ANN has classified correctly the patterns, but the global minimum is not reached and the perfect mapping with the desired output is not achieved. This leads to the conclusion that these algorithms should be used for classification problems or for problems that do not require extreme accuracy. However, the algorithms become effective for problems that require fast training and more specifically it can be used for large scale classification problems.

We have also trained a one hidden layer feedforward ANN with other databases. These databases are from Murphy and Aha (1992). They are called IRIS and WINE and they are recognition tasks. The result of the training with the affine scaling algorithm is shown in Tables 5 and 6. For each database, we run the algorithm 100 times.

Table 6. IRIS and WINE training error with Affine scaling algorithm.

Database Name	*Number of hidden neurons*	*Number of weights*	*Number of patterns*	*Min training error*	*Max training error*	*Mean training error*	*Standard deviation*
IRIS	20	120	150	8.16252	47.2299	11.6692	5.60538
WINE	25	375	178	21.3445	70.1	43.7049	7.11221

7. CONCLUSIONS AND FUTURE RESEARCH

In this paper we have developed two quadratic optimization algorithms to train a one hidden layer ANN using an affine scaling and a trust region method for convex quadratic programming. Our numerical simulations using C++ show significant reductions in learning time in comparison with the backpropagation algorithm for medical diagnosis problems. There is a great potential for improvements and future research. We are currently investigating a Quasi-Newton logarithmic barrier approach (Trafalis and Tutunji 1994) and a Primal-dual interior point method (Trafalis and Couellan 1994). We are also testing the algorithm for larger problems. The combination of interior point methods and backpropagation shows good promise for training of large scale ANN in real time.

ACKNOWLEDGMENT

The present work has been supported by the NSF grant ECS-9212003, Research Initiation Award. We thank Paul Werbos for his encouragement during the work.

APPENDIX

Trust region technique algorithm for quadratic programming :

STEP 1 : Determine the definiteness of Q. If Q is positive definite, go to Step 2, otherwise go to Step 3.

STEP 2 : Determine $x^+ = - Q^{-1}c$; if x^+ is feasible, then $x^* = x^+$, $\mu^* = 0$, STOP; otherwise go to Step 4.

STEP 3 : If $\beta<0$ then $\mu^* = - \lambda_1$, $x^* = x_{min}$; otherwise go to Step 4.

STEP 4 : Determine the unique root μ^* of $\phi(\mu)$ in the interval :

$[\max(0,-\lambda_1)\,,\ \|c\| + n.\max_{i,j}|q_{ij}|\ ]$ and $x^* = x(\mu^*)$.

where :

λ_1 is the least eigenvalue of Q, P the matrix of eigenvectors of Q

$$\phi(\mu) = \sum_{i=1}^{n} \frac{\bar{c}_i^2}{(\lambda_i + \mu)^2} - 1 \quad \text{where } \bar{c} = P'c\,,$$

$$\beta = \lim_{\mu \downarrow -\lambda_1} \Phi(\mu)$$

x_{min} is the minimal-norm solution of $(Q + \mu^* I)\,x = -c$

A Backpropagation Affine Scaling Quadratic Optimization Algorithm :

STEP 1 : $V^2 = D$ (D scaled such that $v_i^2(j) < 0.5$)

STEP 2 : Initialize the weights :

$w_{ij}^1 \in [\,-1\,,\,1\,]$

$w_{ij}^2 \in [\,-1/L\,,\,1/L\,]$

STEP 3 : Compute $E_o(W^1,W^2) = tr\,((D - Y^2)'\,(D - Y^2))$

where $Y^2 = F(U^{2\prime}W^2)$ $\quad U^{2\prime} = [\,Y^1\ \mathbf{1}\,]$

$Y^1 = \Phi(U^{1\prime}W^1)$ U^1 is given

If $E_o < \varepsilon$ then STOP.

else go to STEP 4.

STEP 4 : Compute V^1 using the following formula :

$$v_i^1(j) = f^{-1}\left(y_i^1(j) + \sum_{h=1}^{L+1} coeff.d_{hj}\,w_{ij}^2 \right)$$

$d_h(j)$ is given by the backpropagation scheme:

$d_h(j) = (d_h(j) - y_h^2(j)).(1 - y_h^2(j)^2)$ when using f = tanh

with: $\bar{w} = \max_{i,j}\{w_{ij}^2\}$

if $\bar{w} > 1/L$ then coeff $= L \times \bar{w}$

else coeff = 1

STEP 5 : Define $Q_1 = U^1 U^{1\prime}$

$h_i^1 = 2\,v_i^{1\prime}\,U^{1\prime}$

$$c_i^1 = v_i^{1'} v_i^1 \qquad 1 \le i \le n$$

Minimize the quadratics $w_i^{1'} Q_1 w_i^1 - h_i^1 w_i^1 + c_i^1$
using the affine scaling algorithm for quadratic optimization.
Update W^1.

STEP 6 : $U^{2'} = [\, Y^1 \; \mathbf{1} \,]$ where $Y^1 = \Phi(U^{1'} W^1)$

STEP 7 : Define $Q_2 = U^{2'} U^{2'}$

$$h_i^2 = 2\, v_i^{2'} U^{2'}$$

$$c_i^2 = v_i^{2'} v_i^2 \qquad 1 \le i \le n$$

Minimize n quadratics $w_i^{2'} Q_2 w_i^2 - h_i^2 w_i^2 + c_i^2$
using the affine scaling algorithm for quadratic optimization.
Update W^2.

STEP **8** : Go to STEP 3

REFERENCES

Anstreicher K. M., Hertog, D. Den, Roos, C. & Terlaky, T., (1990). A long step barrier method for convex quadratic programming. Technical Report 90-53: Delft University of Technology, Delft, Netherlands.

Barnard E., (1992). Optimization for training neural nets. *IEEE Transactions on Neural Networks,* **3**,(2), 232-240.

Battiti R., (1992). First and second-order methods for learning between steepest descent and newton's method. *Neural Computation,* **4**, 141-166.

Bazaraa, M.S., Sherali, H.D., & Shetty, C.M., (1993). *Nonlinear programming theory and algorithms*. NY: Wiley.

Bennett, K.P., & Mangasarian, O.L. (1992). Neural network training via linear programming. In P.M. Pardalos (ed), *Advances in Optimization and Parallel Computing*, North Holland, Amsterdam.

Fang S., & Puthenpura, S., (1993). *Linear optimization and extensions: theory and algorithms*. Prentice Hall.

Flippo, O.E., & Jansen, B., (1992). Duality and sensitivity in quadratic optimization over a sphere. Report 92-65. Falculty of Technical Mathematics and Informatics, Delft University of Technology.

Glover, F. (1990). Improved linear programming models for discriminant analysis. *Decision Sciences*, **21**, 4, 771-785.

Glover, F., Keene, S. and Duea, B. (1988). A new class of models for the discriminant problem. *Decision Sciences*, **19**, 269-280.

Gonzaga, C. C. (1991). Large step path-following methods for linear programming, Parts 1 & 2. *SIAM Journal of Optimization* , **1**, 268-280.

Gonzaga, C. C. (1992). Path-following methods for linear programming. *SIAM REVIEW*, **34**, (2,) 167-224.

Hertz, J., Krogh, A., & Palmer, R. G., (1991). Introduction to the theory of neural computation. Redwood City, CA: Addison-Wesley.

Judd, J.S., (1987). Learning in networks is hard. *Proceedings of the First International Conference on Neural Networks* (685-692). San Diego, CA: IEEE.

Judd, J.S., (1990). *Neural network design and the complexity of learning.* Cambridge, MA: MIT Press.

Lustig, I. J., Marsten, R. E., & Shanno, D. F., (1994). Interior point methods for linear programming: computational state of the art. *ORSA Journal on Computing,* **6**(1), Winter 1994, 1-14.

Mangasarian, O.L. (1965). Linear and nonlinear separation of patterns by linear programming. *Operations Research*, **13**,444-452.

Mangasarian, O.L., Setiono, R. and Wolberg, W.H. (1990). Pattern recognition via linear programming: Theory and application to medical diagnosis. In Coleman, T.F., and Li. Y. (Eds.) *Large -scale numerical optimization*, Philadelphia: SIAM, 22-30.

Mangasarian, O.L., (1993). Mathematical programming in neural networks. *ORSA Journal on Computing*, **5**, (4), 349-360.

Murphy, P.M. & Aha, D.W., (1992). UCI repository of machine learning databases. Department of Information and Computer Science, University of California, Irvine, California, http://www.ics.uci.edu/AI/ML/MLDBRepository.html.

Roy, A., Govil, S. and Miranda, R., (1995). An algorithm to generate radial basis function (RBF)-like nets for classification problems. *Neural networks*, **8**, (1), 179-201.

Rumelhart, D. E., Hinton, G. E., & Williams, R., (1986). Learning internal representations by error propagation. In D.E. Rumelhart and J.L. McClelland (Eds*), Parallel distributed processing: explorations in the microstruture of cognition*, **1**, Foundations. Cambridge, MA: MIT Press.

Sartori, M. A., Antsaklis, P. J., (1992). Neural network training via quadratic optimization. *IEEE International Symposium on Circuits and Systems*, 49-52.

Shepanski, J., (1988). Fast learning in artificial neural networks: multilayer perceptron training using optimal estimation. *Proceedings of the IEEE International Conference on Neural Networks*, **1**, 465-472, San Diego: IEEE.

Smith, F.W. (1968). Pattern classifier design by linear programming. *IEEE Transactions on Computers*, C-17, 4, 367-372.

Trafalis, T. B. & Sieger, D. B., (1993). Training of multilayer feedforward artificial neural networks by a logarithmic barrier function adaptation scheme. In C.H. Dagli, L.I. Burke, B.R. Fernandez & J. Ghosh (Eds.), *Intelligent Engineering Systems Through Artificial Neural Networks*, **3**, 167-173.

Trafalis T.B. & Couellan, N.P., (1994). Neural networks training via primal-dual interior point method for linear programming. San Diego: *Proceedings of WCNN*, June 5-9, 1994.

Trafalis, T.B. & Tutunji, T., (1994). A quasi-newton barrier function algorithm for artificial network araining with bounded weights. In C. Dagli, L. Burke, B. Fernandez & J. Ghosh (Eds.), *Intelligent engineering systems through artificial neural networks*, **4**, 161-173.

Wasserman, P.D., (1989). *Neural computing: theory and practice*. NY: Van Nostrand Reinhold.

Werbos, P., (1974). Beyond regression: new tools for prediction and analysis in the behavioral sciences. PhD. Thesis, Committee on Applied Mathematics, Harvard University, Cambridge, MA. Reprinted in P. Werbos, The roots of backpropagation: from ordered derivatives to neural networks and political forecasting. NY: Wiley (1993).

Ye Y., (1992). On affine scaling algorithms for nonconvex quadratic programming. *Mathematical Programming*, **56**,285-300.

5

A NEURAL NETWORK MODEL FOR PREDICTING ATLANTIC HURRICANE ACTIVITY

Ohseok Kwon
Bruce Golden
and Edward Wasil*

College of Business and Management
University of Maryland
College Park, Maryland 20742

**Kogod College of Business Administration*
American University
Washington, D.C. 20016

ABSTRACT

Modeling techniques such as linear regression have been used to predict hurricane activity many months in advance of the start of the hurricane season with some success. In this paper, we construct feedforward neural networks to model Atlantic basin hurricane activity and compare the predictions of our neural network models to the predictions produced by statistical models found in the weather forecasting literature. We find that our neural network models produce reasonably accurate predictions that, for the most part, compare favorably to the predictions of statistical models.

1. INTRODUCTION

Over the last 10 years or so, there has been a great deal of research devoted to building models that predict Atlantic seasonal tropical cyclone activity for the period that starts on June 1 and ends on November 30. Most notably, William Gray and his colleagues (Christopher Landsea, Paul Mielke, and Kenneth Berry) at Colorado State University have constructed linear regression models that predict activity in advance of the start of the hurricane season. When the season is two months old, they use up-to-date information to predict activity for the August to October time period which is the most active portion of the hurricane season. The predictions produced by the Gray, Landsea, Mielke, and Berry (GLMB) models are of great interest to government officials and decision makers in the insurance industry (see Stevens [7] for additional details).

Using the data sets that GLMB have published in the weather forecasting literature (see [2], [3], and [4]), we set out to construct neural network models to predict Atlantic hurricane activity and then compare the predictions produced by the neural network models to the predictions produced by the regression models. In Section 2, we review the published data sets and the GLMB regression procedure. In Section 3, we describe our neural network modeling effort and compare the predictions produced by our models to the predictions of the GLMB models. In Section 4, we present our conclusions.

2. BACKGROUND

In this section, we review the types of models, data sets, and model-building methodology that GLMB have used to predict hurricane activity.

2.1 Types of Models and Data Sets

GLMB built models of the form

$$Y = f(\text{Wind, Rain, Other Factors}) + e$$

where Y is the dependent variable such as the number of hurricanes, Wind, Rain, and Other Factors are composite functions of independent variables, and e is the random error. The definitions of the dependent variables are given in Table 1 and the definitions of the independent variables are given in Table 2.

Variable	*Name*	*Definition**
H	Number of Hurricanes	A hurricane is a tropical cyclone with sustained low-level winds of 74 miles per hour or greater
HD	Number of Hurricane Days	A hurricane day is four 6-hour periods during which a tropical cyclone is observed or estimated to have hurricane-intensity winds.
HD	Number of Hurricane Days	A hurricane day is four 6-hour periods during which a tropical cyclone is observed or estimated to have hurricane-intensity winds.
NS	Number of Named Storms	A named storm (e.g., Tropical Storm Alberto, Hurricane Andrew) can be a tropical storm (maximum sustained winds between 39 and 73 miles per hour) or a hurricane.
NSD	Number of Named Storm Days	A named storm day is four 6-hour periods during which a tropical cyclone is observed or estimated to have attained tropical-storm or hurricane-intensity winds
IH	Number of Intense Hurricanes	An intense hurricane is a hurricane that reaches a sustained low-level wind of at least 111 miles per hour during its lifetime.
IHD	Number of Intense Hurricane Days	An intense hurricane day is four 6-hour periods during which a hurricane is at least a category 3 on the Saffir-Simpson scale (see [6]). A category 1 hurricane is the weakest, while a category 5 hurricane is the most intense.
HDP	Hurricane Destruction Potential	This is a measure that takes into account a hurricane's potential for wind and storm-surge destruction. HDP is the sum of the square of a hurricane's maximum wind speed for each 6-hour period of its existence.
NTC	Net Tropical Cyclone Activity	NTC = (%NS + %H + %IH + %NSD + %HD + %IHD)/6 where each season's percentage value from the long-term mean is used for the six measures of seasonal activity

* Variable names and definitions from GLMB [2], [3], and [4]

Table 1. Definitions of dependent variables.

Variable	*Definition**
U_{50}	Extrapolated zonal winds near 10°N at 50 millibars (about 20 kilometers altitude)
U_{30}	Extrapolated zonal winds near 10°N at 30 millibars
$\lvert U_{50} - U_{30}\rvert$	Absolute shear between the two levels of zonal winds near 10°N
R_s	Rainfall occurring throughout the western Sahel region of West Africa
R_g	Rainfall occurring along West Africa's Gulf of Guinea coast
D_xP	Interior standardized sea level pressure anomalies minus coastal standardized sea level pressure anomalies in West Africa
D_xT	Coastal standardized temperature anomalies minus the interior standardized temperature anomalies in West Africa
SLPA	Caribbean basin sea level pressure anomalies
ZWA	Caribbean basin 200-millibar zonal wind anomalies
SST, SSTA	Equatorial eastern Pacific sea surface temperature anomalies
D_tSST	Change in SST from January and February to April and May
SOI	Standardized Tahiti minus Darwin sea level pressure index, that is, the Southern Oscillation Index
D_tSOI	Change in SOI from January and February to April and May

* Variable names and definitions from GLMB [2], [3], and [4]

Table 2. Definitions of independent variables.

Gray and his colleagues constructed regression models for three different prediction periods and these models are shown in Table 3. The model denoted by GLMB1 is a long-range model that generates predictions by December 1 -- six months prior to the June 1 start of the hurricane season. The model denoted by GLMB2 uses data available by August 1 to make predictions for the peak period of hurricane activity from August through October. The model denoted by GLMB3 uses data available by June 1 to predict activity for the entire season from June to November. GLMB1 was published in the literature in 1992 (see [2]), followed by GLMB2 in 1993 (see [3]), and GLMB3 in 1994 (see [4]).

GLMB1 uses five independent variables and GLMB2 uses nine independent variables to generate predictions for seven dependent variables (H, HD, NS, NSD, IH, IHD, and HDP). In building GLMB1 and GLMB2, Gray and his colleagues used 41 years of data from 1950 to 1990. GLMB3 uses thirteen independent variables to generate predictions for eight dependent variables (H, HD, NS, NSD, IH, IHD, HDP, and NTC). In building GLMB3, Gray and his colleagues used 42 years of data from 1950 to 1991. The data sets that were used to build these three regression models are published in [2], [3], and [4].

Timeframe	Model
December 1	GLMB1
$Y = b_0 + b_1W + b_2R + e$	
$W = a_1U_{50} + a_2U_{30} + a_3\lvert U_{50} - U_{30}\rvert$	
$R = a_4R_s + a_5R_g$	
U_{50}, U_{30}, $\lvert U_{50} - U_{30}\rvert$	Prior 10 months used to make an extrapolation to September
R_s	Rainfall in August and September of the current year
R_g	Rainfall from August through November of the current year
August 1	GLMB2
$Y = b_0 + b_1W + b_2R + b_3E + e$	
$W = a_1U_{50} + a_2U_{30} + a_3\lvert U_{50} - U_{30}\rvert$	
$R = a_4R_s + a_5R_g$	
$E = a_6SLPA + a_7ZWA + a_8SOI + a_9SSTA$	
U_{50}, U_{30}, $\lvert U_{50} - U_{30}\rvert$	July data used to make an extrapolation to September
R_s	Rainfall in June and July of the current year
R_g	Rainfall from August through November of the prior year
SLPA, ZWA, SSTA, SOI	Data taken in June and July
June 1 GLMB3	
$Y = b_0 + b_1W + b_2R + b_3E + e$	
$W = a_1U_{50} + a_2U_{30} + a_3\lvert U_{50} - U_{30}\rvert$	
$R = a_4R_s + a_5R_g + a_6D_xP + a_7D_xT$	
$E = a_8SLPA + a_9ZWA + a_{10}SST + a_{11}D_tSST + a_{12}SOI + a_{13}D_tSOI$	
U_{50}, U_{30}, $\lvert U_{50} - U_{30}\rvert$	May data used to make an extrapolation to September
R_s	Rainfall in August and September of the prior year
R_g	Rainfall from August through November of the prior year
D_xP, D_xT	Current year February - May data
SLPA, ZWA, SST, SOI	Data taken in April and May
D_tSST, D_tSOI	Change measured from January and February to April and May

Table 3. Regression models.

2.2 Model-building Methodology

To build the three models, GLMB used least-absolute deviation (LAD) regression (a special case of linear programming) and a jackknife solution procedure. We describe their methodology for the GLMB2 model (see Table 3).

For each of the seven dependent variables, GLMB determined the values of a_1 through a_9 empirically using LAD regression and a jackknife procedure. The jackknife procedure is a cross-validation method in which a model is built to forecast a single year. With 41 years of data, GLMB withheld one year, built a model on the remaining 40 years of data, and predicted the value of the withheld year. The procedure was repeated 41 times and produced 41 predicted values for each of the seven dependent variables. The predicted values and the observed values for each of the 41 years are then compared via an agreement coefficient. Essentially, the agreement coefficient (a number between 0 and 1) measures how close observed and predicted value pairs come to falling on a line of unit slope that passes through the origin.

To predict future results, GLMB generated estimated values for b_0, b_1, b_2, and b_3 for each of the seven dependent variables. They calculated the regression coefficients using the jackknife solution weights for a_1 to a_9, LAD regression, and a nonjackknife solution procedure (that is, all 41 years of data were used). To illustrate, the prediction equation for the seasonal number of intense hurricanes was

$$\begin{aligned} IH = 3.459 &+ 0.041(1.0U_{50} + 0.614U_{30} - 0.467|U_{50} - U_{30}|) \\ &+ 1.142(1.0R_s + 0.591R_g) - 0.677(1.0SLPA + 0.051ZWA \\ &+ 0.474SOI + 0.630SSTA). \end{aligned}$$

GLMB developed six other prediction equations corresponding to the remaining dependent variables (see [3] for complete details).

3. NEURAL NETWORK MODELING EFFORT

In this section, we construct neural network models that predict hurricane activity (our first modeling effort in this area is described in [5]). We compare the results of our neural network models to the results of the more recent GLMB models (that is, the GLMB2 model built on 41 years of data for August 1 predictions and the GLMB3 model built on 42 years of data for June 1 predictions). Note that we did not experiment with GLMB1.

3.1 Neural Network Configurations and Parameter Settings

The neural network model that we use is a three layer feedforward structure with an input layer, a hidden layer, and an output layer. We build two neural network models: (1) the model denoted by NNA uses data available by August 1 to predict activity for the most active portion of the hurricane season and (2) the model denoted by NNB uses data available by June 1 to predict activity for June to November. The configurations and parameter settings for NNA and NNB are given in Table 4.

	Neural Network Model	
Configuration	*NNA*	*NNB*
Network Type	Feedforward	Feedforward
Learning Algorithm	Backpropagation*	Backpropagation*
Input Layer	9 Variables	13 Variables
Hidden Layer	3 Nodes	3 Nodes
Output Layer	1 of 7 Variables	1 of 8 Variables

* We use a variant known as quick backpropagation

Parameter	*Setting for NNA and NNB*
Learning Rate	0.20
Sigmoid Function Slope	0.40
Momentum Term	None
Training Length	100 Iterations

Table 4. Configurations and parameter settings for neural network models.

We coded a computer program in the C language to build our neural network models. Our program runs on a Sun 4/370 workstation and uses a backpropagation learning algorithm. Based upon previous experience, we end the training session when the number of iterations reaches 100 in order to avoid overtraining the network.

3.2 Training and Testing the Neural Network Models

In the applied neural network literature, the most common way of evaluating a model is to train a neural network on one set of data (known as the *training set*), use the resulting model to make predictions on a different set of data (known as the *test set*), and then evaluate the performance of the model by comparing the observed values to the predicted values for the test set data. However, the small sizes of the two hurricane data sets (41 and 42 observations) make it difficult to split them into useful training and test sets. Instead, we use a jackknife procedure (as did GLMB) to train our neural network and gain some insight into the overall viability of neural networks in modeling hurricane activity. Given a data set of size N, we train our model on N - 1 years of data and test our model on the remaining one year of data. We repeat the training procedure N times. This yields N predicted values for each of the dependent variables. We evaluate each model's performance by calculating the mean absolute error (that is, MAE = ð |Observed Value - Predicted Value|/N) and the agreement coefficient, denoted by AC (a detailed discussion of the agreement coefficient is given in [2]). We point out that it is not possible to directly compare the results of the jackknife models developed by GLMB to our jackknife models as the GLMB jackknife models are not published in the literature. Instead, we develop nonjackknifed neural network models using all years of data and compare the predictions of these models to the predictions of the GLMB nonjackknifed models.

3.3 Results of Neural Network Models

We develop seven nonjackknifed neural network models (denoted by NNA) using all 41 years of data from 1950 to 1990 and compare the predictions of these models to the predictions of the GLMB2 models reported by Gray et al. [3].

Table 5 gives the MAE values and Table 6 gives the agreement coefficient values for the neural network models and the regression models. In examining Tables 5 and 6, we see that for each dependent variable, NNA always produces better results than GLMB2, that is, NNA produces the lowest MAE value and the highest AC value for each dependent variable.

In 1994, GLMB published adjusted data for three dependent variables (IH, IHD, and HDP) and introduced a new variable NTC (see [4] for details). We develop eight nonjackknifed neural network models (denoted by NNA) using all 41 years of adjusted data from 1950 to 1990 and compare the predictions of these models to the predictions of the GLMB2 models reported by Gray et al. [4]. Table 7 gives the MAE values and Table 8 gives the AC values. We point out that the MAE values and the AC values for the first four columns in Tables 7 and 8 are the same as the values shown in Tables 5 and 6.

In examining Tables 7 and 8, we see that for each dependent variable, NNA always produces better results than GLMB2, that is, NNA produces the lowest MAE value and the highest AC value for each dependent variable.

We also develop eight nonjackknifed neural network models (denoted by NNB) using all 42 years of data from 1950 to 1991 and compare the predictions of these models to the predictions of the GLMB3 models reported by Gray et al. [4]. Table 9 gives the MAE values and Table 10 gives the agreement coefficient values for the neural network models and the regression models.

In examining Tables 9 and 10, we see that for each dependent variable, NNB always produces better results than GLMB3, that is, NNB produces the lowest MAE value and the highest AC value for each dependent variable.

3.4 Predictions for the 1993 and 1994 Atlantic Hurricane Seasons

We develop predictions for the 1993 Atlantic hurricane season using our NNA model and compare the neural network predictions to the predictions produced

	Dependent Variable						
Model	*NS*	*NSD*	*H*	*HD*	*IH*	*IHD*	*HDP*
GLMB2	1.573	8.491	1.141	6.708	0.695	2.073	21.250
NNA	1.033	7.295	0.909	5.033	0.551	1.615	14.960

Table 5. MAE values for models predicting by August 1.

	Dependent Variable						
Model	*NS*	*NSD*	*H*	*HD*	*IH*	*IHD*	*HDP*
GLMB2	0.447	0.608	0.472	0.516	0.618	0.617	0.557
NNA	0.654	0.643	0.592	0.643	0.696	0.701	0.701

Table 6. AC values for models predicting by August 1.

	Dependent Variable							
Model	*NS*	*NSD*	*H*	*HD*	*IH*	*IHD*	*HDP*	*NTC*
GLMB2	1.573	8.491	1.141	6.708	0.630	2.047	22.490	22.880
NNA	1.033	7.295	0.909	5.033	0.496	1.528	14.100	14.610

Table 7. MAE values for models predicting by August 1 using adjusted data.

	Dependent Variable							
Model	*NS*	*NSD*	*H*	*HD*	*IH*	*IHD*	*HDP*	*NTC*
GLMB2	0.447	0.608	0.472	0.516	0.568	0.541	0.449	0.511
NNA	0.654	0.643	0.592	0.643	0.670	0.657	0.699	0.712

Table 8. AC values for models predicting by August 1 using adjusted data.

	Dependent Variable							
Model	*NS*	*NSD*	*H*	*HD*	*IH*	*IHD*	*HDP*	*NTC*
GLMB3	1.448	7.239	0.910	4.530	0.582	1.809	14.380	14.770
NNB	1.034	5.407	0.771	3.230	0.517	1.173	11.100	11.370

Table 9. MAE values for models predicting by June 1.

	Dependent Variable							
Model	*NS*	*NSD*	*H*	*HD*	*IH*	*IHD*	*HDP*	*NTC*
GLMB3	0.513	0.659	0.616	0.694	0.636	0.612	0.708	0.717
NNB	0.651	0.740	0.650	0.772	0.644	0.734	0.761	0.772

Table 10. AC values for models predicting by June 1.

by the GLMB2 model. The predicted values generated by each model are given in Table 11. In examining this table, we see that NNA and GLMB2 always produced predicted values that were larger than the observed values. In four cases (NSD, H, HD, and IH), the neural network predictions were closer to the observed values than were the regression model predictions.

We develop predictions for the 1994 Atlantic hurricane season using our NNB model and compare the neural network predictions to the predictions produced by the GLMB3 model. The predicted values generated by each model are given in Table 12. In examining this table, we see that GLMB3 produced predicted values that were larger than the observed values in eight cases, while the predictions of NNB were larger in seven cases. Both GLMB3 and NNB overestimate hurricane activity in 1994. Gray and his colleagues attribute the lack of 1994 hurricane activity in part to the continued warm equatorial Pacific El Niño conditions (see Gray [1] and Stevens [7]).

4. Conclusions

In this application, our neural network models were easy to construct, fit the data reasonably well, produced fairly good predictions for the 1993 and 1994 hurricane seasons, and are easy to replicate. In addition, our neural network models showed a slight edge in accuracy over the regression models. It appears to us that neural network models are good alternatives to traditional statistical models. Neural networks should be given serious consideration when trying to build models that predict Atlantic hurricane activity.

	Dependent Variable						
Model	*NS*	*NSD*	*H*	*HD*	*IH*	*IHD*	*HDP*
GLMB2	9.8	52.1	6.5	23.9	1.7	1.1	50.4
NNA	10.1	40.9	5.1	16.2	1.1	4.3	53.0
OV	8	30	4	10	1	0.75	23

OV Observed Value

Table 11. Predicted and observed values for the 1993 Atlantic hurricane season.

			Dependent Variable					
Model	NS	NSD	H	HD	IH	IHD	HDP	NTC
GLMB3	8.9	28.0	4.7	12.2	0.6	0.2	30.2	0.65
NNB	7.0	20.4	5.8	8.5	1.3	1.8	21.4	0.52
OV	7	28	3	7	0	0	15	0.37

OV Observed Value

Table 12. Predicted and observed values for the 1994 Atlantic hurricane season.

REFERENCES

1. W. M. Gray, "Forecast of Atlantic Seasonal Hurricane Activity for 1994," Technical Report, Department of Atmospheric Sciences, Colorado State University, Fort Collins, Colorado (1994).

2. W. M. Gray, C. W. Landsea, P. W. Mielke, and K. J. Berry, "Predicting Atlantic Seasonal Hurricane Activity 6-11 Months in Advance," *Weather and Forecasting*, *7*, 440-455 (1992).

3. W. M. Gray, C. W. Landsea, P. W. Mielke, and K. J. Berry, "Predicting Atlantic Basin Seasonal Tropical Cyclone Activity by 1 August," *Weather and Forecasting*, *8*, 73-86 (1993).

4. W. M. Gray, C. W. Landsea, P. W. Mielke, and K. J. Berry, "Predicting Atlantic Basin Seasonal Tropical Cyclone Activity by 1 June," *Weather and Forecasting*, *9*, 103-115 (1994).

5. O. Kwon, B. Golden, E. Wasil, and R. Gordon, "Modeling Hurricane Activity Using Neural Networks," in *Proceedings of the International Conference on Neural Information Processing Volume 2* (edited by M-W. Kim and S-Y. Lee), Seoul, Korea, 1005-1009, (1994).

6. R. H. Simpson, "The Hurricane Disaster-Potential Scale," *Weatherwise*, *27*, 169 and 186 (1974).

7. W. K. Stevens, " '95 Hurricane Season Seen as One of the Fiercest in the Last 20 Years," *The New York Times*, Vol. CXLIV, No. 50105 (June 27), C1 and C7 (1995).

6

AN EFFICIENT DUAL SIMPLEX OPTIMIZER FOR GENERALIZED NETWORKS

Jeffery L. Kennington*, and Riad A. Mohamed**

** Southern Methodist University*
Dallas, Texas 75275-0122
***SABRE Decision Technologies*
DFW Airport, Texas 75261-9616

ABSTRACT

This Chapter describes a specialization of the dual simplex algorithm for the generalized network problem. We use a dual two phase method along with efficient dual partial pricing schemes and specialized routines for the dual ratio test. In comparison with CPLEX 3.0 on a set of ten benchmark problems, we found that our specialized dual code performed 20 times faster than the best CPLEX optimizer. Problems having 10 to 20 thousand arcs are routinely solved in under 10 seconds on a 60MHz DECStation 5000/260 with our rapid dual generalized network optimizer.

1 INTRODUCTION

The mathematical problem of finding a minimal cost flow through a capacitated network is a fundamental problem in both operations research and computer science. A generalization of this model which allows for either gains or losses as flows pass along an arc is known as the *generalized network problem*. Generalized networks have been used to model a wide array of applications in many engineering and economic areas. Applications that involve either gains or losses include electric power carried on transmission lines, cash management that involves the time value of money, and manufacturing processes of varying efficiencies. Other applications allow for flow conversion from one unit to another. Further discussion of a wide variety of applications can be found in Ahuja, Magnanti, and Orlin [1] , Glover, Klingman, and Phillips [11], Glover *et al.* [10], and Mulvey and Zenios [19].

1.1 Problem Description

Let $e_k = (i, j)$ denote an arc from node i to node j having multiplier a_k associated with node i, and b_k associated with node j. If $i = j$, then arc e_k has one end which is incident on node i, with a single multiplier a_k. Let $\mathcal{V} = \{1, 2, \ldots, m\}$ denote a set of m nodes, and let $\mathcal{A} = \{e_1, e_2, \ldots, e_n\}$ denote a set of n arcs. The corresponding generalized network can be represented on a directed graph $< \mathcal{V}, \mathcal{A} >$. For $e_k = (i, j)$ let

$$g_\ell^k = \begin{cases} a_k, & \text{for } \ell = i; \\ b_k, & \text{for } \ell = j; \\ 0, & \text{otherwise.} \end{cases}$$

Let $G = [g^1|g^2|\ldots|g^n]$, be an $m \times n$ matrix called the *node-arc coefficient* matrix. Let the n-component vectors c, l, u, and x denote the arc costs, arc lower bounds ,arc capacities, and arc flows; respectively. Let r be an m-component vector representing the right-hand-side. Let $\mathcal{G} = \{x \in \Re^n : Gx = r, l \leq x \leq u\}$ denote the set of feasible solutions. The generalized network problem can be stated mathematically as $\min_x \{cx : x \in \mathcal{G}\}$.

1.2 Survey of Literature

Generalized networks have the special structure that the basis can be represented graphically as a collection of trees and quasi-trees. In 1973, Glover, Klingman, and Stutz [12], developed the first specialized primal simplex code (NETG), that exploits the special graphical structure using the extended augmented predecessor index procedure. In 1987, Glover *et al.* [10] enhanced NETG, by investigating various rules for the starting strategy, the pivot selection criteria, and degeneracy handling. In 1984, Brown and McBride [6] presented a detailed description of a complete implementation of an efficient primal simplex system specialized for the generalized network problem. The system uses a preorder traversal method in addition to the predecessor, depth, and cycle factor to represent the basis. To enhance the algorithm performance, Brown and McBride included a basis aggregation technique, a dynamic queue scheme for selecting the entering variable, and a big-M starting strategy. In 1985, Engquist and Chang [8] presented a brief description for implementing the primal simplex code GRNET for generalized networks that is based on the labeling procedures of Barr, Glover, and Klingman [3]. Mulvey and Zenios [19] investigated the performance efficiency of the primal simplex generalized network code LPNETG, when using different internal programming tactics such as alternative pivot strategies, column normalization, and the big-M starting method. In 1988, Nulty and Trick [22] developed the first primal simplex code written in the C language. They use the predecessor, thread, reverse thread, and the level node labels presented in Kennington and Helgason [15] to represent the basis. In 1988, Bertsekas and Tseng [5] proposed a new class of algorithms, that adopt the nonlinear programming relaxation methods, to solve the linear cost generalized network problem. The algorithm, is based on the iterative improvement of the dual cost while maintaining a flow that satisfies complementary slackness. In 1992, Clark *et al.* [7] developed and empirically tested two primal simplex parallel algorithms, GENFLO, and GRNET2, for solving generalized networks. Both codes use a gradient penalty method to find a starting feasible solution.

1.3 Objective of Investigation

Although several papers have been presented that discuss the development of specialized algorithms and software implementations for solving generalized networks, none of these are based on the dual simplex method. The objective of this investigation is to develop and perform an empirical analysis of a specialization of a dual simplex algorithm for the generalized network problem. There are two key issues in the development of a dual procedure: (i) the creation of an effective dual pricing scheme, and (ii) the creation of an efficient scheme for computing an updated row of the coefficient matrix which exploits the underlying network structure.

2 THE DUAL SIMPLEX ALGORITHM

Duality theory in linear programming is well known. Each variable in the primal corresponds to a constraint in the dual, and each constraint in the primal is associated with a dual variable. The dual simplex method has received limited attention, and its role has been limited to sensitivity analysis, parametric programming, and the solution of integer programming problems (see Nemhauser and Wolsey [21] and Parker and Rardin [23]).

There are some instances where the dual method has an advantage over the primal method, see Ali, Padman, and Thiagarajan [2]. The reported results of their dual simplex code, specialized for pure network problems, illustrates improved performance on a subset of the benchmark of standard NETGEN problems (Klingman, Napier, and Stutz [16]), and an additional set of larger transportation problems.

Under the non-degeneracy assumption, the primal simplex algorithm starts with a basic primal feasible solution and pivots to a new one with an improved objective function. These pivots are repeated until primal optimality is achieved, i.e. obtaining a dual feasible solution. The dual method starts with a dual feasible solution and pivots to a new one with an improved objective function, while attempting to reduce primal infeasibilities. This is repeated until dual optimality (primal feasibility) is achieved. In this section we present the dual simplex algorithm, pricing techniques for selecting the leaving variable, and a specialization for generalized network problems.

Let π_i denote the dual variable associated with the ith conservation of flow constraint, and let λ_j denote the dual variable associated with the lower bound constraint on the jth variable, and let μ_j denote the dual variable associated with the upper bound constraint on the jth variable. Let $\mathcal{D} = \{(\pi \in \Re^m, \lambda \in \Re^n, \mu \in \Re^n) : G^T\pi + \lambda - \mu = c,\ \lambda \geq 0,\ \mu \geq 0\}$, denote the set of dual feasible solutions. Therefore, the dual problem for the generalized network problem is $\max_{\pi,\lambda,\mu} \{r\pi + l\lambda - u\mu : (\pi, \lambda, \mu) \in \mathcal{D}\}$. For any $\hat{q} \in \mathcal{V}$, let $q = A(\hat{q})$ denote the index of the associated basic variable x_q, and let $e^{\hat{q}}$ denote the corresponding column of an $m \times m$ identity matrix. Recall that G is an $m \times n$ matrix with full row rank, $g^j \in \Re^m$ denotes the jth column of G, B denotes a basis. Let $\bar{x}$ be an extreme point and let the flow variables be partitioned into *basic* and *non-basic* variables. A non-basic variable must assume one of it's bounds, i.e. $\bar{x}_k = l_k$ or $\bar{x}_k = u_k$. Let $\mathcal{N}_l = \{k : \bar{x}_k = l_u, e_k \in \mathcal{A}\}$, denote the set of indices of the non-basic variables at lower bound, let $\mathcal{N}_u = \{k : \bar{x}_k = u_k, e_k \in \mathcal{A}\}$, denote the set of indices of the non-basic variables at upper bound, and let $\mathcal{N} = \mathcal{N}_l \cup \mathcal{N}_u$. Let $x^N \in \Re^{n-m}$ denote the vector of non-basic variables; let $x^B \in \Re^m$ denote the vector of basic variables; and let $\mathcal{B} = \mathcal{A} \backslash \mathcal{N}$ denote the set of indices of the basic variables. Recall that $\pi \in \Re^m$ denotes the vector of duals associated with the flow conservation constraints $Gx = r$, and that $A(i)$ is the index k, such that e_k is the basic variable associated with constraint i. Let $\sigma_k = c_k - \pi g^k,\ k \in \mathcal{A}$, denote the reduced cost for arc e_k. The dual simplex algorithm can be stated as follows:

Procedure *Dual*
Input : $c, G, l, r, u, \mathcal{B}, \mathcal{N}_l, \mathcal{N}_u$.
Output : $\bar{x}^B, \bar{x}^N, v^*$.
begin
$\bar{\pi} \leftarrow c^B B^{-1};\ \sigma_k \leftarrow c_k - \bar{\pi} g^k,\ k \in \mathcal{N};$
$\bar{r} \leftarrow r - \sum_{k \in \mathcal{N}_u} g^k u_k - \sum_{k \in \mathcal{N}_l} g^k l_k;\ \ \bar{x}^B \leftarrow B^{-1}\bar{r};$

$v^* \leftarrow \sum_{k\in\mathcal{N}_u} c_k u_k + \sum_{k\in\mathcal{N}_l} c_k l_k + \sum_{k\in\mathcal{B}} c_k \bar{x}_k^B$;
$\mathcal{I} \leftarrow \{i \in \mathcal{V} : (\bar{x}_i^{\mathcal{B}} < l_i^{\mathcal{B}}) \text{ or } (\bar{x}_i^{\mathcal{B}} > u_i^{\mathcal{B}})\}$;
while($\mathcal{I} \neq \phi$)**do**
  select $\hat{q} \in \mathcal{I}$; $q \leftarrow A(\hat{q})$; /* leaving variable */
  $z \leftarrow e^{\hat{q}} B^{-1}$; /* row $\hat{q}$ of basis inverse */
  $w \leftarrow \text{sign}(u_q - \bar{x}_q) z G$; /* updated row */
  $\mathcal{E} \leftarrow \{j \in \mathcal{N}_l : w_j < 0\} \cup \{j \in \mathcal{N}_u : w_j > 0\}$;
  if($\mathcal{E} = \phi$)**then** exit('unbounded dual problem');
  $\begin{bmatrix} \delta \\ s \end{bmatrix} \leftarrow \begin{matrix} \max_{j\in\mathcal{E}} \\ \text{argmax}_{j\in\mathcal{E}} \end{matrix} \left\{ \frac{\sigma_j}{w_j} \right\}$; /* ratio test */
  $y \leftarrow B^{-1} g^s$; $\mathcal{P} \leftarrow \{i : y_i \neq 0\}$; /* updated column */
  $\Delta \leftarrow \left\{ \begin{matrix} \frac{(\bar{x}_q - l_q)}{y_{\hat{q}}} & \text{if } \bar{x}_q < l_q \\ \frac{(\bar{x}_q - u_q)}{y_{\hat{q}}} & \text{if } \bar{x}_q > u_q \end{matrix} \right\}$;
  $\bar{x}_i^B \leftarrow \bar{x}_i^B - \Delta y_i,\ i \in \mathcal{P}$; $\bar{x}_s \leftarrow \bar{x}_s + \Delta$; /* flow update */
  if($\delta < 0$)**then**
    $\sigma_k \leftarrow \sigma_k - \delta w_k,\ k \in \mathcal{N}$; $\sigma_q \leftarrow \left\{ \begin{matrix} \delta & \text{if } \bar{x}_q = l_q \\ -\delta & \text{if } \bar{x}_q = u_q \end{matrix} \right\}$;
    $\bar{\pi} \leftarrow \bar{\pi} + \delta z$; $v^* \leftarrow v^* + \sigma_s \Delta$; /* duals and objective update */
  endif
  $\mathcal{B} \leftarrow \mathcal{B} \cup \{s\} \backslash \{q\}$; $\mathcal{N}_l \leftarrow \mathcal{N}_l \backslash \{s\}$; $\mathcal{N}_u \leftarrow \mathcal{N}_u \backslash \{s\}$; /* basis update */
  if($\bar{x}_q = l_q$)**then** $\mathcal{N}_l \leftarrow \mathcal{N}_l \cup \{q\}$;
  if($\bar{x}_q = u_q$)**then** $\mathcal{N}_u \leftarrow \mathcal{N}_u \cup \{q\}$;
  $\mathcal{I} \leftarrow \{i \in \mathcal{V} : (\bar{x}_i^{\mathcal{B}} < l_i^{\mathcal{B}}) \text{ or } (\bar{x}_i^{\mathcal{B}} > u_i^{\mathcal{B}})\}$;
endwhile
end.

2.1 A Specialized Implementation for Generalized Networks

The main operations of a dual simplex pivot are selecting the leaving variable, calculating the updated row, determining the entering variable, and updating the flows, the reduced cost, and the inverse of the basis. For generalized networks, most of these operations can be carried out on the tree representing the basis, and the inverse of the basis is never calculated.

Calculating a row of the basis inverse, $z = e^{\hat{q}} B^{-1}$, can be considered as a special case of the dual calculations where the vector c is replaced by $e^{\hat{q}}$ with a single nonzero entry in position $\hat{q}$. For more details on the specialized dual calculations see Mohamed [18]. Let $\mathcal{Z} = \{i : z_i \neq 0\}$, denote the set of indices for the nonzero entries in z. Let B^q be the basis partition corresponding to the augmented forest component that contains node $\hat{q}$, and let z^q and $e^{\tilde{q}}$, be the matching partitions of z and $e^{\hat{q}}$. Suppose that B^q corresponds to a rooted-tree, T^r. The system $z^q B^q = e^{\tilde{q}}$ can be solved using back substitution, and $\mathcal{Z}$ will only contain the nodes in the subtree component rooted at node $\hat{q}$. Suppose that B^q corresponds to a one-tree, T^o and consider only the cycle of T^o. For notation purpose, assume that the cycle nodes are numbered $1, \ldots, \hat{m}$. First, the z^q values for the cycle nodes are obtained using

$$\begin{aligned} z_1^q &= \frac{1}{\hat{a}_1(1 - w_1 w_2 \cdots w_{\hat{m}})}, \text{ and} \\ z_{i+1}^q &= \frac{-\hat{a}_i}{\hat{b}_i} z_i, \; i = 1, \ldots, \hat{m} - 1. \end{aligned}$$

Then, the z^q values for the remaining nodes in the one-tree are calculated using back substitution. In this case $\mathcal{Z}$ will contain all nodes in the one-tree T^o.

The calculation of the the updated row, $w = zG$, is a dominant operation in the dual pivot. Preliminary profile investigation revealed that, when using standard matrix operations, approximately 50% of the time is spent in calculating w with a complexity of $O(mn)$. This time can be drastically reduced by exploiting the sparsity of z and the special structure of the network problems. Let G^i denote the ith row of G. To exploit the sparsity of z, the updated row calculation is carried out as follows:

$$w = \sum_{i \in \mathcal{Z}} z_i G^i.$$

This reduces the complexity of calculating w to $O(|\mathcal{Z}|n)$, where $|\mathcal{Z}| \ll m$ in most cases. Let $\mathcal{F}^i = \{k : e_k = (i, v) \in \mathcal{A}, v \in \mathcal{V}\}$, the forward star of node i, and let $\mathcal{T}^i = \{k : e_k = (v, i) \in \mathcal{A}, v \in \mathcal{V}\}$, the backward star of node i. Let $\mathcal{W} = \{j : w_j \neq 0\}$. For network problems, G^i can be represented using the forward and the backward stars of node i. This special structure of the network problem has a significant influence on the sparsity of w, that is $\mathcal{W} = \cup_{i \in \mathcal{Z}} (\mathcal{F}^i \cup \mathcal{T}^i)$. This reduces the complexity of calculating the updated row to $O(|\mathcal{Z}|\tilde{n})$, where $\tilde{n}$ is the average number of arcs incident to a node. In practice $\tilde{n} \ll n$.

The entering variable, x_s, and the change in the reduced cost, δ, are determined in the ratio test, which is conducted on the nonbasic variables in $\mathcal{E} \subseteq \mathcal{W}$. The updated column, y, is calculated using the specialized procedure described in Helgason and Kennington [15]. The flow update is carried out only on the basic arcs associated with the nodes in $\mathcal{P}$. For additional information see Mohamed [18]. This is followed by updating the reduced cost of the nonbasic variables in $\mathcal{W}$, updating the node labels, updating $\mathcal{B}$ and $\mathcal{N}$ to obtain the new basis, and updating the duals.

Let k denote the kth iteration of the dual simplex algorithm. Let $\pi^{(k)}$ be the vector of duals at iteration k, let $B^{(k)}$ be the basis matrix at iteration k, and let $D^{(k)} = [B^{(k)}]^{-1}$. Let $q \in \mathcal{B}$ be the leaving variable, and let $\hat{q}$ denote it's position in B. Therefore, the dual vector at iteration $k+1$ is defined as follows:

$$\pi^{(k+1)} = c^{B^{(k+1)}} D^{(k+1)}, \tag{6.1}$$

where

$$D^{(k+1)} = \left\{ I - \frac{1}{y_p}(y - e^{\hat{q}})e^{\hat{q}^T} \right\} D^{(k)}, \tag{6.2}$$

and

$$c^{B^{(k+1)}} = c^{B^{(k)}} \left\{ I - \left(\frac{c_q - c_s}{c_q} \right) e^{\hat{q}} e^{\hat{q}^T} \right\}. \tag{6.3}$$

Substituting (6.2) and (6.3) into (6.1) we obtain

$$\begin{aligned}
\pi^{(k+1)} &= \left\{ c^{B^{(k)}} - c^{B^{(k)}} \left(\frac{c_q - c_s}{c_q} \right) e^{\hat{q}} e^{\hat{q}^T} \right\} \left\{ D^{(k)} - \frac{1}{y_q}(y - e^{\hat{q}})e^{\hat{q}^T} D^{(k)} \right\} \\
&= c^{B^{(k)}} D^{(k)} - c^{B^{(k)}} \left(\frac{c_q - c_s}{c_q} \right) e^{\hat{q}} e^{\hat{q}^T} D^{(k)} - c^{B^{(k)}} \frac{(y - e^{\hat{q}})}{y_q} e^{\hat{q}^T} D^{(k)} \\
&\quad + c^{B^{(k)}} \left(\frac{c_q - c_s}{c_q} \right) \frac{e^{\hat{q}} e^{\hat{q}^T}}{y_q} (y - e^{\hat{q}}) e^{\hat{q}^T} D^{(k)} \\
&= \pi^k + \frac{1}{y_q} \left\{ c^{B^{(k)}} y_q \left(\frac{c_s - c_q}{c_q} \right) e^{\hat{q}} - c^{B^{(k)}} (y - e^{\hat{q}}) \right. \\
&\quad \left. + c^{B^{(k)}} \left(\frac{c_q - c_s}{c_q} \right) e^{\hat{q}} e^{\hat{q}^T} (y - e^{\hat{q}}) \right\} e^{\hat{q}^T} D^{(k)} \\
&= \pi^k + \frac{1}{y_q} \left\{ c^{B^{(k)}} y_q \left(\frac{c_s - c_q}{c_q} \right) e^{\hat{q}} + c^{B^{(k)}} (e^{\hat{q}} - y) \right. \\
&\quad \left. + c^{B^{(k)}} \left(\frac{c_q - c_s}{c_q} \right) e^{\hat{q}} (y_q - 1) \right\} e^{\hat{q}^T} D^{(k)} \\
&= \pi^k + \frac{1}{y_q} \left\{ c^{B^{(k)}} (e^{\hat{q}} - y) - c^{B^{(k)}} e^{\hat{q}} \left(\frac{c_q - c_s}{c_q} \right) \right\} e^{\hat{q}^T} D^{(k)} \\
&= \pi^k + \frac{1}{y_q} \left\{ c_q - c^{B^{(k)}} y - c_q \left(\frac{c_q - c_s}{c_q} \right) \right\} e^{\hat{q}^T} D^{(k)} \\
&= \pi^k + \left(\frac{c_s - c^{B^{(k)}} y}{y_q} \right) e^{\hat{q}^T} D^{(k)}
\end{aligned}$$

$$= \pi^{(k)} + \left(\frac{\sigma_s}{y_q}\right) e^{\hat{q}^T} D^{(k)}$$

Therefore, the dual vector can be updated using $\pi^{(k+1)} = \pi^{(k)} + \left(\frac{\sigma_s}{y_q}\right) z$, and the sparsity of z can be exploited in this update.

2.2 Dual Pricing Strategies

Since all nonbasic variables assume one of their bounds, primal infeasibility occurs only when the basic variables violate the bounds. The dual simplex pricing strategy to select the leaving variable x_q has received very little attention in the literature. Although the selection of the leaving variable appears to be a simple operation, the exact strategy used is very important in the development of an efficient generalized network code.

The standard approach for choosing the leaving variable is to scan the basic variables and determine the variable having maximum bound violation. The index of the leaving variable selected using the *maximum-infeasibility* strategy is

$$q = \text{argmax}_{k \in \mathcal{B}} \left\{ \begin{array}{ll} l_k - \bar{x}_k & \text{if } \bar{x}_k < l_k \\ \bar{x}_k - u_k & \text{if } \bar{x}_k > u_k \end{array} \right\}.$$

Using this approach, every basic variable is scanned at each iteration, which requires a node length search. This section presents techniques which improve this search by performing a partial pricing where only a subset of the basic variables are scanned.

Let $\rho \leq |\mathcal{V}|$ denote the page length. The *fixed-page pricing* technique scans a page of basic variables at a time. If one prices favorable, then it is selected; otherwise, another page is scanned. This continues until either a favorable variable is found or all basic variables have been priced. The list of basic variables are treated in a wrap-around fashion. Let ℓ denote the index of the most recently priced entry, with ℓ initially set to one. The fixed-page pricing technique can be described mathematically as follows:

Procedure : *D_Fixed_Page_Pricing*
Input : ℓ.
Output : ℓ, q.
begin
  $j \leftarrow q \leftarrow \Delta_q \leftarrow 0$;
  while($j \leq |\mathcal{V}|$ & $q = 0$)**do**
    $i \leftarrow 1$; $\hat{\rho} \leftarrow \min\{\rho, |\mathcal{V}|\}$;

while($i \leq \hat{\rho}$)**do** $/ *$ price $\hat{\rho}$ basic arcs $* /$

$k \leftarrow A(\ell);$ $/ *$ get basic arc index associated with node $\ell * /$

$$\Delta_k \leftarrow \begin{cases} l_k - \bar{x}_k & \text{if } \bar{x}_k < l_k; \\ \bar{x}_k - u_k & \text{if } \bar{x}_k > u_k; \\ 0 & \text{otherwise}; \end{cases} \quad / * \text{ get the bound violation value } * /$$

$$\begin{bmatrix} \Delta_q \\ q \end{bmatrix} \leftarrow \begin{matrix} \max \\ \text{argmax} \end{matrix} \{\Delta_k, \Delta_q\};$$

$i \leftarrow i + 1;\ j \leftarrow j + 1;\ \ell \leftarrow \text{mod}(\ell, |\mathcal{V}|) + 1;$

endwhile

endwhile

end.

The fixed-page pricing strategy can be enhanced by maintaining multiple pricing candidates. A candidate-queue, which is a cyclic list of varying length, is used to keep the indices of the favorable variables (variables violating their bounds). The ***candidate-queue pricing*** technique starts with a general scan pricing a set of basic variables to select the most favorable variable (the variable having the maximum bound violation), while placing other favorable variables into the queue. At each iteration a subset of the queue entries are priced. The most favorable variable becomes the leaving variable, and other variables that price favorable are placed back in the queue. A general scan is used to replenish an exhausted queue, where each scan starts where the previous one has ended.

Let Q, called the ***candidate-queue***, be a sequence of entries, and let ω denote the maximum allowable size of the candidate-queue. Let $Q[1]$ denote the first entry of the candidate-queue, and let $Q[i \dots j]$ denote entries i through j. Let $Q\&\&(t)$ denote the concatenation of (t) to the end of Q. Let ρ denote the number of queue candidates to be priced in each iteration. Let ℓ denote the index of the node associated with the most recently priced basic variable, with ℓ initialized to one. The candidate-queue dual pricing technique can be represented as follows:

Procedure : *D_Candidate_Queue_Pricing*

Input : ℓ, Q.

Output : ℓ, q, Q.

begin

$q \leftarrow \Delta_q \leftarrow 0;$

while($Q \neq \phi$ & $q = 0$)**do**

$i \leftarrow 1;\ \hat{\rho} \leftarrow \min\{\rho, |Q|\};$

while($i \leq \hat{\rho}$)**do** $/ *$ price $\hat{\rho}$ queue entries $* /$

$j \leftarrow Q[1];\ \ Q \leftarrow Q[2 \dots |Q|];$ $/ *$ get the first queue entry $* /$

$k \leftarrow A(j);$ $\quad$ / * get basic arc index associated with node j * /

$$\Delta_k \leftarrow \begin{cases} l_k - \bar{x}_k & \text{if } \bar{x}_k < l_k; \\ \bar{x}_k - u_k & \text{if } \bar{x}_k > u_k; \\ 0 & \text{otherwise;} \end{cases} \quad \text{/ * get the bound violation value * /}$$

$$\begin{bmatrix} q \\ k \end{bmatrix} \leftarrow \begin{matrix} \text{argmax} \\ \text{argmin} \end{matrix} \{\Delta_k, \Delta_q\};$$

if($\Delta_k > 0$)**then** $Q \leftarrow Q$ && (k);

$i \leftarrow i + 1;$

endwhile

endwhile

if($q = 0$)**then** *D_General_scan*(ℓ, q, Q); / * replenish the empty queue * /

end.

The mechanism for replenishing an empty queue, is to perform a general pricing scan on κ basic variables, and place the favorable arcs into the queue. The general scan procedure can be stated as follows:

Procedure: *D_General_scan*

Input : ℓ, Q.

Output : ℓ, q, Q.

begin

$q \leftarrow \Delta_q \leftarrow 0;\ j \leftarrow 1;$

while($j \leq |\mathcal{V}|$ & $q = 0$)**do**

$i \leftarrow 1;$

while($i \leq \kappa$ & $|Q| \leq \omega$)**do**

$k \leftarrow A(\ell);$ $\quad$ / * get basic arc index associated with node ℓ * /

$$\Delta_k \leftarrow \begin{cases} l_k - \bar{x}_k & \text{if } \bar{x}_k < l_k; \\ \bar{x}_k - u_k & \text{if } \bar{x}_k > u_k; \\ 0 & \text{otherwise;} \end{cases} \quad \text{/ * get the bound violation value * /}$$

$$\begin{bmatrix} q \\ k \end{bmatrix} \leftarrow \begin{matrix} \text{argmax} \\ \text{argmin} \end{matrix} \{\Delta_k, \Delta_q\};$$

if($\Delta_k > 0$)**then** $Q \leftarrow Q$ && (k);

$i \leftarrow i + 1; j \leftarrow j + 1;\ \ell \leftarrow \text{mod}(\ell, |\mathcal{V}|) + 1;$

endwhile

endwhile

end.

The above pricing strategies (maximum-infeasibility, fixed-page, candidate-queue) are similar to strategies that have been used in the primal algorithm. For problems having 4000 or more nodes, the pricing strategy can affect the performance of the dual algorithm.

2.3 Steepest Edge Dual Simplex Algorithm

In the simplex method, we proceed from one vertex to an adjacent one along a selected edge. When using standard pricing, the edge is selected such that the objective function improves the most along that edge. In the standard dual simplex method, the size of the bound violation is taken directly to indicate the gradient improvement. This can be regarded as measuring the gradient in the framework of the current basic variables. In the steepest edge method, the selected edge is the one along which the improvement of the objective function is the highest with respect to the original solution space. This involves taking into account the variables that change along the edge. That is, the selected edge is the one that makes the largest, most obtuse angle with the gradient of the objective function.

It is easy to show that the edge directions for a particular basic dual solution are given by the rows of the corresponding basis inverse, B^{-1}. To select edges that are row-length independent, edge normalization was introduced, where weighting factors are used for estimating the edge-length. The steepest-edge pricing is one of a variety of methods for normalized pricing, in which the bound violations are row scaled. However, it is not practical to calculate the edge-lengths at each iteration. Harris [14], used a dynamic subset of the original solution space as a working framework. An approximation of the weighting factors are obtained using the Euclidean norm of the sub-vector consisting of just the components in the current working framework. That is, the Harris' variant of the steepest edge, called *Devex*, chooses the edges that are only approximately steepest. Goldfarb and Reid [13], developed a recurrence formula for the squares of the Euclidean norms of the edge direction when pivoting from one feasible basis to a new one. Forrest and Goldfarb [9], introduced several new steepest edge algorithms for both the primal and the dual simplex methods.

Let $\eta_i = \|e^i B^{-1}\|^2, i = 1, \ldots, m$, denote the length squared of row i of the basis inverse. Following the previously described dual simplex terminology, the steepest edge dual simplex algorithm, using the Goldfarb and Reid recurrence formula, for a given dual feasible solution defined by $\mathcal{B}, \mathcal{N}_l$, and $\mathcal{N}_u$ can be stated as follows:

Procedure : *Dual_Steepest_Edge*
Input : $c, G, l, r, u, \mathcal{B}, \mathcal{N}_l, \mathcal{N}_u$.
Output : $\bar{x}^B, \bar{x}^N, v^*$.
begin

$\bar{\pi} \leftarrow c^B B^{-1}; \quad \sigma_k \leftarrow c_k - \bar{\pi} g^k,\ k \in \mathcal{N};$

$\bar{r} \leftarrow r - \sum_{k \in \mathcal{N}_l} g^k l_k - \sum_{k \in \mathcal{N}_u} g^k u_k;\ \bar{x}^B \leftarrow B^{-1}\bar{r};$
$v^* \leftarrow \sum_{k \in \mathcal{N}_u} c_k u_k + \sum_{k \in \mathcal{N}_l} c_k l_k + \sum_{k \in \mathcal{B}} c_k \bar{x}_k^B;$
for($i \in \mathcal{B}$)**do** $\quad z \leftarrow e^i B^{-1};\ \eta_i \leftarrow zz;$ $\qquad$ /∗ initiate the row norms ∗/
while($\exists i \in \mathcal{B} : (\bar{x}_i < l_i)$ or $(\bar{x}_i > u_i)$)**do**

$$\hat{q} \leftarrow \operatorname{argmax} \left\{ \begin{array}{ll} \frac{l_{A(k)} - \bar{x}_k^B}{\eta_k} & \text{if } \bar{x}_k^B < l_k \\ \frac{\bar{x}_k^B - u_{A(k)}}{\eta_k} & \text{if } \bar{x}_k^B > u_k \end{array} \right\},\ k = 1, \ldots, m;$$

$q \leftarrow A(\hat{q});\ z \leftarrow e^{\hat{q}} B^{-1};$
$w \leftarrow \operatorname{sign}(u_q - \bar{x}_q) z G;$ $\qquad$ /∗ updated row ∗/
$\mathcal{E} \leftarrow \{j \in \mathcal{N}_l : w_j < 0\} \cup \{j \in \mathcal{N}_u : w_j > 0\};$

$$\begin{bmatrix} \delta \\ s \end{bmatrix} \leftarrow \begin{array}{c} \max_{j \in \mathcal{E}} \\ \operatorname{argmax}_{j \in \mathcal{E}} \end{array} \left\{ \frac{\sigma_j}{w_j} \right\};$$ /∗ Ratio Test ∗/

$y \leftarrow B^{-1} g^s;\ \mathcal{P} = \{i : y_i \neq 0\};$ $\qquad$ /∗ updated column ∗/

$$\Delta \leftarrow \left\{ \begin{array}{ll} \frac{(\bar{x}_q - l_q)}{y_{\hat{q}}} & \text{if } \bar{x}_q < l_q \\ \frac{(\bar{x}_q - u_q)}{y_{\hat{q}}} & \text{if } \bar{x}_q > u_q \end{array} \right\};$$

$\bar{x}_i^B \leftarrow \bar{x}_i^B - \Delta y_i, i \in \mathcal{P};\ \bar{x}_s \leftarrow \bar{x}_s + \Delta;$ $\qquad$ /∗ flow update ∗/

$$\sigma_k \leftarrow \sigma_k - \delta w_k,\ k \in \mathcal{N};\ \sigma_q \leftarrow \left\{ \begin{array}{ll} \delta & \text{if } \bar{x}_q = l_q \\ -\delta & \text{if } \bar{x}_q = u_q \end{array} \right\};$$

$v^* \leftarrow v^* + \sigma_s \Delta;$ $\qquad$ /∗ update the objective value ∗/
$\bar{\pi} \leftarrow \bar{\pi} - \delta z;\quad t \leftarrow B^{-1} z;$
for($i = 1, \ldots, m : i \neq \hat{q}$)**do** $\qquad$ /∗ update the norms ∗/

$$\eta_i \leftarrow \eta_i - 2\left(\frac{y_i}{y_q}\right) t_i + \left(\frac{y_i}{y_q}\right)^2 \eta_{\hat{q}};$$

$$\eta_i \leftarrow \max\left(\eta_i, 1 + \left(\frac{w_i}{w_q}\right)^2\right);$$

endfor

$$\eta_{\hat{q}} \leftarrow \left(\frac{1}{w_q}\right)^2 \eta_{\hat{q}};$$

$\mathcal{B} \leftarrow \mathcal{B} \cup \{s\} \backslash \{q\};\ \mathcal{N}_l \leftarrow \mathcal{N}_l \backslash \{s\};\ \mathcal{N}_u \leftarrow \mathcal{N}_u \backslash \{s\};$ /∗ basis update ∗/
if($\bar{x}_q = l_q$)**then** $\mathcal{N}_l \leftarrow \mathcal{N}_l \cup \{q\};$
if($\bar{x}_q = u_q$)**then** $\mathcal{N}_u \leftarrow \mathcal{N}_u \cup \{q\};$
endwhile
end.

In the above dual steepest edge procedure, the standard maximum-infeasibility strategy is used for selecting the leaving variable. The partial pricing techniques were also extended to be used in variants of the steepest edge algorithm.

2.4 An Initial Dual Feasible Solution

A solution $\bar{x}$, defined by $\mathcal{B}, \mathcal{N}_l, \mathcal{N}_u$, is said to be dual feasible if the corresponding dual variables $(\pi, \lambda, \mu) \in \mathcal{D}$. Thus far, we assumed that the dual simplex algorithms are initiated with a dual feasible solution. The strategy needed to obtain a starting dual feasible solution was not evident. A starting basis B for the generalized network problem is obtained by adding an artificial variable $(\zeta_i, i = 1, \ldots, m)$ with zero bounds, to each of the constraints $Gx = r$. Therefore, the set of dual feasible solutions $\mathcal{D}^A = \{\pi \in \Re^m, \lambda^x \in \Re^n, \lambda^\zeta \in \Re^m, \mu^x \in \Re^n, \mu^\zeta \in \Re^m : G^T\pi + \lambda^x - \mu^x = c^x,\ \pi + \lambda^y - \mu^y = c^\zeta,\ \lambda^x \geq 0, \mu^x \geq 0,\ \lambda^y \geq 0, \mu^y \geq 0\}$. In order to assure that $\mathcal{D}^A$ is equivalent to $\mathcal{D}$, the artificial cost vector c^ζ is set to zero. Therefore, for this starting basis, $\pi = 0$, the arc reduced costs are simply the arc costs, and any feasible solution of $\mathcal{D}$, say (π, λ^x, μ^x), has a corresponding feasible solution $(\pi, \lambda^x, \lambda^\zeta, \mu^x, \mu^\zeta)$ where $\lambda^\zeta - \mu^\zeta = \pi$. In the following sections, we introduce techniques for obtaining a dual feasible solution from any starting basis.

Big-M Method

In this approach, the initial dual feasible solution is obtained by setting the original variables $(x_i, i = 1, \ldots, n)$ to be nonbasic at the appropriate bound based on the sign of the corresponding reduced costs. That is, $\mathcal{N}_l = \{j \notin \mathcal{B} : \sigma_j \geq 0\}$, and $\mathcal{N}_u = \{j \notin \mathcal{B} : \sigma_j < 0\}$. Then, the artificial variables $\zeta = \bar{r}$, where $\bar{r}$ is the updated right-hand-side. An infinite upper bound is assigned to the original unbounded variables, where infinity is approximated by a large positive constant, usually called big-M. The dual problem becomes $\max_{(\pi,\lambda,\mu)\in\mathcal{D}^A} \{r\pi + l\lambda - u\mu\}$, and optimality is achieved when all variables satisfy primal feasibility. Since the artificial variables have zero bounds, the corresponding flows will be forced to zero at optimality. Also, variables with big-M upper bounds will be driven to a smaller value. Other big-M approaches have been suggested to obtain a dual feasible starting basis (Murty [20]).

Two Phase Method

The big-M approach for setting the initial basis is not recommended because of difficulties in making the appropriate choice of M. To avoid this critical setting of the big-M value, a two phase approach is usually used. In phase 1 a new problem associated with the original dual problem is defined whose optimal solution is dual feasible for the original problem. The initial solution is obtained by setting the original variables $(x_i, i = 1, \ldots, n)$ to be nonbasic at the appropriate bound based on the sign of the corresponding reduced costs. Unbounded variables with negative reduced costs are assigned to their lower bounds, resulting in a dual infeasibility equal to the sum of the corresponding reduced cost. Therefore, $\mathcal{N}_u = \{j \notin \mathcal{B} : \sigma_j < 0, u_k \text{ is finite}\}$, $\mathcal{I} = \{j \notin \mathcal{B} : \sigma_j < 0, u_k \text{ is undefined}\}$, and $\mathcal{N}_l = \{j \notin \mathcal{B} : \sigma_j \geq 0\} \cup \mathcal{I}$. The phase 1 objective function is obtained by temporarily setting the lower and upper bounds to zero for all variables and introducing a new right-hand-side, $\tilde{r} = \sum_{k \in \mathcal{I}}\{-g^k\}$. Therefore, the artificial variables $\bar{\zeta} = \tilde{r}$, and the phase 1 objective is to $\max_{(\pi,\lambda,\mu)\in\mathcal{D}^A}\{\tilde{r}\pi\}$. This approach is similar to and influenced by the two phase method developed by Professor Bixby in CPLEX 3.0. Let σ_j denote the current reduced cost for $j \in \mathcal{I}$. The contribution to dual infeasibility by x_j is $(0 - (\sigma_j - \delta w_j))$. Thus, the total sum of infeasibility is $\sum_{j\in\mathcal{I}}(\delta w_j - \sigma_j)$. Differentiating with respect to δ, the rate of change of the sum of dual infeasibilities is $\sum_{j\in\mathcal{I}}\{w_j\} = \sum_{j\in\mathcal{I}}\{-zg^j\} = z\sum_{j\in\mathcal{I}}\{-g^j\}$. Let x_q be the leaving variable. Therefore, the rate of change of the objective function is $x_q = e^{\hat{q}}\bar{x} = e^{\hat{q}}B^{-1}\bar{r} = z(r - \sum_{j\in\mathcal{N}_u} g^j u_j - \sum_{j\in\mathcal{N}_l} g^j l_j)$. Setting the upper and lower bounds on all variables to zero, and setting the right-hand-side to be defined in terms of the variables that violate dual feasibility, results in having the rate of change of the objective function to be the same as the rate of change of the sum of dual infeasibilities. This approach can be applied to any basis, and it is not restricted to an all artificial starting basis. Another intuitive interpretation of this setting is that for any $j \in \mathcal{I}$ the reduced cost $\sigma_j = c_j - \pi g^j < 0$. The objective function to drive σ_j to be nonnegative is $\min\{\pi g^j\}$. Therefore, for all the dual infeasible variables the objective function is $\min\{\pi \sum_{j\in\mathcal{I}} g^j\}$.

The computational steps of the dual simplex algorithm in phase 1 are modified to incorporate dual infeasibilities. The pivoting rules are as follows:

1. the leaving variable assumes the closest of the upper or lower bound, and the reduced cost will assume the correct sign. Unbounded variables are allowed to move only to the lower bound.

2. the basic variables remain dual feasible after updating the basis.

3. the entering variable assumes a dual feasible reduced cost.

In this approach the total number of infeasibilities is reduced without restricting the total amount of infeasibility. Other methods that reduce the total amount of infeasibility require fewer iterations, but need some extra computational effort. Belling-Seib [4] presents similar ideas for the primal simplex method.

Following the previous terminology, the phase 1 dual simplex algorithm can be stated as follows:

Procedure *Dual_Phase 1*
Input : $c, G, l, r, u, \mathcal{B}, \mathcal{N}$.
Output : $\bar{x}^B, \bar{x}^N, v^*$.
begin

$\bar{\pi} \leftarrow c^B B^{-1}; \quad \sigma_k \leftarrow c_k - \bar{\pi} g^k,\ k \in \mathcal{N};$

$\mathcal{N}_u \leftarrow \{k \in \mathcal{N} : \sigma_k < 0, u_k \neq \infty\}; \quad \mathcal{I} \leftarrow \{k \in \mathcal{N} : \sigma_k < 0, u_k = \infty\};$

$\mathcal{N}_l \leftarrow \{k \in \mathcal{N} : \sigma_k \geq 0\} \cup \mathcal{I};$

$\tilde{r} \leftarrow \sum_{k \in \mathcal{I}} \{-g^k\}; \quad \bar{x} \leftarrow B^{-1}\tilde{r}; \quad v^* \leftarrow \sum_{k \in \mathcal{I}} \sigma_k;$

while($\mathcal{I} \neq \phi$)**do**

$\hat{q} \leftarrow \mathrm{argmax}_{k \in \mathcal{V}} \left\{ \begin{array}{ll} -\bar{x}_k^{\mathcal{B}} & \text{if } \bar{x}_k^{\mathcal{B}} < 0 \\ \bar{x}_k^{\mathcal{B}} & \text{if } \bar{x}_k^{\mathcal{B}} > 0, u_k^{\mathcal{B}} \neq \infty \end{array} \right\};$ / * leaving variable * /

$q \leftarrow A(\hat{q}); \quad z \leftarrow e^{\hat{q}} B^{-1};$ / * row $\hat{q}$ of basis inverse * /

$w \leftarrow \mathrm{sign}(-\bar{x}_q) z G;$ / * updated row * /

$\mathcal{E} \leftarrow \{j \in \mathcal{N}_l : w_j < 0\ ,\ \sigma_j \geq 0\} \cup \{j \in \mathcal{N}_u : w_j > 0\ ,\ \sigma_j < 0\};$

if($\mathcal{E} = \phi$)**then** $\mathcal{E} \leftarrow \{j \in \mathcal{I} : w_j > 0\};$

$\begin{bmatrix} \delta \\ s \end{bmatrix} \leftarrow \begin{array}{c} \max_{j \in \mathcal{E}} \\ \mathrm{argmax}_{j \in \mathcal{E}} \end{array} \left\{ \frac{\sigma_j}{w_j} \right\};$ / * ratio test * /

$y \leftarrow B^{-1} g^s; \ \mathcal{P} \leftarrow \{i : y_i \neq 0\};$ / * updated column * /

if($\bar{x}_q < 0$)**then** $\mathcal{N}_l \leftarrow \mathcal{N}_l \cup \{q\};$

if($\bar{x}_q > 0$)**then** $\mathcal{N}_u \leftarrow \mathcal{N}_u \cup \{q\};$

$\Delta \leftarrow \frac{\bar{x}_q}{y_{\hat{q}}}; \ \bar{x}_i^B \leftarrow \bar{x}_i^B - \Delta y_i,\ i \in \mathcal{P}; \ \bar{x}_s \leftarrow \Delta;$ / * flow update * /

if($\delta > 0$)**then** $\sigma_k \leftarrow \sigma_k - \delta w_k,\ k \in \mathcal{N}; \ \sigma_q \leftarrow \mathrm{sign}(-\bar{x}_q)\delta;$

$\mathcal{B} \leftarrow \mathcal{B} \cup \{s\} \backslash \{q\}; \ \mathcal{N}_l \leftarrow \mathcal{N}_l \backslash \{s\}; \ \mathcal{N}_u \leftarrow \mathcal{N}_u \backslash \{s\};$ / * basis update * /

for($k \in \mathcal{I}$ & $\sigma_k \geq 0$)**do**

$\bar{x}^B \leftarrow \bar{x}^B - B^{-1} g^k;$ / * adjust the flow * /

$\mathcal{I} \leftarrow \mathcal{I} \backslash \{k\};$ / * variable k became feasible * /

endfor

endwhile

dual$(c, G, l, r, u, \mathcal{B}, \mathcal{N}, \bar{x}^B, \bar{x}^N, v^*);$

end.

The search for the leaving variable in phase 1 of the dual simplex method is enhanced by exploiting the infrequency (small number) of the dual infeasibilities and the basis structure of generalized networks. The search is limited to small parts of the basis components associated with the nodes in $\hat{\mathcal{I}} = \{i \in \mathcal{V} : g_i^j \neq 0, j \in \mathcal{I}\}$; precisely the path from a node in $\hat{\mathcal{I}}$ to the root of the corresponding rooted-tree component, and the nodes in the cycle along with the nodes in the path from a node in $\hat{\mathcal{I}}$ to the cycle of the corresponding one-tree component.

3 BENCHMARK PROBLEMS AND TESTING ENVIRONMENT

A set of generalized network test problems were randomly generated using a modified version of NETGEN by Klingman *et al.* [16]. The modified code, generates a variety of generalized network problems based on the user controlled problem description. The main inputs are the number of nodes, the number of arcs, the number of source nodes, the number of sink nodes, the number of transshipment sources, and the number of transshipment sinks. In addition, the minimum supply and the maximum supply specifications bounds the total supply. The % arcs with high cost parameter specifies the number of arcs having the maximum cost, and the remaining arc costs are randomly generated on a user specified interval. Similarly, the % arcs with bounds parameter specifies the percentage of bounded arcs, where the generated arc bounds are uniformly distributed on a given interval. A seed is supplied to a random number generator that allows one to reproduce the problems.

Our modified code can be divided into two main segments. In the first segment, a skeleton network that guarantees the problem connectivity and feasibility is generated. Initially, the total supply is randomly distributed among the source nodes, including the transshipment source nodes. Then, for each source node a chain is created to carry the flow through a series of distinct transshipment nodes, to be randomly allocated among a randomly chosen subset of sink nodes. These chains are mutually exclusive except for the sink nodes. In the second segment, randomly generated arcs are appended to the network to fill out the total required arcs. The final network will contain approximately the required number of arcs. For more detail about generating the arc costs and bounds see Klingman *et al.* [16]. For the skeleton arcs, the from and the to multipliers are set to 1 and -1, respectively. For each of the remaining arcs, the two multipliers are randomly selected from the continuous interval $[-1, 1)$, while insuring that the first multiplier is nonzero.

The comparison and examination of the efficiency for different solution methods is of practical importance, and the selection of the test set used in the comparison is very crucial. Therefore, we generated ten test problems of various sizes and structures. The number of nodes ranged from 1,000 to 15,000, and the required number of arcs ranged from 3,000 to 90,000. The characteristics of the test problems are given in Tables 1 and 2. The naming convention is $G\_\hat{m}\_\hat{n}$ for a generalized problem having $\hat{m}$K nodes, and $\hat{n}$K arcs. All testing was performed on a 60 MHz DECStation 5000/260 running the Ultrix 4.3a operating system. The reported performance measures are the number of pivots

Table 1 Characteristics of test problems one through five.

Specification	**Problem**				
	G_01_03	G_01_06	G_02_06	G_02_12	G_04_12
Seed	85	173	319	51	920
Nodes	1K	1K	2K	2K	4K
Arcs	3K	6K	6K	12K	12K
Sources	100	100	200	200	200
Sinks	100	100	200	200	200
Trans. Sources	50	50	100	100	100
Trans. Sinks	50	50	100	100	100
Chain Length	8	8	10	20	25
Min Cost	-1K	-2K	-1K	-3K	-4K
Max Cost	10K	10K	10K	10K	10K
Min Supply	10K	10K	50K	50K	50K
Max Supply	20K	20K	50K	50K	70K
Min Arc Bounds	10	10	10	10	10
Max Arc Bounds	1K	1K	1K	1K	1K
High Cost Arcs	58%	40%	29%	95%	78%
Bounded Arcs	70%	99%	87%	73%	92%

Table 2 Characteristics of test problems six through ten.

Specification	Problem				
	G_04_20	G_09_20	G_09_45	G_15_45	G_15_90
Seed	11	194	239	357	421
Nodes	4K	9K	9K	15K	15K
Arcs	20K	20K	45K	45K	90K
Sources	200	200	500	2000	2000
Sinks	200	200	500	2000	2000
Trans. Sources	100	100	250	1000	1000
Trans. Sinks	100	100	250	1000	1000
Chain Length	25	50	100	200	400
Min Cost	-2K	-1K	-3K	-3K	-3K
Max Cost	10K	30K	30K	30K	30K
Min Supply	70K	80K	90K	90K	90K
Max Supply	80K	90K	95K	95K	99K
Min Arc Bounds	10	10	100	100	100
Max Arc Bounds	1K	1K	1K	2K	3K
High Cost Arcs	73%	20%	84%	75%	80%
Bounded Arcs	60%	30%	30%	60%	80%

(or iterations) and the user central processing time rounded to two decimal places. The user cpu time is obtained using the standard Ultrix function *getrusage*, that returns information describing the system resource utilization. This information includes the cumulative user cpu time, measured in microseconds, consumed to the point of the function call. In the implementation of the simplex algorithm, getrusage is called immediately preceding and immediately following the call to the optimizer. The user cpu time consumed is obtained by subtracting the first call returned time from the second call returned time.

4 COMPUTATIONAL EXPERIENCE WITH CPLEX 3.0

To provide a comparison base, the test problems were solved using the four optimization algorithms available in CPLEX 3.0 (a state-of-the-art solver [24].) This version includes efficient implementations of the primal simplex, the dual simplex, the network simplex, and the barrier algorithms. All CPLEX runs were made using the default parameter settings, which activate the presolver and the aggregator problem preprocessors. The presolver helps simplify, reduce, and eliminate redundancies in the problem presentation. The aggregator attempts to make simplifying substitutions that reduce the basis size.

In the CPLEX dual simplex optimizer, the dual problem is solved using the primal simplex presentation, where all the linear algebra is carried out on the associated primal basis. The network optimizer, recognizes and exploits the embedded network structure of the problem, and uses an efficient specialized algorithm on the pure network portion to obtain an advanced start for the dual optimizer. The barrier optimizer, is an efficient implementation of the primal-dual method fully integrated with the predictor corrector basis crossover of Megiddo [17].

The empirical results for solving the test problems using the four CPLEX solvers are presented in Table 3, where the numbers in parenthesis represent the number of phase I pivots. From the first four problems, it is clear that the barrier solver performs poorly for generalized network problems, and it was not run for the remaining ones. The CPLEX dual simplex solver is approximately two times faster than the primal simplex solver for all problems except G_01_03, G_15_45 and G_15_90. The over all time performance is 1.4 times faster than the primal algorithm. Using the default settings, the CPLEX network solver failed to extract any network component for three of the test problems, and it terminated with an unbounded network for problems G_09_45 and G_15_45.

5 EMPIRICAL ANALYSIS

The specialized big-M dual simplex algorithm is implemented in C, and the performance in solving the ten test problems, using the three pricing techniques, is shown in Tables 4, 5, and 6. For all testing M is set to 10^9. Comparing the big-M standard (non-normalized) dual RAMSES, using the three pricing strategies, and the CPLEX 3.0 dual solver, it is clear that the dual RAMSES with a candidate-queue pricing is the superior. For the G_15_90, it performed approximately 30 times faster than CPLEX, 13 times faster than RAMSES with a maximum-infeasibility pricing strategy, and 28% faster than RAMSES with a fixed-page pricing strategy. On average, for the ten problems it performed 20 times faster than CPLEX, 9 times faster than RAMSES with the maximum-infeasibility pricing strategy, and 25% faster than RAMSES with fixed-page pricing strategy. The fixed-page pricing strategy resulted in only a 2% increase in the total number of pivots, while the candidate-queue pricing strategy resulted in a 19% increase in the total number of pivots over the maximum-infeasibility pricing strategy.

For the three pricing strategies, the standard (non-normalized) RAMSES performed consistently much faster, and required fewer pivots than the steepest edge RAMSES. This can be ascribed to the generated problem characteristics of having arc multipliers in the range [-1,1], and uniformly distributing the arcs among the network nodes, which results in basis inverse rows having approximately the same length.

The specialized two phase dual simplex algorithm is implemented in C and the performance in solving the ten test problems, using the three pricing techniques, is shown in Tables 7, 8, and 9. It is clear that the two phase dual RAMSES with a candidate-queue pricing strategy is the superior implementation. For the G_15_90, it performed approximately 31 times faster than CPLEX, 13 times faster than RAMSES with a maximum-infeasibility pricing strategy, and 29% faster than RAMSES with a fixed-page pricing strategy. On average, for all test problems it performed 22 times faster than CPLEX, 10 times faster than RAMSES with a maximum-infeasibility pricing strategy, and 22% faster than RAMSES with a fixed-page pricing strategy.

Table 3 Solving generalized networks using CPLEX version 3.0

Prob. Name	Primal		Dual		Network		Barrier		Objective Value
	Pivots	Time	Pivots	Time	Pivots	Time	Iter	Time	
G_01_03	975(295)	2.90	647(34)	2.73	647	2.53	20	45	1.0676511e08
G_01_06	2162(311)	7.70	1088(13)	4.67	N/A[a]	N/A	16	151	3.6451339e07
G_02_06	2408(765)	11.70	1526(43)	8.85	1527	8.78	16	373	5.9484944e08
G_02_12	3399(608)	21.95	1294(34)	11.68	N/A[a]	N/A	19	1955	7.8177439e08
G_04_12	4968(1678)	79.15	2748(68)	33.10	2744	37.82	-	-	1.3194873e09
G_04_20	5692(1213)	60.20	2124(243)	29.52	2125	32.52	-	-	1.1445576e09
G_09_20	6394(2538)	176.53	3212(94)	76.53	3194	76.08	-	-	5.2429920e09
G_09_45	7694(2205)	249.03	3958(186)	107.97	238[b]	8.63	-	-	1.4578699e10
G_15_45	16061(5941)	999.37	13106(236)	786.02	1797[b]	10.95	-	-	2.2251216e11
G_15_90	22636(8101)	1576.32	15055(94)	1167.02	15107	1277.27	-	-	2.2564331e11
Totals	72389(23655)	3184.85	44758(1045)	2226.09	-	-	-	-	-

[a]Small or non-existent network
[b]Unbounded network

Table 4 Results for big-M dual RAMSES with maximum-infeasibility pricing.

Prob. Name	CPLEX (DUAL)		Standard			Time Ratio	Steepest Edge			Time Ratio
	Pivots	Time	Pivots		Time		Pivots		Time	
	Total	[1]	Total	Deg.	[2]	[1]/[2]	Total	Deg.	[3]	[1]/[3]
G_01_03	647(34)	2.73	830	107	0.69	4.0	971	122	0.98	2.8
G_01_06	1088(13)	4.67	1340	85	1.29	3.6	1381	89	1.68	2.8
G_02_06	1526(43)	8.85	1977	195	3.35	2.6	2266	230	4.82	1.8
G_02_12	1294(34)	11.68	1660	465	3.98	2.9	1901	571	4.72	2.5
G_04_12	2748(68)	33.10	3389	377	11.49	2.9	3867	418	16.59	2.0
G_04_20	2124(243)	29.52	3145	677	12.14	2.4	3247	645	12.78	2.3
G_09_20	3212(94)	76.53	4900	386	47.87	1.6	5388	317	61.66	1.2
G_09_45	3958(186)	107.97	6195	2244	63.88	1.7	6725	2315	75.52	1.4
G_15_45	13106(236)	786.02	20978	6788	371.47	2.1	25083	7724	530.99	1.5
G_15_90	15055(94)	1167.02	25566	13230	492.77	2.4	25540	11229	559.72	2.1
Totals	44740(1045)	2226.09	69980	24554	1008.93	2.2	76369	23660	1269.46	1.8

Table 5 Results for big-M dual RAMSES with fixed-page pricing.

Prob. Name	CPLEX (DUAL)		Standard				Steepest Edge			
	Pivots	Time	Pivots		Time	Time Ratio	Pivots		Time	Time Ratio
	Total	[1]	Total	Deg.	[2]	[1]/[2]	Total	Deg.	[3]	[1]/[3]
G_01_03	647(34)	2.73	948	114	0.43	6.3	1001	115	0.62	4.4
G_01_06	1088(13)	4.67	1444	86	0.80	5.8	1444	89	1.02	4.6
G_02_06	1526(43)	8.85	2502	197	1.50	5.9	2509	206	1.87	4.7
G_02_12	1294(34)	11.68	2185	545	2.61	4.5	2083	563	2.43	4.8
G_04_12	2748(68)	33.10	4500	416	4.95	6.7	4624	380	8.99	3.7
G_04_20	2124(243)	29.52	4108	755	4.88	6.0	4097	722	6.98	4.2
G_09_20	3212(94)	76.53	6693	408	13.77	5.6	6449	392	21.70	3.5
G_09_45	3958(186)	107.97	7684	2509	15.43	7.0	7687	2498	23.94	4.5
G_15_45	13106(236)	786.02	23198	7017	42.66	18.4	24640	7425	56.54	13.9
G_15_90	15055(94)	1167.02	25738	11187	49.74	23.5	26034	11243	58.35	20.0
Totals	44740(1045)	2226.09	79000	23234	136.77	16.3	80568	23633	182.44	12.3

Table 6 Results for big-M dual RAMSES with candidate-queue pricing.

Prob.	CPLEX (DUAL)		Standard			Time	Steepest Edge			Time
Name	Pivots	Time	Pivots		Time	Ratio	Pivots		Time	Ratio
	Total	[1]	Total	Deg.	[2]	[1]/[2]	Total	Deg.	[3]	[1]/[3]
G_01_03	647(34)	2.73	981	121	0.44	6.2	1042	123	0.64	4.3
G_01_06	1088(13)	4.67	1620	84	1.04	4.5	1600	96	1.27	3.7
G_02_06	1526(43)	8.85	2474	236	1.17	7.6	2599	237	1.75	5.1
G_02_12	1294(34)	11.68	2249	600	1.91	6.1	2371	660	2.53	4.6
G_04_12	2748(68)	33.10	4561	411	3.45	9.6	4637	410	5.99	5.5
G_04_20	2124(243)	29.52	3811	742	3.35	8.8	3887	735	4.30	6.9
G_09_20	3212(94)	76.53	6407	421	11.26	6.8	6467	413	28.73	2.7
G_09_45	3958(186)	107.97	7644	2613	11.00	9.8	7753	2706	16.56	6.5
G_15_45	13106(236)	786.02	26115	8149	33.80	23.3	26271	8154	51.99	15.1
G_15_90	15055(94)	1167.02	27282	10992	38.97	30.0	27832	11206	55.88	20.9
Totals	44740(1045)	2226.09	83144	24369	106.39	20.9	84459	24740	169.64	13.1

Table 7 Results for two phase dual RAMSES with maximum-infeasibility pricing.

Prob. Name	CPLEX (DUAL) Pivots Total	CPLEX (DUAL) Time [1]	Standard Pivots Total	Standard Pivots Deg.	Standard Time [2]	[1] — [2]	Steepest Edge Pivots Total	Steepest Edge Pivots Deg.	Steepest Edge Time [3]	[1] — [3]
G_01_03	647(34)	2.73	813(39)	103	0.68	4.0	938(40)	115	0.93	2.9
G_01_06	1088(13)	4.67	1337(13)	85	1.23	3.8	1376(12)	87	1.76	2.7
G_02_06	1526(43)	8.85	1948(50)	193	3.31	2.7	2224(51)	226	4.26	2.1
G_02_12	1294(34)	11.68	1662(36)	446	4.00	2.9	1923(36)	579	4.51	2.6
G_04_12	2748(68)	33.10	3410(72)	387	12.56	2.6	3905(72)	406	18.30	1.8
G_04_20	2124(243)	29.52	3012(288)	558	10.51	2.8	3206(288)	530	14.85	2.0
G_09_20	3212(94)	76.53	4869(173)	361	40.69	1.9	5356(173)	297	62.08	1.2
G_09_45	3958(186)	107.97	5939(242)	2049	62.26	1.7	6565(245)	2122	76.23	1.4
G_15_45	13106(236)	786.02	20572(309)	6657	371.17	2.1	24088(312)	7330	537.78	1.5
G_15_90	15055(94)	1167.02	24648(119)	12469	496.94	2.3	24552(119)	10500	533.64	2.2
Totals	44740(1045)	2226.09	68210(1341)	23308	1003.35	2.2	74133(1348)	22192	1254.34	1.8

Table 8 Results for two phase dual RAMSES with fixed-page pricing.

Prob.	CPLEX (DUAL)		Standard			[1]	Steepest Edge			[1]
Name	Pivots	Time	Pivots		Time	—	Pivots		Time	—
	Total	[1]	Total	Deg.	[2]	[2]	Total	Deg.	[3]	[3]
G_01_03	647(34)	2.73	975(39)	110	0.42	6.5	1017(40)	117	0.58	4.7
G_01_06	1088(13)	4.67	1445(13)	88	0.84	5.6	1446(13)	91	1.01	4.6
G_02_06	1526(43)	8.85	2502(50)	198	1.50	5.9	2548(51)	204	1.93	4.6
G_02_12	1294(34)	11.68	2252(36)	546	2.50	4.7	1988(36)	521	2.70	4.3
G_04_12	2748(68)	33.10	4368(72)	415	4.59	7.2	4496(72)	369	9.69	3.4
G_04_20	2124(243)	29.52	4072(288)	762	5.23	5.6	3949(283)	605	8.11	3.6
G_09_20	3212(94)	76.53	6646(173)	383	13.11	5.8	6963(173)	401	30.94	2.5
G_09_45	3958(186)	107.97	7506(242)	2387	16.32	6.6	7466(245)	2357	24.48	4.4
G_15_45	13106(236)	786.02	22778(312)	8014	40.37	19.5	25099(312)	7553	71.05	11.1
G_15_90	15055(94)	1167.02	25258(119)	10929	49.24	23.7	26027(119)	10962	63.53	18.4
Totals	44740(1045)	2226.09	77802(1344)	23832	134.12	16.6	80999(1344)	23180	214.02	10.4

Table 9 Results for two phase dual RAMSES with candidate-queue pricing.

Prob. Name	CPLEX (DUAL)		Standard			[1]	Steepest Edge			[1]
	Pivots	Time	Pivots		Time	—	Pivots		Time	—
	Total	[1]	Total	Deg.	[2]	[2]	Total	Deg.	[3]	[3]
G_01_03	647(34)	2.73	956(39)	120	0.44	6.2	1036(39)	123	0.66	4.1
G_01_06	1088(13)	4.67	1554(13)	85	1.05	4.4	1577(13)	92	1.29	3.6
G_02_06	1526(43)	8.85	2461(50)	237	1.13	7.8	2519(50)	238	1.61	5.5
G_02_12	1294(34)	11.68	2216(36)	628	1.88	6.2	2337(36)	664	2.58	4.5
G_04_12	2748(68)	33.10	4571(72)	419	3.59	9.2	4799(72)	423	7.57	4.4
G_04_20	2124(243)	29.52	3591(288)	613	3.01	9.8	3612(283)	606	4.28	6.9
G_09_20	3212(94)	76.53	6387(173)	401	10.90	7.0	6450(173)	387	27.24	2.8
G_09_45	3958(186)	107.97	7447(242)	2467	10.48	10.3	7465(245)	2474	16.14	6.7
G_15_45	13106(236)	786.02	25826(312)	8014	33.29	23.6	26099(312)	8114	58.94	13.3
G_15_90	15055(94)	1167.02	26846(119)	10786	38.09	30.6	27283(119)	10873	59.96	19.5
Totals	44740(1045)	2226.09	81855(1344)	23770	103.86	21.5	83177(1342)	23994	180.27	12.4

6 SUMMARY AND CONCLUSIONS

Generalized networks are used to model a diversity of practical problems that include class scheduling, machine loading, manpower planning, resource distribution systems, and currency exchange. This paper describes the development of efficient techniques for solving generalized network problems. As in the primal simplex algorithm, the dual simplex method is initiated by an all artificial basis. In Section 2.4 we introduce two approaches to obtain a dual feasible solution from any starting basis. In the two approaches, the nonbasic variables are assigned to the appropriate bound based on the the sign of the corresponding reduced costs. In the first approach, a big-M value is used to approximate the upper bound for unbounded variables. For the three dual pricing strategies, the big-M dual RAMSES code with a candidate-queue is superior. It achieves a performance 30 times faster than the CPLEX 3.0 dual optimizer for problem G_15_90, and 20 times faster for all the benchmark problems. However, the success of the big-M method is contingent upon the selection of a proper big-M value. In the second approach, this critical selection is avoided by using a two phase method, where dual infeasibilities are temporarily permitted in phase 1 for unbounded variables with negative reduced cost. For the three dual pricing strategies, the two phase dual algorithm with a candidate-queue is superior. It achieves a performance 31 times faster than the CPLEX dual optimizer for problem G_15_90, and 22 times faster for the ten test problems.

Acknowledgements

This work was partially supported by the Office of Naval Research under Grant Number N00014-95-1-0645.

REFERENCES

[1] R. Ahuja, T. Magnanti, and J. Orlin. *Network Flows: Theory, Algorithms, and Applications.* Prentice-Hall, Inc., Englewood Cliffs, NJ., 1993.

[2] A. Ali, R. Padman, and H. Thiagarajan. Dual Algorithms for Pure Network Problems. *Operations Research*, 37(1):159–179, 1989.

[3] R. Barr, F. Glover, and D. Klingman. Enhancement of Spanning Tree Labeling Procedures for Network Optimization. *INFORS*, 17:66–85, 1979.

[4] K. Belling-Seib. An Improved General Phase-1 Method in Linear Programming. *European Journal of Operational Research*, 36:101–106, 1988.

[5] D. Bertsekas and P. Tseng. Relaxation Methods for Minimum Cost Ordinary and Generalized Network Flow Problems. *Operations Research*, 36(1):93–114, 1988.

[6] G. Brown and R. McBride. Solving Generalized Networks. *Management Science*, 30(12):1497–1523, 1984.

[7] R. Clark, J. Kennington, R. Meyer, and M. Ramamurti. Generalized Networks: Parallel Algorithms and an Empirical Analysis. *ORSA Journal on Computing*, 4(2):132–145, 1992.

[8] M. Engquist and M. Chang. New Labeling Procedures for The Basis Graph in Generalized Networks. *Operations Research Letters*, 4(4):151–155, 1985.

[9] J. Forrest and D. Goldfarb. Steepest-edge Simplex Algorithm for Linear Programming. *Mathematical Programming*, 57:341–374, 1992.

[10] F. Glover, J. Hultz, D. Klingman, and J. Stutz. Generalized Networks: A Fundamental Computer-Based Planning Tool. *Management Science*, 24(12):1209–1220, 1978.

[11] F. Glover, D. Klingman, and N. Phillips. *Network Models in Optimization and Their Applications in Practice*. John Wiley & Sons, New York, 1992.

[12] F. Glover, D. Klingman, and J. Stutz. Extensions of the Augmented Predecessor Index Method to Generalized Network Problems. *Transportation Science*, 7(4):377–384, 1973.

[13] D. Goldfarb and J. Reid. A Practicable Steepest-Edge Simplex Algorithm. *Mathematical Programming*, 12:361–371, 1977.

[14] P. Harris. Pivot Selection Methods of the DEVEX LP Code. *Mathematical Programming*, 5:1–28, 1973.

[15] J. Kennington and R. Helgason. *Algorithms for Network Programming*. John Wiley & Sons, New York, 1980.

[16] D. Klingman, A. Napier, and J. Stutz. A Program for Generating Large Scale Capacitated Assignment, Transportation, and Minimum Cost Flow Network Problems. *Management Science*, 5(5):814–821, 1974.

[17] N. Megiddo. On Finding Primal- and Dual-Optimal Bases. *ORSA Journal on Computing*, 3(1):63–65, 1991.

[18] R. Mohamed. Efficient Dual Simplex Optimizers for Generalized Network Models. unpublished dissertation, Department of Computer Science and Engineering, Southern Methodist University, Dallas, TX 75275, 1995.

[19] J. Mulvey and S. Zenios. Solving Large Scale Generalized Networks. *Journal of Information and Optimization Sciences*, 6(1):95–112, 1985.

[20] K. Murty. *Linear and Combinatorial Programming.* Robert E. Krieger Publishing Company, Malabar, Florida, 1985.

[21] G. Nemhauser and L. Wolsey. *Integer and Combinatorial Optimization.* John Wiley & Sons, New York, 1988.

[22] W. Nulty and M. Trick. GNO/PC Generalized Network Optimization System. *Operations Research Letters*, 7(2):101–102, 1988.

[23] R. Parker and R. Rardin. *Discrete Optimization.* Academic Press, New York, 1988.

[24] Using the CPLEX Callable Library Version 3.0. Incline Village, Nevada 89451-9436, 1994.

7

SOLVING LARGE SCALE CREW SCHEDULING PROBLEMS

Hai D. Chu
Eric Gelman
Ellis L. Johnson*

Sabre Decision Technologies, Dallas-Fort Worth, Texas

**Georgia Tech, Atlanta, Georgia*

Abstract

The crew pairing problem is posed as a set partitioning zero-one integer program. Variables are generated as legal pairings meeting all work rules. Dual values obtained from solving successive large linear program relaxations are used to prune the search tree. In this paper we present a graph based branching heuristic applied to a restricted set partitioning problem representing a collection of "best" pairings. The algorithm exploits the natural integer properties of the crew pairing problem. Computational results are presented to show realized crew cost savings.

1. Introduction

Recent work at Sabre Decision Technologies (SDT) in crew pairing optimization has been described in two papers [1, 2]. The problem being solved is to partition a set of flight segments into pairings, where a pairing is a sequence of segments that begin and end at a crew base. The objective function is to minimize excess crew cost associated with operating a given schedule. The principal cost component is pay and credit. It is defined to be guaranteed hours of pay minus the hours actually flown. The ratio between the pay and credit time and the actual flying time is known as the percent of pay and credit. The quality of a solution can be determined by the percent of pay and credit. Since a solution is highly dependent on the flight schedule this percentage may necessarily be large. However, for American Airlines' largest fleet, the MD-80, we are able to achieve pay and credit of about 0.3%.

The crew pairing problem can be represented as a set partitioning problem, where each column is a pairing and each row is a flight segment. Each segment is required to be covered by exactly one pairing in the final solution. The cost of each column is determined by the excess crew cost of the pairing associated with that column.

The set partitioning formulation puts the crew pairing problem into a class of combinatorial optimization problem that is nondeterministic polynomial in complexity (NP-complete). Solving this problem for large flight schedules becomes a difficult task. For example, an American Airlines MD-80 daily flight schedule of 1,000 segments may produce on the order of 2^{1000} pairings as potential columns in the set partitioning matrix. However, the number of columns can be reduced to the order of 10^{12} to 10^{15} by imposing constraints such as reasonable connection time and three day pairings. Nevertheless, it is impossible to fully represent the constraint matrix for large schedules. Techniques, however, are employed to generate only "good" columns that have a higher chance of appearing in the optimal solution.

One tractable way of obtaining crew pairing solutions is the subproblem methodology described by [1]. This methodology is robust, easy to implement, and produces feasible solutions. The subproblem approach generates columns for randomly chosen subsets of flight segments from the whole schedule. To solve the subproblems, the (0,1) restrictions on the columns are relaxed to nonnegativity constraints. If the size of the subproblems are small then the LP solutions to the relaxed problems are typically integer solutions for the original problems. The obvious limitations of the subproblem methodology are the size of the subproblems and the number of subproblems one can solve in a reasonable time frame. The limitations of the subproblem approach are that local optima can prevent further improvements and the quality of solutions is hard to measure against any known LP bound for the entire schedule.

An alternative approach to address the limitations of the subproblem methodology has been described in a joint SDT-IBM study [2]. The approach is a large scale methodology where successive LP's of 2 million columns are solved to determine "good" columns for inclusion in the set partitioning problem. The results produced by this large scale method are significantly better than the results of the subproblem methodology. American Airlines, using this large scale optimizer, was able to save approximately 1.2 million dollars from mid 1991 to mid 1992. The gap, however, between the lower bound obtained from the LP solution and the IP solution for a typical crew pairing problem was between \$2,000 to \$5,000. This gap was related to the fact that at the IP stage the set partitioning formulation was relaxed to a set packing formulation.

SDT has recently been successful at solving large set partitioning problems to obtain integer solutions with objective values very close to the LP objective value. This has resulted in additional savings for American Airlines. American Airlines confirmed that record savings are being observed with projected annual savings of \$2 million beyond the savings already reported. The gaps between the LPs and the IPs are now typically less than \$1000 for the MD-80 fleet. The MD-80 excess crew cost obtained is about \$50,000 per day, including hotel, per diem, and pay and credit. The percent of pay and credit has been reduced to about 0.3%.

This paper describes the current large scale crew pairing optimization methodology used at SDT. The solution procedure consists of three stages. In the first stage, linear programs are successively solved and new columns generated until 20 million columns have been generated. At the end of this stage, an optimum linear programming solution is obtained on a matrix with 2 million columns selected from the total of 20 million generated. The first stage is described in section 2. The next stage is a branching heuristic applied to 15,000 of the "best" columns from stage 1. The branching heuristic is based on a follow-on logic described in [2] with improvements added to solve the 15,000 columns set partitioning problem. The second stage is described in section 3. Section 4 contains computational results. The final section, section 5, is our conclusions.

2. LP Methodology Used in Crew_Solve.

The mathematical formulation of the crew pairing problem, as mentioned in the introduction leads to a set partitioning problem.

$$\min \sum_{j \in M} c_j x_j$$
$$\sum_{j \in M} a_{ij} x_j = 1, \quad \forall i \in N \qquad \text{(SSP)}$$
$$x_j = 0,1 \quad \forall j \in M$$

where:

c_j - cost of a pairing *j*;
N - set of all flights in the schedule;
M - set of all pairings generated;
The constraint matrix **A** is defined as:

$$a_{ij} = \begin{cases} 1 & \text{if flight } i \text{ participates in pairing } j \\ 0 & \text{otherwise} \end{cases}$$

Typically, the number of rows (flight segments) in these problems varies between 300 - 1500 for the cockpit crew. Crew_Solve algorithm jointly developed by SDT and IBM are used to solve these crew scheduling problems.

The algorithm includes a matrix representation module known as the Pairing Generator. In the paper this module will be referred to as Col_Gen. The module produces the constraint matrix that is used as input to the optimization modules.

The LP optimization group modules contains four main modules. The first - Sprint_Solve_LP is used for faster solution of large scale LP problems. Auxiliary modules - LP_Column_Choose and IP_Column_Choose are used in the LP and IP solution process respectively for selection of the "best" reduced cost columns from a given large scale LP. After the LP solution is found, IP_Branch is used to efficiently search through the IP branching tree. All the modules are called in conjunction with IBM's OSL subroutines.

The Crew_Solve algorithm as well as Sprint_Solve_LP will be described next, while the IP_Branch algorithm will be described in the next section.
The following definitions will be used in further discussions:

Definition 1. The original large scale problem LP consists of a set of "smaller" problems, that are chosen in the course of the solution process. This could be described as follows:

$LP = \{ LP_1, LP_2, ...LP_i, ... \}$

Similarly

$LP_i = \{ LP_{i1}, LP_{i2} ,..., LP_{ij},...\}$

For example, LP could be a 20 million column problem; LP_1, LP_2, LP_i could be 2 million column problems each; while LP_{i1}, LP_{i2} , LP_{ij}, could be 5,000 columns problems each.

Definition 2. Optimal solution for the problem LP will be defined by $\mathbf{x}$; similarly for LP_i the optimal solution will be $\mathbf{x_i}$ and for LP_{ij} the optimal solution will be $\mathbf{x_{ij}}$.

Definition 3. The procedure **LP_Columns_Choose(LP_{i-1}, LP_i)** means choose the "best" reduced cost columns from LP_{i-1} and place them into LP_i . Similarly the meaning of **IP_Columns_Choose(LP_i, IP_i)** procedure is choose the "best" reduced cost columns from LP_i and place them into IP_i .

The pseudo-code and a short explanation of the Crew_Solve algorithm follows:

```
Crew_Solve:
  0. i = 0,  LP0 = {∅}
  1. i = i + 1
  2. call Col_Gen LPi : { K million columns ( pairings)}
  3. call Sprint_Solve_LP (LPi)
  4. Check stopping criterion:
     if || Σ LPi cjxj - Σ LPi-1 cjxj || ≤ ε
        then do
          call IP_Columns_Choose(LPi, IPi)
          call IP_Branch
          call OSL_IP_Solve
        end do
     else
         do
          call LP_Columns_Choose (LPi, LPi+1)
          go to 1.
         end do
     end if
end Crew_Solve.
```

For our application K = 2 has been selected, i.e. the LP_i problems contain 2 million columns. The column generator is being used to produce successive LP_i problems. The stopping criterion of Crew_Solve has been set to sense minor fluctuations in the objective function between two successive Sprint_Solve_LP calls with the threshold $\varepsilon = 10^{-1}$. The "best" 15,000 columns are chosen by IP_Columns_Choose for further processing by IP_Branch. The integer solution is found by IP_Branch or by OSL_IP_Solve. More detailed discussion of the IP_Branch will be given in the next section.

If the current solution is not optimal according to the stopping criterion, LP_Columns_Choose module is used to select 30-40% of the columns to be passed into the next Sprint_Solve_LP iteration. Then, additional columns are generated as needed (see step 2 of the Crew_Solve algorithm) and passed to the next iteration of Sprint_Solve_LP.

The pseudo-code of the Sprint_Solve_LP procedure follows:

```
Sprint_Solve_LP ( LPi ) Procedure:
  0. j = 0,  LPi0 = {∅},  S = 5,000
```

1. j = j + 1
2. place the optimal solution of $LP_{i(j-1)}$ into LP_{ij}
3. call LP_Columns_Choose (LP_i, LP_{ij}) such that $|LP_{ij}| = S$
4. call OSL_LP_Solve(LP_{ij}):
   $\mathbf{x_{ij}}$ optimal solution for LP_{ij}
5. if $\mathbf{x_{ij}}$ is not optimal for LP_i
   go to 1.
6. optimal solution $\mathbf{x_i}$ found for LP_i

end Sprint_Solve_LP

SPRINT approach has been introduced by J. Forrest and was described in [2]. It has been successfully used for solution of large scale crew scheduling LP's. The main idea behind the algorithm is to generate the optimal solution of a large scale LP by solving a sequence of considerably smaller LP's. The optimal solution to the large scale LP typically is generated after 10-20 small LP's are solved. This constitutes about 10% of the large scale LP. Recent enhancements to Sprint_Solve_LP resulted in improved and more robust performance in solving these large scale LP's. One of the improvements was slight perturbation of the RHS in the original large scale LP; since the solution of the perturbed problem is "almost always" still feasible for the original problem it is also optimal for the original LP. Furthermore the performance of Sprint_Solve_LP has been improved as a result of a more precise selection of "best" reduced cost columns in both LP_Columns_Choose and IP_Columns_Choose modules.

3. IP_Branch Heuristic

Once Sprint_Solve_LP has found an LP bound a set of "best" 15,000 columns is collected from the large LP to be solved by IP_Branch and OSL_IP_Solve. The set partitioning problem is solved by modeling the crew pairing problem as a problem of finding a set of node-disjoint directed cycles spanning the nodes of a graph. The nodes are the flight segments and the directed arcs are the legal connections between flight segments.

Let N = { 1, 2, ..., n } be the rows and M = { 1, 2, ..., m } be the columns of the set partitioning matrix A. Each row of A is a flight segment and each column of A is a pairing covering a sequence of flight segments. Let G = { N, E } be a directed graph on N = { 1, 2, ..., n } and M = { 1, 2, ..., m } such that N is the nodes of G and a directed arc $e = (i, j) \varepsilon E$ exists if and only if there is a column in M where flight segment **i** is followed by flight segment *j*.

Let $G^0 = \{ N, E^0 \}$ be the extension of G to include artificial arcs directed from the last flight segment to the first flight segment of a pairing (Figure 1). Conceptually, the artificial arcs allow for ending a pairing at the pairing's crew base. For a directed arc $e = (i, j) \in E^0$ we say node *i* is a *predecessor* of node *j* and node *j* is a *successor* of node *i*. Let an arc $e = (i, j) \in E^0$ correspond to a set of columns say $C(i, j) \Pi M$, so that the predecessor-successor pair (i, j) belongs in any column in C(i, j). Thus, $e = (i, j) \in E^0$ if and only if $|C(i, j)| \neq 0$. For the example given in Figure 1, $C(i, j) = \{c_4, c_5\}$ and $|C(i, j)| = 2$.

Each column of A is a directed cycle in G^0. The converse is true only if a cycle contains exactly one artificial arc as it defines the crew base of the pairing associated with that cycle. From here on we define a cycle in G^0 to be a directed cycle containing exactly one artificial arc. Thus, assuming A contains no duplicate columns, there is a one-to-one correspondence between columns in A and cycles in G^0.

$N = \{ i, j, k, r, l, p\}$, $M = \{ c_1, c_2, c_3, c_4, c_5 \}$

A =

	c_1	c_2	c_3	c_4	c_5
i = (BOS →DFW)	0	0	1	1	1
j = (DFW→BOS)	1	0	0	1	1
k = (DFW→BOS)	0	0	1	0	1
r = (BOS→ORD)	1	1	0	0	1
l = (ORD→DFW)	1	0	0	0	1
p = (ORD→BOS)	0	1	0	0	0

Note: $G^0 = G + \{$ **Artificial arcs**$\}$

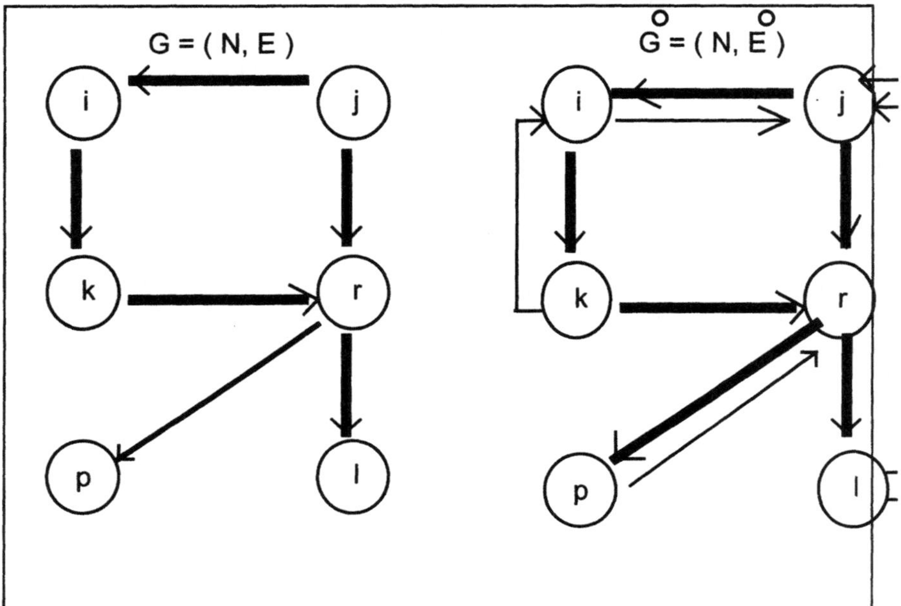

Figure 1.Example of matrix A and corresponding graphs G and G^0

Let IS be a valid solution to the set partitioning problem SPP. Since every segment is covered exactly once by IS $\Rightarrow$ there is a node-disjoint set of cycles in G^0 say $C(IS) = \{ C^1, C^2, ..., C^{|IS|} \}$ which span the nodes of G^0. By our definition of cycles on G^0 the converse is also true and we denote this equivalency between the existence of IS and the representation C(IS) as IS $\equiv$ C(IS).

Remark 1. IS is a valid solution to problem SPP $\Leftrightarrow$ there is a corresponding set of node-disjoint cycles C(IS) in G^0 such that IS $\equiv$ C(IS).

In Figure 1, node r is a *unique predecessor* of l if arc $(r, l) \varepsilon E^0$ and there is no other arc $(a, l) \varepsilon E^0$ where $a \neq r$. Thus, node r = segment (BOS-ORD) is a unique predecessor of node l = segment (ORD-DFW). Similarly, node r is a *unique successor* of p if arc $(p, r) \varepsilon E^0$ and there is no other arc $(p, z) \varepsilon E^0$ where $z \neq r$. Thus, node r = segment (BOS-ORD) is a unique successor of node p = segment (ORD-BOS). In both cases, any valid solution must have a cycle containing arc (r, l) and arc (p, r) since node l can only be covered by a directed arc from node r, and node p can only be covered by a directed arc from node r. Clearly, if node r is a unique predecessor of more than one node, say node l and node p, then there is no valid solution **IS** since there exists no node-disjoint cycles covering both 1 and p. Flight (BOS-ORD) is a unique predecessor of flight (ORD-DFW) and flight (ORD-BOS) in separate pairings. There is no way to cover flight (BOS-ORD) without leaving either flight (ORD-DFW) or flight (ORD-BOS) uncovered. The following remark states a condition on G^0 under which a valid solution to problem SPP does not exist.

Remark 2. If there exist a unique predecessor or unique successor of two or more nodes in G^0 then there exists no valid solution to SPP.

Remark 1 stipulates that each flight segment must be flown by exactly one crew and Remark 2 determines flight segments that compete for the same crew as well as flight segment that compete for different crews. These observations suggest a branching heuristic that either finds a representation C(IS) in G^0 which corresponds to a valid solution to problem SPP as stated in Remark 1, or preconditions SPP before using the branch-and-bound method as proposed in Remark 2.

The following definitions will be used in the description of the branching heuristic.

Definition 4. The procedure $Succ(x) = \{ y : (x, y) \in E^i \}$ returns the successors of x. The procedure $Pred(y) = \{ x : (x, y) \in E^i \}$ returns the predecessors of y.

Definition 5. The procedure Min_Succ(E^i) returns node list { a, ..., i } such that $1 < | Succ(a) | = ... = | Succ(i) | < | Succ(v) |$ for $v \in \{ N - \{ a, ..., i \} \}$. The procedure Min_Pred(E^i) returns node list { b, ..., j } such that $1 < | Succ(b) | = ... = | Succ(j) | < | Succ(v) |$ for $v \in \{ N - \{ b, ..., j \} \}$.

The pseudo-code and short explanation of the IP_Branch algorithm follows.

IP_Branch:

```
1.   i = 0, G^i = ( N, E^i ). Assume there is a valid solution in G^0.
2.   for all (r ε N that have unique predecessor or successor)
  do
     if ( r unique successor of x ) set
          | C (k, r) | = 0 and E^i = E^i - { (k, r) } for all k ≠ x, (k, r) ε E^i
     if ( r unique predecessor of y ) set
          | C (r, k) | = 0 and E^i = E^i - { (r, k) }for all k ≠ y, (r, k) ε E^i
     enddo
3.   solve LP^i : x^i optimal
4.   if (x^i is integer) stop
5.   if (LP^i feasible) then
```

```
    do
        compute arc weights
            w(r, p) = Σk  C (r, p) xk for all (r, p) ε E^i.
        if | Min_Succ(E^i) | < | Min_Pred(E^i) | then
            w(x, y) = max (v, w) { w(v, w) : v ε Min_Succ(E^i), w ε Succ(v) }
        else
            w(x, y) = max (v, w) { w(v, w) : w ε Min_Pred(E^i), v ε Pred(w) }
        update graph:
                E^{i+1} = E^i - { (x, v) ∪ C (w, y) : v ε Succ(x), w ε Pred(y) }
                G^{i+1} = ( N, E^{i+1} )
                i = i + 1
        update matrix:
                for all v ≠ y and w ≠ x, set | C (x, v) | = 0 and | C (w, y) | = 0
        enddo
    endif
8.  if (LP^i infeasible) or (G^i infeasible by remark 2) then
    do
        fathom arc e = (x, y)
        G^i = ( N, E^{i-1} )
    enddo
  endif
9.  if (matrix columns > P) goto 2
end IP_Branch
```

For our application P = 3000 columns have been selected before IP_Branch terminates and OSL_IP_Solve is called. The strategy that is suggested in IP_Branch algorithm is to branch on a predecessor or a successor of a given node in G^0, i.e. for a given flight segment branch on a preceding or succeeding flight segment. At each branch stage a pair of segments is locked by fixing a directed arc from a node to its predecessor or to its successor. Preference is given to nodes with a minimum number of predecessors or successors. Conceptually, in a "hub and spoke" system typical of domestic flight patterns we first branch on segments involving spoke stations before considering the hubs.

Once a pair of nodes (x, y) is determined to be locked all arcs (x, z) and (w, y) is deleted from the edge set of the graph to make (x, y) a unique predecessor-successor pair. Since each arc in the graph determines a set of columns in the matrix A of SPP, the deletion of an arc implies the deletion of a set of cycles in the graph. Using this method, cycles are removed from the graph in an iterative manner to produce a sequence of graphs G^0, G^1, G^2, ..., G^k thereby reducing the number of cycles in the graph. At each iteration **i** the feasibility of the reduced graph G^i is verified by checking the condition of remark 2 and by solving the linear programming problem

LP^i corresponding to it. The predecessor-successor that is chosen to be branched on is determined by a maximum arc-weight computed as a sum of primal values on columns associated with that arc in the current graph. Let $w(i, j) = \Sigma_{k\ C(i,j)}\ x_k$, be the arc-weight of (i, j) where C(i, j) has been defined as a set of columns having (i, j) as a predecessor-successor pair and x_k the primal value of such a column. Note that for a given node **j**, $\Sigma_i\ w(i, j) = \Sigma_k\ w(j, k) = 1$, and locking (i, j) and (j, k) implies that $w(i, j) = w(j, k) = 1$. In this manner entire sets of columns may become inactive thereby significantly reducing the size of the search tree. As the nodes are fathomed the search tree can be efficiently probed to detect infeasibility as the SPP problem is preconditioned for further branch-and-bound processing.

4. Computational Results.

The code of the Crew_Solve algorithm was programmed in FORTRAN and uses IBM Optimization Subroutine Library. All the test and production runs were made on IBM RS6000-550. The system described in this paper has been successfully used by American Airlines Crew Planning Department since mid 1992. Over and above the savings produced by its predecessor, that has been installed in 1991 (see [1]), this system produces additional 2 million dollars in annual savings. A sample of three production runs is presented in Table 1. All 3 runs were performed for American's largest MD80 fleet. In each run a 20 million column LP was solved using Sprint_Solve_LP technique with successive calls to OSL. After the LP solution was found IP_Branch has been used to find an integer solution for a significantly smaller IP. The gap between the IP solution and the LP bound in all cases was under 2%. Summary of the results as well as the problem dimensions, run times and pay and credit are given in the following table.

Table 1. Computational Results for AA Crew Pairing Problems

	SEP01	*OCT01*	*NOV01*
# ROWS	1,152	1,256	1,145
# COLUMNS	20 million	20 million	20 million
LP BOUND	47,520	48,391	47,908
IP	48,069	49,103	48,410
% GAP	1.1%	1.5%	1.0%
%PAY&CREDIT	0.30%	0.32%	0.35%
RUN TIME	6.3 hours	7.1 hours	7.8 hours

Significant reduction in run time was achieved by removing the duplicate columns that were generated by the Col_Gen module. The effect was about factor of 2 for the Sprint_Solve_LP and about 30-40% for the IP_Branch and IP_Solve phases of the Crew_Solve algorithm. The methodology as well as the results will be described in our future papers that are currently in the process of preparation.

5. Conclusions.

A comprehensive description of Crew_Solve algorithm has been given in this paper. Current advances in the computer technology in conjunction with our implementation of the Crew_Solve algorithm enabled the solution of large scale linear programs as well as large scale integer programs in the crew scheduling area. These problems as a whole were intractable in the large scale sense only 3-4 years ago. The implementation of the system can be viewed not only as a technological breakthrough. It can be viewed as an important business decision support tool in a competitive airline industry. For example, the benefits of using the system in crew planning at American Airlines were reflected in both increased crew utilization and significant annual savings produced by lower pay and credit. In fact with the most recent enhancements to the system the pay and credit for the largest of American's fleets MD80 is lower than 0.2%. It translates into about 2-4 hours of "synthetic" pay and credit time out of a total 2,500 hours flown daily. As the airline business undergoes a difficult transition to a lower operational cost with more efficient utilization of human and machine resources, the current results achieved at American Airlines crew planning are hard to improve on. However, we are constantly looking for ways to enhance the current system to keep pace with the growing complexity of the crew scheduling problem.

References

[1] Anbil, R., Gelman, E., Patty, B., and Tanga, R., 1991, "Recent advances in crew-pairing optimization at Amemrican Airlines", *Interfaces*, Vol. 21, No. 1, pp. 62-74.

[2] Anbil, R.,Johnson, E., and Tanga, R., 1991, "A global approach to crew-pairing optimization", Research Report #77509, IBM Research Division.

8

HOURIA III: A SOLVER FOR HIERARCHICAL SYSTEMS OF FUNCTIONAL CONSTRAINTS

M. Bouzoubaa, B. Neveu and G. Hasle*

INRIA-CERMICS,B.P.93 06902 Sophia-Antipolis Cedex, France

**SINTEF, Applied Mathematics,P.O.BOX 124 Blindern Oslo, Norway*

ABSTRACT

Constraints allow programmers and users to state declaratively a relation that should be maintained, rather than requiring them to write procedures to maintain the relation themselves. They are thus useful in many applications such as user interface and simulation packages. In many situations, it is desirable to be able to state both hard and soft constraints. The hard constraints must hold, and the soft constraints are satisfied as much as possible, depending on the criterion used. A constraint hierarchy consists of a set of constraints, each labeled as either hard or soft at some strength, and each soft constraint is weighted by a real weight. The existing solvers for satisfying constraint hierarchy impose that every class of soft constraints labeled by a strength contains constraints with the same weight. This paper surmounts this restriction by presenting an incremental solver that proposes extended definitions of constraint hierarchy. This solver handles different classes of soft constraints labeled with strengths where each class may contain weighted constraints. The criterion of comparison used in this solver is global. It allows the comparison of some valuations, which are not comparable by local criteria used in some existing solvers. This criterion is the generalization of the global criterion used in other solvers.

1 INTRODUCTION

Local Propagation is an efficient constraint satisfaction algorithm that takes advantage of potential locality of constraint systems [1]. Garnet [2] is based on local propagation and handles one-way constraints. A one-way constraint always outputs a value to a certain variable. However, one-way constraints are often insufficient because they cannot change dependencies among variables.

Multi-way constraints are proposed by Borning in [3]. A multi-way constraint has multiple methods for one constraint. A system of multi-way constraints is solved as follows: one method is selected from each constraint, a solution graph is generated out of the system so that the graph has no conflicts and no cycles. Then, local propagation is applied to the solution graph. Multi-way constraints also embody the problem that output variables are not determined uniquely and it would easily result in over-constrained systems.

A solution graph is a constraint graph extended so that it dictates how constraints will be solved. A correct solution graph produces correct solutions, and satisfies the following two properties: the value of each variable must be determined by at most one constraint, (i.e. the graph must have no conflicts) and all the constraints must be partially ordered (i.e. the graph must have no cycles). Applying local propagation to solution graph is, in short, equivalent to solving necessary constraints in a topological order.

In [4, 5], Borning et al. proposed constraint hierarchies to cope with the problem of over-constrained systems. A constraint hierarchy is a system of constraints with hierarchical strengths. If the system is over-constrained, it is solved so that there are as many satisfied strong constraints as possible. For example the constraints $v_2 = 1$ and $v_2 = 2$ are in conflict. However, if $v_2 = 1$ and $v_2 = 2$ are, respectively labeled with strength *strong* and *weak*. The constraint system is solved by satisfying only the constraint $v_2 = 1$.

The existing solvers (Blue, DeltaBlue, SkyBlue [6, 7, 8]), for over-constrained systems, are based on local criteria: valuation comparison. According to these local criteria, many valuations might not be possible to compare (i.e two different valuations satisfying two disjoint constraint sets are not comparable) and regarding the semantics of the hierarchy, the best valuation can not be obtained. The solvers for over-constrained systems described in [9, 10] are based on global criteria. These solvers find a valuation which satisfies more constraints than the one produced by solvers based on local criteria for the same over-constrained problems. However, existing incremental solvers for satisfying constraint hierarchy have the property that every class of soft constraints labeled by a strength contains constraints with the same weight. This paper surmounts this restriction by presenting a incremental solver based on global criteria. This solver is the extension ! of the one described in [10].

2 THEORY OF FUNCTIONAL CONSTRAINT HIERARCHIES

Each functional constraint has a set of methods that can be invoked to satisfy the constraint. For example, the constraint $v_1 = v_2 - v_3$ has three methods: $v_1 \leftarrow v_2 - v_3$ (i.e. calculates the value of v_1 from the values of v_2 and v_3), $v_2 \leftarrow v_1 + v_3$ and $v_3 \leftarrow v_2 - v_1$.

In [11], the problem of constraint satisfaction in functional constraints hierarchies is a triplet *(V,D,C)* defined by:

- A set of n variables: $V = \{v_1, v_2, ..., v_n\}$.
- Each variable v_i ranges over a domain d_i. The set of domains d_i is noted by $D : D = \{d_1, d_2, ..., d_n\}$.
- A constraint system C. A constraint is an n-ary relation among a subset of V. Each constraint has some methods. A method uses some of the constraint variables as inputs called ***antecedent*** and computes the remainder as outputs called ***consequent***. A method may only be executed when all its input variables are determined by at least one other constraint, and none of its outputs have been determined by other constraints. C is partitioned into sets $C_0, C_1, C_2..., C_m$ where C_i contains the constraints labeled with strength i. C_0 contains the hard (i.e. required) constraints. $C_1, C_2...C_m$ contains the soft constraints. C_1 contains the most strongly preferred constraints. C_2 the next weaker level, and so forth through C_m, where m is the number of distinct non-required strengths. Each constraint in C_i where: $0 < i \leq m$, is associated with a real weight w_j where: $0 \leq w_j \leq 1$.

Solutions to a constraint hierarchy are defined as a set S of valuations, each valuation in S must be such that it satisfies all constraints in C_0. In addition, we desire each valuation in S to be such that it satisfies the soft (i.e. non-required) constraints $C_1, C_2, ..., C_m$ as well as possible, with respect to their relative strengths and weights.

To formally define this set of solutions, we first present the set S_0 of valuations such that each valuation in this set satisfies C_0. Then, using S_0, we present the desired set S by eliminating all potential valuations that are worse than some other potential valuation using the comparator ***predicate better***. (*Sat* is a boolean predicate, its value is true when the valuation θ satisfies every constraint in C_0, otherwise its value is false).

$S_0 = \{\theta : Sat(\theta, C_0)\}$ and $S = \{\theta : \theta \in S_0 \wedge \forall \eta \in S_0 \ \neg better_C(\eta, \theta)\}$

Many alternative definitions for comparators are given in [11, 12, 4], *better* is irreflexive and transitive. However, *better* will not provide a total ordering, so, there may exist θ and η in S such that $\neg better_C(\theta, \eta)$ and $\neg better_C(\eta, \theta)$.

In [11, 13, 8, 4] several different comparators are defined. The error function $e(c\theta)$ is used. This error function returns a non-negative real number indicating how nearly constraint c is satisfied for a valuation θ. This function has the property that $e(c\theta) = 0$ if and only if $Sat(\theta, c)$. For any domain D, the trivial error function that returns 0 if the constraint is satisfied and 1 if it is not, can be used. The first of the comparators, *locally-better*, considers each constraint in C individually. The definition of this local comparator is :

Definition 1: A valuation θ is *locally-better* than another valuation η if, for each of the constraints through some level $k - 1$, the error after applying θ is equal to that after applying η , and at level k the error is strictly less for at least one constraint and less than or equal for all the rest.

$locally - better(\theta, \eta, C) \Leftrightarrow \exists k > 0 \, such \, that \, \forall i \in 1...k - 1 \ \forall p \in C_i$
$(e(p\theta) = e(p\eta)) \wedge (\exists q \in C_k e(q\theta) < e(q\eta)) \, (\wedge \forall r \in C_k e(r\theta) \leq e(r\eta))$.

Next, the schema *globally-better* for global comparators is parameterized by a function g that combines the errors of all the constraints C_i at a given level. The definition of this global comparator is :

Definition 2: A valuation θ is *globally-better* than another valuation η if, for each level through some level $k - 1$, the combined error of the constraints after applying θ is equal to that after applying η, and at level k it is strictly less.

$globally{-}better(\theta, \eta, C, g) \Leftrightarrow \exists k > 0 \, such \, that \, \forall i \in 1...k{-}1 \ g(\theta, C_i) = g(\eta, C_i) \wedge$
$g(\theta, C_k) < g(\eta, C_k)$.

Using globally-better, we present two global comparators defined in [4] , by using different combining functions g. The weight for constraint c is denoted by $\bar{c}$.

$weighted - sum - better(\theta, \eta, C) \Leftrightarrow globally - better(\theta, \eta, C, g)$ where
$g(\eta, C_i) \equiv (\sum_{c \in C_i} \bar{c} e(c\eta))$

$least - squares - better(\theta, \eta, C) \Leftrightarrow globally - better(\theta, \eta, C, g)$ where $g(\eta, C_i) \equiv (\sum_{c \in C_i} \bar{c}e(c\eta)^2)$

The predicate versions of the two comparators are respectively ***weighted-sum-predicate-better*** and ***least-squares-predicate-better***. The global comparator ***least-squares-predicate-better*** is not particularly useful, because it always gives the same results as ***weighted-sum-predicate-better*** (since $1^2 = 1$).

Solvers as SkyBlue or DeltaBlue use the ***locally-predicate-better*** comparator, variation of ***locally-better*** comparator (i.e. ***locally-better*** using the error function that return 0 if the constraint is satisfied and 1 if it is not). It may produce a very large set S of valuations. This criterion cannot compare (i.e. order) two valuations that satisfy two disjoint constraints set at a given level. Generally, a solver based on a local criterion, cannot always find the best valuation which satisfies the soft (i.e. non-required) constraints. In order to handle a global comparator, the solver described in [10] uses the ***unsatisfied-count-better*** comparator. This comparator is a special case of ***weighted-sum-predicate-better***, using weights of 1 on each soft constraint, it counts the number of satisfied constraints in making its comparisons. In this paper, we propose a generalization of the solver described in Houria [10]. It aims to use the *!* ***weighted-sum-predicate-better*** c

3 HOURIA III SYSTEM

3.1 Overview

Based on the theory in section 2, Houria III solves constraint hierarchies with the following restriction:

- All constraints in a constraint hierarchy are independent and unique. For example, a hierarchy must not contain the constraints $(strong, 0.9)\, x = 3$ and $(medium, 0.9)\, x = 3$, or the constraints $(strong, 0.9)\, x = 3$ and $(strong, 0.8)\, x = 3$.
- For each constraint, any of their methods can be invoked to satisfy this constraint.

3.2 The Criterion of Comparison used

The solver presented in this article is based on the *weighted-sum-predicate-better* comparator. This comparator uses the sum of weights of errors in each level of the hierarchy. This comparator is global and finds solutions at reasonable computational cost. The size of the set of valuations produced using this comparator is generally smaller than the one produced by the local criteria. Its definition is:

Definition 3: A valuation θ is *weighted-sum-predicate-better* to another valuation η , if and only if, for each level through some level $k-1$, the sum of weights of non-satisfied constraints after applying θ is equal to the sum of weights of non-satisfied constraints after applying η , and at level k it is strictly less.

In the following paragraphs we define how to obtain the valuation(s) in the set S, by constructing first a graph with no conflict and no cycle (called *Lexicographic-Weighted-Sum-Better-Graph*) and executing all the methods in this graph by applying the local propagation algorithm.

3.3 Lexicographic-Weighted-Sum-Better-Graph

In graph-theoretic terms, a constraint is considered to be satisfied if it is enforced in the solution graph. A constraint is enforced if it is included in the solution graph. A constraint is unforced, or unsatisfied, if it is not included in the solution graph. A graph is admissible if it enforces all the hard constraints (i.e., the constraints in C_0). A constraint satisfier would like to choose the best of these admissible graphs. In order to obtain the best valuation that satisfies the hierarchy by using the criterion defined in Definition 3. Houria III plans a Lexicographic Weighted Sum Better Graph (or *LWSBG*) and applies the local propagation algorithm to this graph. We present the definition of a lexicographic better graph.

Definition 4: Solution graph s_1 is lexicographic better than another solution graph s_2 if and only if: for each level through some level $k-1$, the sum of weights of constraints unforced in s_1 is equal to the the sum of weights of constraints unforced in s_2, and at level k it is strictly less.

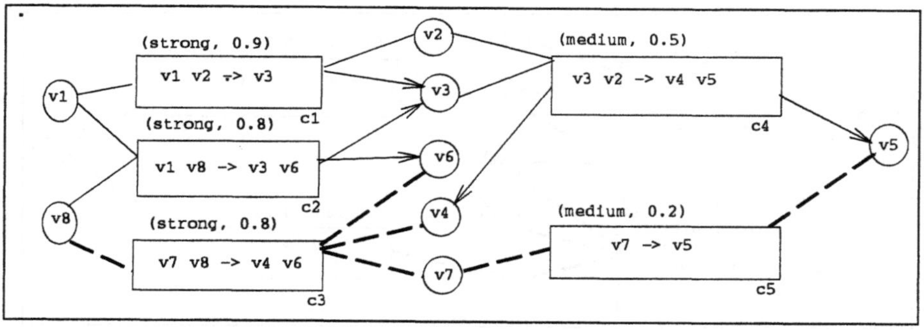

Figure 1 a non-LWSBG Solution Graph

The Houria III uses the constraint weights to construct a *Lexicographic-Weighted-Sum-Better-Graph* (or *LWSBG*) solution graph. A solution graph is *LWSBG* if and only if :

- there are no method conflicts and no cycles, and
- there are no unforced method at any k level, that if it becomes enforced then it generates a lexicographic better solution graph.

Given an over-constrained constraint hierarchy, for obtaining the *LWSBG(s)*, the Houria III performs the following two steps :

- In the same class, it leaves constraints with a small weight unsatisfied (i.e. unforced in the solution graph) in order to satisfy (i.e enforce in the solution graph) the constraint with a large weight.
- Between the classes, it leaves constraint labeled by a weaker strength unsatisfied in order to satisfy the constraint labeled by a higher strength.

For example, consider the solution graph in Figure 1, we suppose that each constraint contains one method.

This solution graph is not *LWSBG*, because the constraints c_1 and c_2 are in conflict at the variable v_3. Since the weight of the constraint c_2 is lower than the weight of the constraint c_1, c_2 can be revoked, producing Figure 2.

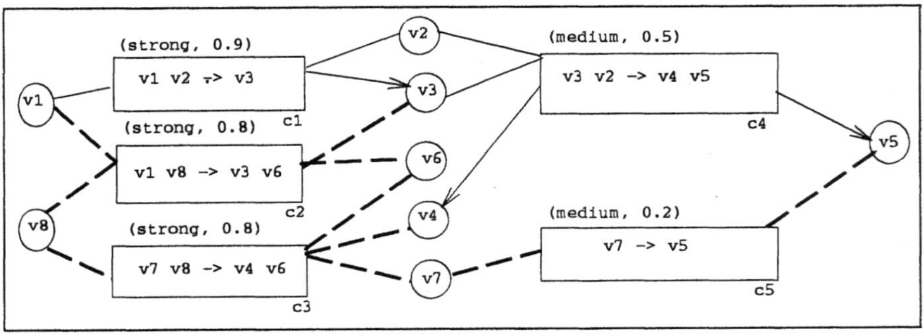

Figure 2 a non-LWSBG Solution Graph ng

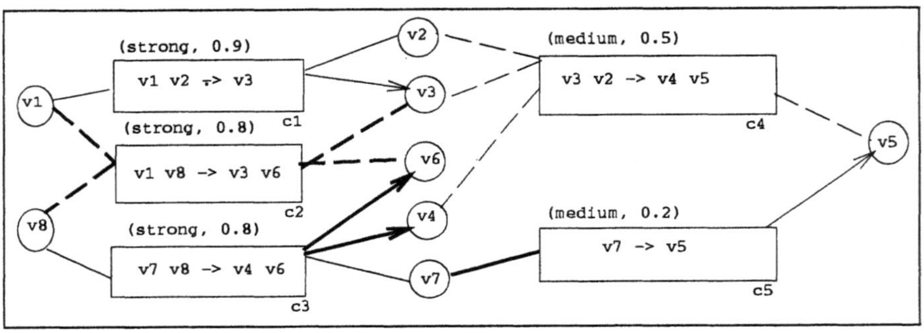

Figure 3 a LWSBG Solution Graph

Actually, this solution graph is not *LWSBG* either, because c_3 could be enforced, by revoking the constraint c_4 and enforcing the constraint c_5, producing Figure 3. The solution graph in Figure 3 is *LWSBG* since the unforced constraints cannot be enforced and produce a better solution graph.

3.4 General description of Houria III

The set S can be seen as the set of *LWSBG solution graphs.* For obtaining a *LWSBG solution graph*, Houria III uses some tools. These tools are defined in paragraph *Houria III tools.* The algorithmic approach for computing the set S is described in the paragraph *The set S computation.* We describe an improvement of this approach. The objective of this improvement is to reduce the number of solutions graphs developed in the set S. We describe another improvement that reduces the cost of *LWSBG solution graphs* computation.

Paragraphs *Add-constraint* and *Remove-constraint*, respectively describe the add constraint and remove constraint procedures.

Houria III tools

Notation: $\bar{c}$ is the pair *(label, weight)* where *label* is the strength of the constraint c and *weight* is the weight of the constraint c. M_c is the set of methods of constraint c. Each m_{jc} in M_c is represented by a couple noted by *lwsg (sg, lw)*, where the *sg* is: *(Ant, Int, Cons)* and *lw* is a list containing the pair *(label, weight)* of the constraint c. *Ant* is a set containing the antecedent variables of m_{jc}. *Int* is a set of internal variables, the internal variables of one method is null. *Cons* is a set containing the consequent variables of m_{jc}.

Example: $c \equiv \{v_1 = v_2 - v_3\}$, $\bar{c} = (strong, 0.9)$
$M_c \equiv \{(v_2 - v_3 \rightarrow v_1), (v_1 + v_3 \rightarrow v_2), (v_2 - v_1 \rightarrow v_3)\}$
$m_{1c} \equiv ((\{v_2 v_3\}, \{\}, \{v_1\}), ((strong, 0.9)))$,
$m_{2c} \equiv ((\{v_1 v_3\}, \{\}, \{v_2\}), ((strong, 0.9)))$,
$m_{3c} \equiv ((\{v_1 v_2\}, \{\}, \{v_3\}), ((strong, 0.9)))$.

A solution graph can be seen as a couple *(sg, lw)*, where *sg* is the representation of the conjunction of the methods enforced in this solution graph, and *lw* is the list of pairs *(label, weight)* of this solution graph. For obtaining this couple, we define the representation of the conjunction of one method and one method set:

Definition 5:
let $m_{jc} \equiv ((Ant(m_{jc}), \{\}, Cons(m_{jc})), (\bar{c}))$ and
let $s \equiv ((Ant(s), \{\}, Cons(s)), (\bar{s}))$

$$m_{jc} \wedge s \equiv [(Ant(m_{jc}) \cup Ant(s) \setminus (Cons(s) \cup Int(s)),$$

$$Int(s) \cup (Ant(m_{jc}) \cap (Cons(s) \cup Int(s))) \cup (Cons(m_{jc}) \cap (Ant(s) \cup Int(s))),$$

$$(Cons(m_{jc}) \cup Cons(s)) \setminus (Ant(s) \cup Int(s))), \ (\bar{c}) \oplus (\bar{s})].$$

The operator $\oplus$ appends two ordered lists of pair *(label, weight)* into one ordered list of pair (label, weight), by respectively adding the weights associated with the same label.

The representation of the conjunction of one method and one method set is a couple *(sg, lw)*. Note that *Int* contains the common variables set found in both parts:

- The antecedent part of a method set and the consequence part of a method.
- The consequence part of a method set and the antecedent part of a method.

Example: $m_{1c} \equiv ((\{v_2 v_3\}, \{\}, \{v_1\}), ((strong, 0.9)))$ and $s \equiv ((\{v_4\}, \{\}, \{v_2\}), ((medium, 0.8))), m_{1c} \wedge s \equiv ((\{v_3 v_4\}, \{v_2\}, \{v_1\}), ((strong, 0.9)(medium, 0.8)))$.

Intuitively, the goal of this representation mode is to handle a set of variables instead of other complex data structure. We can simply compare a set of variables, to know if the conjunction of one method and one method set contains conflicting methods or cycles. To attempt this goal, we define the consistency of one method and one method set. Before presenting the definition, we can make the following remark:

Remark: A representation of the method set s can be seen as the set of the representations of all connected components in s.

We define now the consistency of the representation of one method set and one method :

Definition 6: Let s be the representation of one method set, and C_s, the representation set of connected components in s. Let m be the representation of one method

$$Conflict - Consistent(s, m) \Leftrightarrow (\forall \psi \in C_s)$$
$$(Cons(\psi) \cap Cons(m) = \emptyset) \wedge (Int(\psi) \cap Cons(m) = \emptyset) \wedge (Int(m) \cap Cons(\psi) = \emptyset).$$

$$Cycle - Consistent(s, m) \Leftrightarrow (\forall \psi \in C_s)$$
$$[\neg((Ant(\psi) \cap (Cons(m) \cup Int(m)) \neq \emptyset) \wedge ((Cons(\psi) \cup Int(\psi)) \cap Ant(m) \neq \emptyset))$$
$$\vee ((\forall v_i \in ((Cons(\psi) \cup Int(\psi)) \cap Ant(m))$$
$$\wedge \quad \forall v_j \in ((Cons(m) \cup Int(m)) \cap Ant(\psi))) \Rightarrow \neg Path(v_i, v_j))].$$

$$Consistent(s, m) \Leftrightarrow Conflict - Consistent(s, m) \wedge Cycle - Consistent(s, m).$$

Regarding the first condition of *LWSBG* solution graph in paragraph 3.4, the Definition 6 means that :

- the method *m* is not in conflict with the method set *s* if and only if the predicate *Conflict-Consistent* is true.
- they are no cycles between *m* and *s* if and only if the predicate *Cycle-Consistent* is true.
- *s* is consistent with *m* if and only if both predicates are true (i.e. the conjunction of *s* and *m* is not consistent if and only if it contains conflicting methods or cycles).

The set S computation

S is a set of *lwsg: (sg, lw)* where *sg* is a solution graph and *lw* is the list of pairs *(label, weight)* of this solution graph. Houria III maintains the solution graphs in a set *S*, by constructing an *LWSBG solution graph*, and executing its methods in order to satisfy the enforced constraints. Initially, the set *S* of solution graphs contains the set of solution graphs corresponding to the constraints in C_0 (i.e. the hard constraints). Houria III is invoked by calling two procedures, ***add-constraint*** to add a constraint to each solution graph in *S*, and *remove-constraint* to remove a constraint from each solution graph in *S*. As constraints are added and removed, Houria III incrementally updates each solution graph in *S* and sorts the set *S* to find *LWSBG solution graphs*. The approach of Houria III can be described by the following statements:

- A constraint *c* is added to the hierarchical system.
- If a method *m* of a constraint *c* is consistent with the solution graph in any *lwsg* $\in S$, then Houria III forms a new *lwsg* and adds it to *S*. This new *lwsg* will have the following form:
  (the solution graph in *lwsg* $\wedge$ the method *m* of the constraint *c*, the ordered list of pairs *(label, weight)* obtained by computing $lw \oplus (\bar{c}\ )$).

Houria III tries to add each method in the constraint *c* to each *lwsg* $\in S$. This approach is complete but very costly.

Improvement for reducing the space complexity of the set S

The first improvement to this approach consists in keeping in *S* only the maximal elements (i.e. it does not exist a *lwsg* in *S*, which contains another *lwsg* in

S). When Houria III tries to add m to the *lwsg*, (m is one method of constraint c, *lwsg* is one couple in *S*) we can make the following distinctions:

- When m is consistent with the solution graph in *lwsg*, Houria III updates *lwsg* by adding m to the solution graph in *lwsg*, and adding ($\bar{c}$) to *lw* by using the operator $\oplus$.
- When m is not consistent with the solution graph in *lwsg*, Houria III updates S by executing the following steps:
  - Extract from the solution graph in *lwsg* a set of *lwsg* (*S'*), where m is consistent with the solution graph in each *lwsg* in *S'*.
  - For any $lwsg \in S'$, the solver forms a new *lwsg* by considering the method m.
  - Add this new *lwsg* to S if it is maximal (i.e. if it does not exist another *lwsg* in S, which contains this new *lwsg*).

This improvement reduces the size of S, while still preserving the completeness of the solver.

Improvement for reducing the computational cost of the set S

Another improvement to this approach aims at reducing the cost of the solver, this enhancement uses the list of pairs ***(label, weight)*** of all *lwsg* in S. We illustrate this improvement with the following example: $C = C_0, C_1, C_2$ where C_0 is the class of hard constraints. C_1 is the class of constraints labeled with strength ***strong***. C_2 is the class of constraints labeled with strength ***medium***. C_0 contains two constraints c_{01} and c_{02} such that:

$M_{c_{01}} \equiv \{(v_1 \rightarrow v_2), (v_2 \rightarrow v_1)\}$ and $M_{c_{02}} \equiv \{(v_2 v_3 \rightarrow v_4)\}$.
So, $S = \{lwsg_1^0, lwsg_2^0\}$where $lwsg_1^0 = ((\{v_1 v_3\}, \{v_2\}, \{v_4\}), ((\mathit{required}, \bot)))$ and $lwsg_2^0 = ((\{v_2 v_3\}, \{\}, \{v_4 v_1\}), ((\mathit{required}, \bot)))$.

Now we introduce in the class C_1 the constraintc_{11} weighted by 0.9 where $M_{c_{11}} \equiv \{(v_5 \rightarrow v_1)\}S$ is updated and its value is the following :

$S = \{lwsg_1^1, lwsg_2^1\}$where
$lwsg_1^1 = ((\{v_5 v_3\}, \{v_2 v_1\}, \{v_4\}), ((\mathit{required}, \bot)(\mathit{strong}, 0.9)))$ and
$lwsg_2^1 = ((\{v_2 v_3\}, \{\}, \{v_4 v_1\}), ((\mathit{required}, \bot)))$.

$\{lwsg_2^0 = lwsg_2^1\}$ because a method of the constraint c_{11} is not consistent with one hard constraint in $\{lwsg_2^0\}$.

Now we introduce in the class C_2 the constraint c_{21} weighted by 0.7 where $M_{c_{02}} \equiv \{(v_6v_7 \rightarrow v_2)\}$ .S is updated and its value is the following:

$S = \{lwsg_1^2, lwsg_2^2\}$ where
$lwsg_1^2 = ((\{v_5v_3\}, \{v_2v_1\}, \{v_4\}), ((required, \perp)(strong, 0.9)))$ and
$lwsg_2^2 = ((\{v_6v_7v_3\}, \{v_2\}, \{v_4v_1\}), ((required, \perp)(medium, 0.7)))$.

$\{lwsg_1^1 = lwsg_1^2\}$ because a method of the constraint c_{21} is not consistent with one required method in $\{lwsg_1^1\}$.

We observe that: when c_{21} is added to the system, the *LWSBG* solution graph is the solution graph in $\{lwsg_1^1\}$ and it was not necessary to compute $\{lwsg_2^2\}$.

Now we introduce in the class C_2 the constraint c_{22} weighted by 0.9 where $M_{c_{22}} \equiv \{(v_7 \rightarrow v_5)\}$. S is updated and its value is the following:

$S = \{lwsg_1^3, lwsg_2^3\}$where
$lwsg_1^3 = ((\{v_7v_3\}, \{v_2v_1v_5\}, \{v_4\}), ((required, \perp)(strong, 0.9)(medium, 0.9)))$ and
$lwsg_2^3 = ((\{v_6v_7v_3\}, \{v_2\}, \{v_4v_1v_5\}), ((required, \perp)(medium, 1.6)))$.

We observe that: when c_{22} is added, the solution graph in $\{lwsg_1^2\}$ is consistent with c_{22}. Since the *lw* of $\{lwsg_1^2\}$ is the strongest in S, it will be sufficient to compute the $\{lwsg_1^3\}$ because the *LWSBG* solution graph is in $\{lwsg_1^3\}$.

We can generalize these observations and formulate this improvement by the following remarks:

S is partitioned into subsets, where each subset contains the *lwsg* that have an equal weight. Each subset is associated with a queue called queue of constraints to add. Each subset has two weights: *current-label-weight* and *potential-label-weight*. The *current-label-weight* of a subset is equal to the *lw* of one *lwsg* in this subset. The *potential-label-weight* of a subset is equal to an ordered list of pair *(label, weight)*. This ordered list is obtained by applying the operator $\oplus$ on the *current-label-weight* and all pairs *(label, weight)* of the constraints in the associated queue of this subset. All subsets in S will be lexicographically ordered on respect to their *potential-label-weight*. The *current-label-weight* and *potential-label-weight* with the first subset in S are always equal. When a constraint$c$ is added to the system, Houria III tries to add the constraint $c!$ to each lwsg in the first subset

The goal of this improvement is to delay the computations as much as possible, and to not enumerate all *LWSBG* solution graphs. This means that Houria III performs the computations only if we are certain that a better solution graph may exist.

Another heuristic is considered by this approach. This heuristic consists to perform on priority the subset which has the highest *current-label-weight* from all subsets which have the same *potential-label-weight*. In average this improvement reduces the complexity of computing the *LWSBG* solution graphs, while still being complete.

The application of this improvement on the precedent example is the following:

$S = \{[\{lwsg_1^0, lwsg_2^0\}, q_1]\}$ where $lwsg_1^0 = ((\{v_1v_3\}, \{v_2\}, \{v_4\}), ((required, \perp)))$ and $lwsg_2^0 = ((\{v_2v_3\}, \{\}, \{v_4v_1\}), ((required, \perp))), q_1 = \emptyset$.

S contains only one subset because all *lwsg* in this subset have equal*potential-label-weight*. q_1 is the queue associated to this subset.

Now we introduce in the class C_1 the constraint c_{11} where $M_{c_{11}} \equiv \{(v_5 \rightarrow v_1)\}$. S is updated and its value is the following :

$S = \{[\{lwsg_1^1\}, q_1], [\{lwsg_2^1\}, q_2]\}, q_1 = \emptyset, q_2 = \emptyset$
$lwsg_1^1 = ((\{v_5v_3\}, \{v_2v_1\}, \{v_4\}), ((required, \perp)(strong, 0.9)))$ and
$lwsg_2^1 = ((\{v_2v_3\}, \{\}, \{v_4v_1\}), ((required, \perp)))$.

$lwsg_2^0 = lwsg_2^1$ because a method of the constraint c_{11} is not consistent with one hard constraint in $\{lwsg_2^0\}$.

Now we introduce in the class C_2 the constraint c_{21} weighted by 0.7 where $M_{c_{21}} \equiv \{(v_6v_7 \rightarrow v_2)\}$. S is updated and its value is the following :

$S = \{[\{lwsg_1^2\}, q_1], [\{lwsg_2^1\}, q_2]\}, q_1 = \emptyset, q_2 = \{c_{21}\}$
$lwsg_1^2 = ((\{v_5v_3\}, \{v_2v_1\}, \{v_4\}), ((required, \perp)(strong, 0.9)))$.

Since the *current-label-weight* of $lwsg_1^2$ is lexicographic strictly greater than the *potential-label-weight* of $lwsg_2^1$, Houria III halts because the *LWSBG* solution graph is computed. Now we introduce in the class C_2 the constraint c_{22} weighted by 0.9 where $M_{c_{22}} \equiv \{(v_7 \rightarrow v_5)\}$ S is updated and its value is the following :

$S = \{[\{lwsg_1^3\}, q_1], [\{lwsg_2^1\}, q_2]\}, q_1 = \emptyset, q_2 = \{c_{21}, c_{22}\}$,
$lwsg_1^3 = ((\{v_7v_3\}, \{v_2v_1v_5\}, \{v_4\}), ((required, \perp)(strong, 0.9)(medium, 0.9)))$.

Houria III halts because it cannot find another solution graph better than the one computed in since :$(lw_2^1 \oplus \bar{q_2}) \leq_{lex} lw_1^3$ (i.e the *potential-label-weight* of the second subset is not greater than the *current-label-weight* of the first subset in S).

In the two following paragraphs, we describe the procedures *Add-constraint* and *Remove-constraint* for adding and removing a constraint c to the hierarchical functional constraint systems. The notations used are:

Notation: $S = ((s_1, q_1), ..., (s_n, q_n))$, the set S is partitioned into n subsets. Each subset s_i contains the *lwsg* which are of equal *current-label-weight*, ($s_i = \{lwsg_1, ..., lwsg_m\}$ such that $lw_1 = ... = lw_m$). Each subset s_i is associated with a queue q_i where $q_i = \{c_0, ..., c_n\}$. The list of pair *(label, weight)* of q_i denoted by $\bar{q_i}$ is equal to $(\bar{c_0} \oplus ... \oplus \bar{c_n})$. The *current-label-weight* of subset s_i is $\bar{s_i}$ and it is equal to one pair *(label, weight)* in one of its *lwsg*. The *potential-label-weight* of the subset s_i is equal to $\bar{s_i} \oplus \bar{q_i}$.

3.5 Add-constraint

When a new constraint is added, the procedure *Add-constraint* puts this new constraint in every queue associated with a subset in S except the first subset. The procedure tries to add the new constraint to the first subset in S by calling the *Add* procedure. The result of this call is sorted with all subsets in S except the first one, and assigned to S. The procedure examines the queue of the first subset in S. When the queue of the first subset is not null (i.e. Houria III is not certain that it has computed the *LWSBG* solution graph). The procedure performs the first subset. The procedure tries to add any constraint in the queue of the first subset to all *lwsg* in this first subset. The resulting subsets are sorted with the subsets in S except the first one, by respecting their *potential-label-weight*. Houria III continues to performs the first subset in S until the queue associated to the first subset in ! S is empty. The procedure re

The effective update is realized by the *Add* procedure. The *Add* procedure tries to add each method of the constraint c to each *lwsg* in the subset s_i. If the method of the constraint c is consistent with the solution graph in the *lwsg*, then the procedure adds the method to the *lwsg*, otherwise the procedure tests the consistency of the required methods in the solution graph in *lwsg* with the method of constraint *c*. The procedure constructs some new *lwsg* by determining the maximal subset of methods from the *lwsg*. Each new *lwsg* is consistent with the method to be added. The new *lwsg* constructed is returned by the procedure if and only if it is maximal.

3.6 Remove-constraint

In order to remove a constraint *c*, Houria III partitions the set *S* into two sets *S'* and*S"*. *S"* contains the subsets which associated queues include the constraint *c* to remove. The procedure removes the constraint *c* by substracting this constraint from any queue of the subset in *S"*. *S'* contains some subsets where a *lwsg* in each subset contains one method of the constraint *c* to be removed. The procedure ***Remove-constraint*** removes all methods of the constraint *c* from all *lwsg* in each subset in *S'*, and keeps each *lwsg* in each subset of *S'* maximal consistent. The two sets *S'* and *S"* are sorted by considering the ***potential-label-weight*** of their subsets . The procedure returns the first subset of *S*, this first subset contains the *LWSBG* solution graph. The actual algorithm is giving in [14].

4 COMPARISON WITH OTHER SOLVERS

QuickPlan is an incremental algorithm for satisfying hierarchies of multi-way dataflow constraints [9]. SkyBlue is a successor of the DeltaBue solver [7]. Like DeltaBlue, SkyBlue solves hierarchies of multi-way constraints using ***Locally-graph-better***, a variation of ***Locally-predicate-better***. The critical differences between Houria III, QuickPlan, DeltaBlue and SkyBlue solver are summarized as follows:

- Skyblue supports constraints with multi-output methods and cycles of constraints by invoking cycle solvers, while Houria III supports only constraints with multi-output methods and eliminates cycles. Houria III can be extended to support cycles by modifying the definition of the consistency and providing subsolvers, which solve constraints in cycles simultaneously.
- If a constraint hierarchy has one or more cyclic solutions, and at least one acyclic solution, Houria III and QuickPlan are capable of guaranteeing an acyclic solution while SkyBlue is incapable.

- Houria III handles incrementally constraint hierarchies where each class in the hierarchy can contain constraints weighted with different weights while SkyBlue, DeltaBlue and QuickPlan handle incrementally constraint hierarchy where each class in the hierarchy contains constraints weighted with the same weight.

- Houria III handles incrementally the set of solutions in a single constraint hierarchy while SkyBlue, DeltaBlue and QuickPlan handle one solution using the Locally-predicate-better criterion. This is particularly interesting, if the user desires to have more than one solution when constraints are added or removed to the system.

- Given an over-constrained problem, the criterion of comparison used in Houria III is more powerful than that of SkyBlue and DeltaBlue. So, the solutions found by Houria III satisfy more constraints than the one produced by DeltaBlue or SkyBlue while respecting the hierarchy. The representation mode used by Houria III is easy to use, because the representation of a solution graph is easy to handle. We can simply compare a set of variables, so it is useful to know if the solution is already in S or not.

The following example illustrates the strength of the criterion used by Houria III compared to the criterion used by SkyBlue. For this example, we suppose that all soft constraints have the same weight, since SkyBlue can not handle constraints weighted by different weights.

$C = \{C_0, C_1, C_2\}, C_0 = \{c_{01}\}, C_1 = \{c_{11}, c_{12}, c_{13}\}, C_2 = \{c_{21}\}.$
$M_{c_{01}} \equiv \{(v_1 v_2 \rightarrow v_3),\}$ $M_{c_{11}} \equiv \{(v_3 v_4 \rightarrow v_5 v_6)\}, M_{c_{12}} \equiv \{(v_5 v_6 \rightarrow v_1 v_2)\},$
$M_{c_{13}} \equiv \{(v_8 v_7 \rightarrow v_5)\}, M_{c_{21}} \equiv \{(v_8 \rightarrow v_6)\}.$

The Locally Graph Better (or *LGB*) constructed by SkyBlue is presented in Figure 4. SkyBlue keeps the constraint c_{13} unforced, because there are no method conflicts and there are no unforced constraints that could be enforced by revoking one or more strongest constraint. SkyBlue keeps the constraint c_{21} unforced, because there is a method conflict with the constraint c_{11}, and the strength of c_{21} is weaker than the strength of c_{11}. The *LWSBG* solution graphs constructed by Houria III is in Figure 5. If we respect the semantics of the hierarchy, the valuation produced by the *LWSBG* in the Figure 4 should be preferred to the one produced by the *LGB* in the Figure 5.

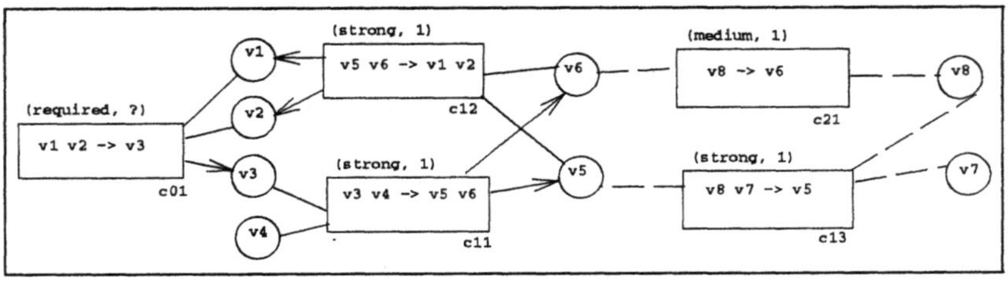

Figure 4 a Locally Graph Better (LPG)

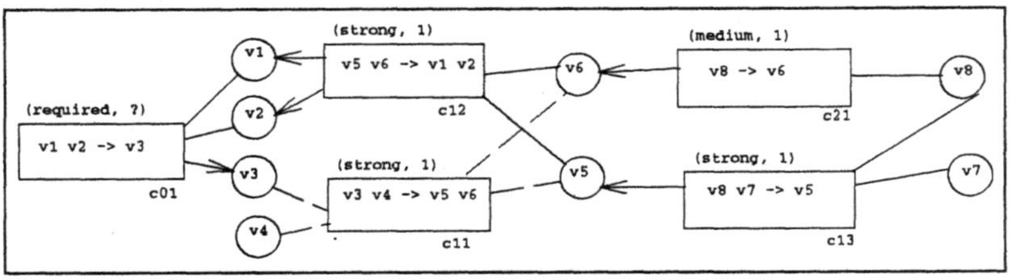

Figure 5 a LWSBG Solution Graph

5 TIME COMPLEXITY IMPLEMENTATION AND PERFORMANCE MEASUREMENTS

In practice, the observed running time of this solver is actually polynomial, and in many cases, linear, in the number of constraints in the constraint system. There are many factors that contribute to this actual-case performance :

- The constraint graph is often divided into several completely disjoint subgraphs. Thus, even if the solver exhibits its worst case behavior, it will only examine a fraction of the constraints in the constraint system.
- The number of constraints inconsistent with a constraint to be enforced is typically only a small fraction of the constraints in the subgraph that is being examined. In many cases this number is one. The solver examines only this constraint, retracts it, and enforces the constraint.

We implemented the Houria III in Lisp. Houria III produces the *LWSBG* solution graphs, and applies local propagation to them. In order to check out

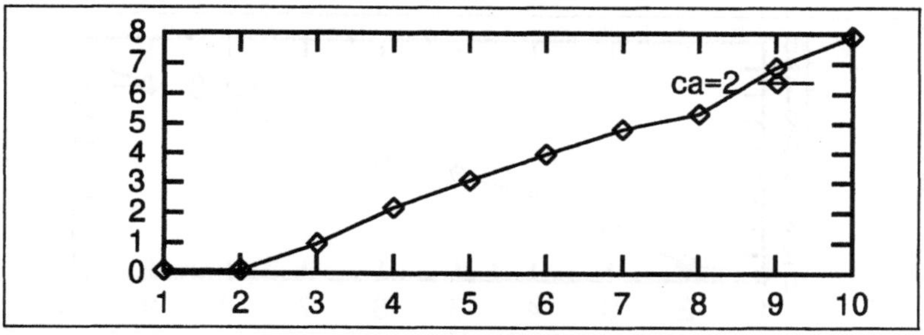

Figure 6 planning time with ca=2

the performance of our solver, we have performed experiments using a Sun SPARC station on some random problems.

We generate a suite of constraint system based on two parameters : the constraint number *cn* and the constraint arity *ca*. We have run two sets of experiments, the first with $ca = 2$ (Figure 6) the second with $ca = 3$ (Figure 7). For each set, we had the constraint number range from 100 to 1000 by steps of 100 and the variable number is equal to the constraint number (then the problems generated are not very dense). Each random generated constraint consists of one method. The pair *(label, weigth)* of each constraint is generated randomly where $label \in \{required, strong, medium, weak\}$ and $weight \in [0.1...1]$. The performance reported for Houria III includes the amount of time in seconds required to find one *LWSBG* solution graph. We have investigated 50 different random over-constrained problems, reporting the average of the results. The results in both figures 6 and 7 can be interpreted as follows :

- With $ca = 2$ or $ca = 3$, the solver time to plan a *LWSBG* solution graph is acceptable.
- When the size of the problem is too large(> 1000), the solver time to plan a *LWSBG* solution graph is long for same applications such as user interface design, document formatting or interactive graphics.

6 CONCLUSION

Houria III can be used to incrementally handle constraint hierarchies where each class in the hierarchy can contain constraints weighted with different weights.

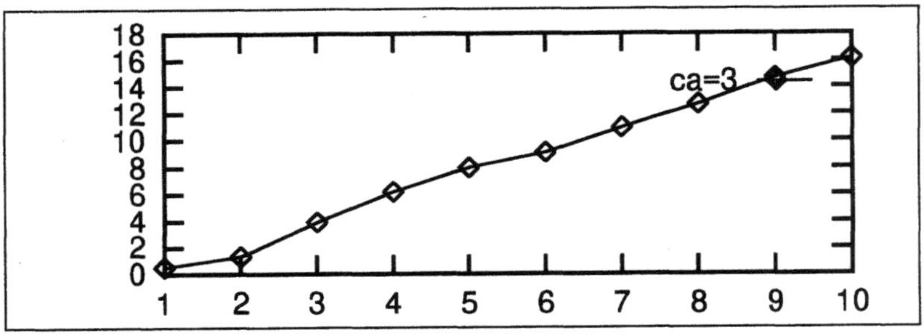

Figure 7 planning time with ca=3

It can be used also in the applications where the user desires to have more than one solution such as geometric layout, physical simulations, user interface algorithm animation [15, 16, 17, 18] design and analysis of mechanical devices and electrical circuits.

7 ACKNOWLEDGMENTS

Thanks for many useful discussions, and comments to early versions of this paper to Alain Borning and Michael Sannella. Thanks to the anonymous referees for useful comments. This project was supported in part by the INRIA-CERMICS in Sophia Antipolis and by SINTEF in Oslo.

REFERENCES

[1] H. Hosobe, K. Miyachita, S. Takahashi, S. Matsuoka, and A. Yonezawa. Locally simultaneous constraint satisfaction lncs 874. In *Principles and Practice of Constraint Programming*, November 1994.

[2] B. A. Myers, D. A. Giuse, R. B. Dannenberg, B. Vander Zanden, A. Mickish D. S. Kosbie, E. Pervin, and P. Marchal. Garnet: Comprehensive support for graphical, highly interactive user interfaces. In *IEEE Computer 23*, pages 71–85, November 1989.

[3] A. Borning. a constraint-oriented simulation laboratory. In *ACM Transactions on Programming Languages and Systems*, number 3. October 1991.

[4] A. Borning, S. Freeman-Benson, and M. Wilson. Constraint hierarchies. In *Lisp and Symbolic Computation*, number 5. 1992.

[5] M. Wilson and A. Borning. Hierarchical constraint logic programing. Tech. rep., University of Washington, May 1993.

[6] G. J. Sussman and G. L. Steele. Constraint language for expressing almost-hierarchical descriptions. In *Artif. Intell.*, number 14-1. 1980.

[7] M. Sannella. The skyblue constraint solver. Tech. rep., University of Washington, February 1992.

[8] M. Sannella, B. Freeman-Benson, J. Maloney, and A. Borning. Multi-way versus one-way constraints in user interfaces: Experience with the deltablue solvery. Tech. rep., University of Washington, 1992.

[9] Brad V. Z. An incremental algorithm for satisfying hierarchies of multi-way. dataflow constraints. Tech. rep., Departement of Computer Science. University of Tennessee, March 1995.

[10] M. Bouzoubaa. The houria constraint solver. In *Fifth Scandinav ian Conference On Artificial Inteligence. IOS press.*, May 1995.

[11] A. Borning, M. Maher, A. Martindale, and M. Wilson. Constraint hierarchies and logic programming. In *the Sixth International Logic Programming Conference*, pages 149–164, April 1989.

[12] B. N. Freeman-Benson. Multiple solutions from constraints hierarchies. Tech. rep., University of Washington, April 1988.

[13] J. H. Maloney, A. Borning, and B. Freman-Benson. Comstraint technology for user-interface construction in thinglabii. In *ACM Conference on Objet-Oriented-Programing Systems Languages and Applications*, pages 381–388, April 1989.

[14] M. Bouzoubaa. Un resolveur de syst me de contraintes fonctionnelles hi rarchique. Tech. rep., CERMICS-INRIA Sophia-Antipolis, November 1994.

[15] M. Sannella and A. Borning. Multi-garnet: Integrating multi-way constraints with garnet. Tech. rep., Departement of Computer Science and Engineering University of Washington, September 1992.

[16] Brad A., Myers, Dario G., Roger B. D., Brad V. Z., David K., Philippe M., and P. Ed. Comprehensive support forgraphical highly-interactive user interfaces: The garnet user interface devlopment environment. In *IEEE Computer 23(11)*, pages 71–85, November 1990.

[17] B. Freeman-Benson and A. Borning. The design and implementation of kaleidoscope'90. a constraint imperative programming language. In *IEEE Computer Society International Conference on Computer Languages*, pages 174–180, April 1992.

[18] G. Misand, B.S. Johansen, G. Hasle, and J. Haukland. Integration of geographical information technology and constraint reasoning. In *Geographical Information Systems*, June 1995.

9

SOME RECENT DEVELOPMENTS OF USING LOGICAL ANALYSIS FOR INFERRING A BOOLEAN FUNCTION WITH FEW CLAUSES

Evangelos Triantaphyllou
Boris Kovalerchuk
and Aniruddha S. Deshpande

ABSTRACT

One of the main areas of great potential in the interface of operations research and computer science is in inductive inference. Inductive inference refers to the extraction of a pattern from observations which belong to different classes. This is the essence of building many intelligent systems with learning capabilities. In this paper we discuss a logical analysis approach to this problem. Given are sets of binary vectors. The main problem is how to extract a Boolean function, in CNF or DNF form, with as few clauses (conjunctions or disjunctions) as possible. Therefore, this is an optimization problem. We present a number of theoretical results and also some possible extensions.

1 INTRODUCTION

A common problem in many scientific and engineering applications is the identification of the characteristics or combination of characteristics which can explain the occurrence of some desirable or undesirable effects. Traditionally, such challenges are examined by statistical and/or artificial intelligence procedures. Among the specific procedures used are: clustering, factor analysis, discriminant analysis, linear regression, decision trees, and neural networks. All this kind of approaches, however, are limited because they rely on some assumptions which are probabilistic in nature (e.g., regarding the existence of a statistical distribution on the data) or geometric in nature (e.g., the existence of separating hyperplanes in some related Euclidean space) [Boros *et al.*, 1994]. These assumptions may severely limit the applicability of the above methods on some real life data.

Moreover, very often the output of these procedures is difficult to be interpreted by field experts. For instance, medical doctors may feel uncomfortable in accepting diagnoses made by a neural network. This situation is increasingly recognized lately and some hybrid systems try to combine rule based systems with neural network techniques [Fu, 1991]. Such hybrid systems produce a set of logical rules which in turn can be used for the classification of new observations. However, the results ares not encouraging because some recent research [Shavlik, 1994] shows that the number of rules generated in this manner can be exponentially large.

2 THE LOGICAL ANALYSIS APPROACH

Logical analysis can be used to extract a set of logical rules (decision rules) which describe patterns of classes of observations. The central hypothesis of using logical analysis [Boros *et al.*, 1994] is that *many phenomena are governed by logical decision rules*. Thus, logical analysis aims at discovering these rules. Logical analysis attempts to capture the cause-effect relationships contained in a data set.

In a typical problem available are collections of examples. Each example is described in terms of a finite set of attributes (also called atoms, variables, characteristics, parameters, or factors). These examples are classified into a finite set of mutually exclusive and exhaustive classes (usually only two classes are considered). Then, the problem is to extract a set of logical rules which describe the patterns implied by these data. The main steps required when applying this kind of data analysis are as follows:

- Definition of the relevant characteristics (i.e., parameters or attributes or factors).
- Extraction of the patterns (i.e., the logical decision rules) which can explain the observations in the various classes.

The above concepts are best described in the following illustrative example.

Suppose that the functioning behavior of a system depends on the values of the five Boolean variables: A_1, A_2, A_3, A_4, and A_5. That is, each such parameter takes on a value which is either true (denoted by 1) or false (denoted by 0). Suppose that the following observations in the set E^+ describe some examples in which the system was functioning correctly, while the observations in the set E^- describe some examples in which the system was malfunctioning:

$$E^+ = \begin{bmatrix} 0 & 1 & 0 & 0 & 0 \\ 0 & 1 & 1 & 1 & 1 \\ 0 & 0 & 1 & 0 & 1 \\ 0 & 1 & 0 & 0 & 1 \\ 1 & 0 & 1 & 0 & 0 \\ 1 & 1 & 1 & 0 & 0 \\ 1 & 1 & 0 & 0 & 0 \\ 1 & 0 & 0 & 1 & 0 \\ 0 & 1 & 1 & 0 & 1 \\ 0 & 0 & 1 & 1 & 0 \\ 1 & 1 & 1 & 1 & 0 \\ 1 & 0 & 1 & 0 & 1 \\ 1 & 1 & 1 & 0 & 1 \end{bmatrix}, \qquad E^- = \begin{bmatrix} 1 & 0 & 0 & 1 & 1 \\ 0 & 0 & 0 & 1 & 1 \\ 1 & 0 & 1 & 1 & 0 \\ 0 & 0 & 0 & 0 & 1 \\ 1 & 1 & 0 & 1 & 1 \\ 1 & 1 & 0 & 0 & 1 \\ 0 & 0 & 0 & 0 & 0 \end{bmatrix}$$

For instance, the first example in E^+ (i.e., [0 1 0 0 0 1 0 1 1 1]) implies that when the values of $(A_1, A_2, A_3, A_4$ *and* $A_5)$ are equal to (*FALSE, TRUE, FALSE, FALSE, FALSE*), respectively, (or equivalently, the values of $(A_1, A_2, A_3, A_4$, *and* $A_5)$ are equal to (*TRUE, FALSE, TRUE, TRUE, TRUE*), respectively), then the system functions correctly (because this example belongs to E^+). The main problem is to use these two classes of observations (or examples of behavior) and describe the conditions on the values of the attributes which cause these examples to belong to either one of these two classes (sets with observations).

Next, consider the following Boolean function defined on the previous five Boolean parameters (or their negations) given in CNF form:

$$(A_1 \vee A_2 \vee A_3) \wedge (\bar{A}_1 \vee A_2 \vee \bar{A}_3 \vee \bar{A}_4) \wedge (\bar{A}_1 \vee A_3 \vee \bar{A}_5) .$$

It can be easily verified that the above expression satisfies the requirements of the previous examples. That is, each positive example makes *each* conjunction to have true value. Furthermore, each negative example is rejected by at *least one* of the conjunctions. In other words, the previous Boolean function returns true value for each one of the positive examples, while it returns false value for each one of the negative examples.

Given the previous E^+ and E^- data, then a new and unclassified example is a positive example (i.e., describes a correct behavior of the system under consideration) if all the previous three rules are satisfied. Otherwise, the example will be classified as a negative one. For the above reasons the terms *"Boolean function"* and *"logical decision rules"* will be used in this paper to mean the same concept.

3. LITERATURE REVIEW

The extraction of a set of logical decision rules from collections of classified data is a type of learning from examples. Complexity issues of this type of learning can be found in [Valiant, 1984 and 1985], [Kearns *et al.*, 1987], and [Pitt and Valiant, 1988]. A considerable amount of related research is today known as the PAC (Probably Approximately Correct) learning theory (see, for instance, [Angluin, 1988] and [Haussler and Warmuth, 1993]). Since the very early days it was recognized that the problem of inferring a Boolean function with a specified number of clauses is NP-complete (see, for instance, [Brayton *et al.*, 1985] and [Gimpel, 1965]).

A learning algorithm is a computational procedure which takes a random sample of positive and negative examples of the target concept c and returns a hypothesis h. In the literature a learning algorithm A is a PAC algorithm if for all positive numbers ε and δ (where: $1 > \varepsilon,\delta > 0$), when A runs and accesses unclassified examples, then it eventually halts and outputs a concept h with probability at least $1 - \delta$ and error at most ε [Angluin, 1992]. A related issue is the one of *properly PAC learnable* concepts. This occurs when the concept c and the hypothesis h are assumed to belong to the same spaces of concepts [Haussler and Warmuth, 1993].

Conjunctive concepts are properly PAC learnable [Valiant, 1984]. However, the class of concepts in the form of the disjunction of two conjunctions is not properly PAC learnable [Pitt and Valiant, 1988]. The same is also true for the class of existential conjunctive concepts on structural instance spaces with two objects [Haussler, 1989]. The classes of k-DNF, k-CNF, and k-decision lists are properly PAC learnable for each fixed k [Kearns *et al.*, 1987], but it is unknown whether the classes of all DNF, or CNF functions are PAC learnable [Haussler and Warmuth, 1993] and [Goldman, 1990]. Note that the present paper deals with Boolean functions in CNF or DNF form. Therefore, the learning algorithms described later in this paper are not known whether they are PAC learnable. Also, in [Mansour, 1992] an $n^{O(\log\log n)}$ algorithm is given for learning DNF formulas (however, *not* of minimal size) under a uniform distribution using membership queries.

There are many reasons why one may be interested in inferring a Boolean function with the minimum (or near minimum) number of terms. In a circuit design environment, a minimum size Boolean representation is the prerequisite for a successful VLSI application. In a learning from examples environment, one may be interested in deriving a *compact set* of decision rules which satisfy the requirements of the input examples [Hammer and Kogan, 1992]. Too many rules may increase the chances for errors. When, for instance, one wishes to use the rules to control the safety of a structure, then a small number of rules will also require a small number of sensors. In medical diagnosis one may wish to derive a small set of rules, because too many rules are difficult to be verified and validated. These are the main reasons why we take an optimization approach in dealing with the problem of inferring a small number of logical decision rules (i.e., a compact Boolean function).

In [Triantaphyllou, Soyster, and Kumara, 1994] the problem of inferring a set of logical clauses is examined. In that work a rather small set of clauses (not necessarily of minimal size) is determined. That approach is based on a branch-and-bound (B&B) algorithm. That B&B algorithm was later significantly enhanced and improved in [Triantaphyllou, 1994].

There is another clause inference approach which can be used to determine a minimal size set of logical clauses. This method, denoted as SAT (for satisfiability), has been proposed in [Kamath *et al.*, 1992]. Then a *clause satisfiability* model is formed as follows. Let M_1 and M_2 be the numbers of examples in the E^+ and E^- sets, respectively. In [Kamath *et al.*, 1992] it is shown that given two collections of positive and negative examples (called the *ON-set* and *OFF-set*, respectively) then, then a DNF system can be inferred to satisfy the requirements of these examples. This is achieved by formulating a satisfiability problem and then using the interior point method developed by Karmakar and his associates [1992] to solve it. This approach pre-assumes the value of *k*; the number of conjunctions in the DNF system. The SAT problem uses the following Boolean variables (Kamath et al. [12]):

if A_i is in the j-th conjunction

if A_i is not in the j-th conjunction

if $\bar{A}_i$ is in the j-th conjunction

if $\bar{A}_i$ is not in the j-th conjunction

if A_i = 1 in the positive example $\alpha \in E^+$

if A_i = 0 in the positive example $\alpha \in E^+$

if the positive example α is accepted by the j-th co
otherwise

Then, the clauses of this SAT problem are as follows:

$$s_{ji} \vee s'_{ji}, \quad \text{for } i = 1, \ldots, n, \text{ and } j = 1, \ldots, k, \tag{1}$$

$$(\bigvee_{i \in P_r} \bar{s}'_{ji}) \vee (\bigvee_{i \in \bar{P}_r} \bar{s}_{ji}), \quad \text{for } j = 1, \ldots, k, \text{ and } r = 1, \ldots, M_2, \tag{2}$$

$$\bigvee_{j=1}^{k} z_j^{\alpha}, \quad \text{for } \alpha = 1, \ldots, M_1, \tag{3}$$

$$\sigma_{ji}^{\alpha} \vee \bar{z}_j^{\alpha}, \quad \text{for } i = 1, \ldots, n, \; j = 1, \ldots, k, \text{ and } \alpha = 1, \ldots, M_1, \tag{4}$$

where P_r is the set of indices of A for which $A_i = 1$ in the negative example $r \in E^-$. Similarly, P_r is the set of indices of A for which $A_i = 0$ in the negative example $r \in E^-$.

Clauses of type (1) ensure that never both A_i and A_i will appear in any conjunction. Clauses of type (2) ensure that each negative example is rejected by all conjunctions. Clauses of type (3) ensure that each positive example is accepted by *at least one* conjunction. Finally, clauses of type (4) ensure that $z_j^{\alpha} = 1$ if and only if the positive example α is accepted by the j-th conjunction. In general, this SAT problem has $k(n(M_1 + 1) + M_2) + M_1$ clauses, and $k(2n + M_1)$ Boolean variables. A detailed example on this formulation can be found in [Kamath *et al.*, 1992].

If the above clause satisfiability problem is feasible, then the conclusion is that it is possible to correctly classify all the examples with k or fewer clauses. If this SAT problem is infeasible, then one must increase k until feasibility is reached. In this manner, the SAT approach can yield a system with the minimum number of clauses. It is very important one to observe at this point that computationally it is much harder to prove that a given SAT problem is infeasible than it is feasible. Therefore, trying to determine a minimum size Boolean function by using the SAT approach may be computationally too difficult.

It should also be emphasized here that it is not very critical whether an inference algorithm determines a CNF or DNF system (i.e., CNF or DNF Boolean function). By applying theorem 1 (described later in section 4.1), either a CNF or DNF system can be derived by using *any algorithm*. Also, some computational results indicate that the B&B approach proposed by the first author in [Triantaphyllou, 1994] is much more efficient than the previous satisfiability based approach.

The SAT results were derived by using a VAX 8700 computer running the 10-th Edition *UNIX*. The code for the SAT tests was written in FORTRAN and in C. The SAT results were originally reported in Kamath *et al.* [1992]. The OCAT results were derived by using an IBM 3090-600S computer running *XA/VM* and the code was written in FORTRAN. On the average, in those tests the OCAT/B&B approach was 5,579 times faster [Triantaphyllou, 1994]. However, it should be stated at this point that the IBM3090-600S machine is, in general, 4-5 times faster than the VAX 8700 computer used in the SAT tests. Thus, due to the different hardware/software platforms and the limited number of test problems, the current results can convey only a flavor on the relative performance of the two methods, and by no means should be considered as a direct comparison of the two approaches.

A related problem is to study the construction of a partially defined Boolean function (or *pdBf*), given disjoint sets of positive and negative examples [Boros *et al.*, 1994]. That is, now it is required that the atoms of the function be grouped according to a given scheme (called a decomposition structure). The problem then is to determine whether a given *pdBf* has an extension when a specific decomposition structure is considered. Typically, a *pdBf* may have exponentially many different extensions. In [Boros *et al.*, 1994] computational and existence issues of this type of problem are examined for many different cases. Note that very often (i.e., depending on the case) these problems are NP-complete. In that treatment Boolean functions were not attempted to be represented by a minimum size CNF or DNF expression (as is the case in this paper).

Other approaches dealing with this type of learning is to use neural networks. However, although a neural network approach may yield encouraging results, it fails to extract the underlying logic in the form of a set of decision rules (i.e., a Boolean function). Some recent efforts have been successful in combining symbolic approaches to neural networks and thus generate rules [Fu, 1994]. However, they are impractical since they may generate an exponential number of rules [Shavlik, 1994]. As result of this, field experts hesitate to use neural network approaches because they do not understand the decision-making process and thus feel uncomfortable.

Mangasarian and his associates [1991] did some considerable work in using linear programming in solving similar classification problems. That approach constructs a family of pairs of separating hyper-planes. Then, new instances are classified by examining where they belong in relation to these planes. They applied their approach in diagnosing breast cancer and also in diagnosing heart disorders. The proposed logical analysis approach, however, besides the promising computational results, it also provides a set of decision rules which can be easily understood by medical doctors. As result, it has the potential to become accepted anywhere other methods have failed to gain the trust of the end users.

4 CURRENT RESEARCH RESULTS

4.1 Some Definitions and Terminology

Let $\{A_1, A_2, A_3, \ldots, A_t\}$ be a set of t Boolean *predicates* or *atoms*. Each atom A_i ($i=1,2,3,\ldots,t$) can be either true (denoted by 1) or false (denoted by 0). Let F be a *Boolean function* over these atoms. That is, F is a mapping from $\{0,1\}^t \rightarrow \{0,1\}$ which determines for each combination of truth values of the arguments $A_1, A_2, A_3, \ldots, A_t$ of F, whether F is true or false (denoted as 1 or 0, respectively). For each Boolean function F, the *positive examples* are the vectors $v \in \{0,1,\}^t$ such that $F(v) = 1$ (i.e., they are *accepted* by F). Similarly, the *negative examples* are the vectors $v \in \{0,1,\}^t$ such that $F(v) = 0$ (i.e., they are *rejected* by F).

In the previous paragraph a Boolean function F is assumed to be of any form. However, in propositional calculus it is convenient to represent Boolean functions using the conjunctive normal form (CNF) or the disjunctive normal form (DNF). See, for instance, [Blair *et al.*, 1985], [Cavalier *et al.*, 1990], [Hooker, 1988a and 1988b], [Jeroslow, 1988 and 1989], and [Williams, 1986]. Peysakh in [1987] describes an algorithm for converting any Boolean expression into CNF. In this paper a set of positive examples will be denoted as E^+. Similarly, a set of negative examples will be denoted as E^-.

Next, define the general form of a CNF and DNF system as (I) and (II), respectively. That is:

$$\bigwedge_{j=1}^{k} \left(\bigvee_{i \in \rho_j} a_i \right), \tag{I}$$

and

$$\bigvee_{j=1}^{k} \left(\bigwedge_{i \in \rho_j} a_i \right), \tag{II}$$

where a_i is either A_i or $\bar{A}_i$ and ρ_j is the superset of the indices of the atoms in the j-th conjunction or disjunction. Furthermore, $\bar{v}$ is defined as the *complement* of example v. For instance, if $v = [1,0,0]$, then $\bar{v} = [0,1,1]$. The following definition introduces the concept of the complement of a set of examples. Let E be a collection of examples (either positive or negative). Then, $\bar{E}$ is defined as the *complement of the collection E*.

The following theorem (proved in [Triantaphyllou and Soyster, 1995a]) states an important property which exists when CNF and DNF systems are inferred from collections of positive and negative examples. It is important to observe here that this theorem is independent of the algorithm used to infer a Boolean expression from positive and negative examples.

Theorem 1: *Let E^+ and E^- be the sets of positive and negative examples, respectively. A CNF system given as (I) satisfies the constraints of the E^+ and E^- sets if and only if the DNF system given as (II) satisfies the constraints of $\bar{E}^-$ (considered as the positive examples) and $\bar{E}^+$ (considered as the negative examples).*

4.2 The One Clause At a Time Approach

In [Triantaphyllou, Soyster, and Kumara, 1994] an algorithm which infers CNF systems from positive and negative examples is developed. In that approach, CNF clauses are generated in a way which attempts to minimize the number of CNF clauses that constitute the recommended CNF system. In this way, a compact CNF system can be derived. The strategy followed there is called the *One Clause At a Time* approach.

The One Clause At a Time (or OCAT) approach is greedy in nature. It uses as input data two collections of positive and negative examples. It determines a set of CNF clauses that, when taken together, rejects all the negative examples and accepts all the positive examples. The OCAT approach is sequential. In the first iteration it determines a single clause (conjunction) which accepts all the positive examples in the E^+ set while it rejects as many negative examples in E^- as possible. This is the greedy aspect of the approach. In the second iteration it performs the same task using the original E^+ set but the revised E^- set has only those negative examples which have not been rejected by any clause (i.e., the first clause) so far. The iterations continue until a set of clauses is constructed which reject all the negative examples in the original E^- set. More on this approach can be found in [Triantaphyllou, Soyster, and Kumara, 1994] and [Triantaphyllou, 1994]. Figure 1 summarizes the iterative nature of the OCAT approach. It is interesting to observe here that the LP approach proposed by Mangasarian and his associates [1991] also uses a greedy approach in the way the hyper-planes are determined.

The core of the OCAT approach is step 2, in figure 1. In [Triantaphyllou, Soyster, and Kumara, 1994] a branch-and-bound based algorithm is presented which solves the problem posed in step 2. That algorithm was later significantly enhanced and improved in [Triantaphyllou, 1994]. The OCAT approach returns the set of desired clauses (i.e., the CNF or DNF system) as set *C*.

A related problem was examined in [Turksen and Zhao, 1993]. In that development it is shown that Quinlan's ID3 algorithm [Quinlan, 1979] is equivalent with the result of applying a Boolean simplification treatment on the rules generated by ID3. Thus, the authors propose a scheme which first derives a set of rules by using the ID3 algorithm, and then this set is simplified (fewer logical clauses are derived) by applying a Boolean simplification procedure. Some experiments contacted by the authors of this paper compared the OCAT approach with the ID3 algorithm in terms of the number of CNF clauses derived. It was shown that the OCAT approach, when it was combined with the heuristic described in [Deshpande and Triantaphyllou, 1996] created by far much less clauses. The logic inference procedures developed by Triantaphyllou and his associates directly aim at deriving a small number of clauses from collections of positive and negative examples.

The procedure proposed by Turksen and Zhao applies a two step approach: In the first step a system is induced by using the ID3 algorithm and in the second step that system is reduced to an equivalent system with (hopefully) fewer terms. That is, the result of the Turksen/Zhao approach may be drastically dependent by the outcome of the ID3 algorithm. However, no experiment results are available to compare the Turksen/Zhao approach with the OCAT approach.

4.3 The Rejectability Graph of Two Collections of Examples

This section presents the motivation and definition of a special graph which can be easily derived from positive and negative examples. The concept of this graph was originally developed in [Triantaphyllou and Soyster, 1994].

> $i = 0$; $C = \phi$;
> **DO WHILE** ($E^- \neq \phi$)
>   **Step 1:** $i \leftarrow i + 1$; /* i stands for the i-th clause */
>   **Step 2:** Find a clause c_i which accepts all members of E^+
>   while it rejects as many members of E^- as possible;
>   **Step 3:** Let $E^-(c_i)$ be the set of members of E^- which are rejected by c_i;
>   **Step 4:** Let $C \leftarrow C \cup c_i$;
>   **Step 5:** Let $E^- \leftarrow E^- - E^-(c_i)$;
> **REPEAT;**

Figure 1. The One Clause At a Time (OCAT) Approach.

To see the motivation for introducing this graph consider a situation with $t = 5$ atoms. Suppose that the vector $v_1 = [1,0,1,0,1]$ is a positive example while the two vectors $v_2 = [1,0,1,1,1]$ and $v_3 = [1,1,1,0,1]$ are negative examples. If $\bar{A}_i$ denotes the negation of the atom A_i, then the positive example v_1 indicates that A_1, $\bar{A}_2$, A_3, $\bar{A}_4$ and A_5 are true (or equivalently, $\bar{A}_1$, A_2, $\bar{A}_3$, A_4 and $\bar{A}_5$ are false in this example). Similar interpretations can easily be derived for the remaining two examples v_2 and v_3.

Let the set $T(v)$ denote the set of the atoms that are true in the example (either positive or negative) v (where $v \in \{1,0,\}^t$, and t is the number of atoms). When this definition of the $T(v)$ set is applied on the previous three examples v_1, v_2, and v_3, then the following sets are derived:

$$T(v_1) = T(\,[1,0,1,0,1]\,) = \{A_1, \bar{A}_2, A_3, \bar{A}_4, A_5\}$$
$$T(v_2) = T(\,[1,0,1,1,1]\,) = \{A_1, \bar{A}_2, A_3, A_4, A_5\}$$
$$T(v_3) = T(\,[1,1,1,0,1]\,) = \{A_1, A_2, A_3, \bar{A}_4, A_5\}.$$

Consider a single CNF clause (conjunction), denoted as C, of the form (where a_i is either A_i or $\bar{A}_i$, for some $N \leq t$): $\bigvee_{i=1}^{N} a_i$. Then, the clause C accepts an example v (i.e., v is a positive example of C) if and only if any of the atoms in the set $T(v)$ is one of the atoms involved in the expression $\bigvee_{i=1}^{N} a_i$. Otherwise, the example v is not accepted (i.e., v is a negative example of C). For instance, if the clause C is defined as: $C = (A_2 \vee A_4)$, then the examples v_1 and v_2 are accepted by C, while the example v_3 is not accepted.

Using the previous three sets $T(v_1)$, $T(v_2)$, and $T(v_3)$ and the notion of acceptance it can be easily verified that there is no single CNF clause which can simultaneously reject both the two negative examples v_2 and v_3, while at the same time, it accepts the positive example v_1. This is true because any clause which simultaneously rejects the two examples v_2 and v_3, should not involve any of the atoms present in the union of the two sets $T(v_2)$ and $T(v_3)$. But if none of the atoms of the set $\{A_1, A_2, \bar{A}_2, A_3, A_4, \bar{A}_4, A_5\}$ $(= T(v_2) \cup T(v_3))$ is present in the clause, then it is impossible to accept the positive example $v_1 = [1,0,1,0,1]$.

Therefore, given the positive example v_1, then the previous two negative examples v_2 and v_3 cannot be rejected by any single CNF clause (conjunction) which also accepts the positive example v_1. In general, given a set of positive examples E^+, then two negative examples v_1 and v_2 are rejectable by a single CNF clause if and only if the condition in the following theorem [Triantaphyllou and Soyster, 1996a] is satisfied:

Theorem 2: *Let E^+ be a set of positive examples and v_1 , v_2 are two negative examples. Let $T(v)$ be the set of the atoms that are true in the (either positive or negative) example v. Then, the two examples v_1 and v_2 are rejectable by a single clause which accepts all the positive examples in E^+ if and only if the following condition is true:*

$$T(v_i) \not\subseteq T(v_1) \cup T(v_2) \textit{ for any positive example } v_i \in E^+.$$

Given two collections of positive and negative examples, denoted as E^+ and E^-, respectively, the above theorem motivates the construction of a graph $G = (V, E)$ as follows:

$V = \{ V_1, V_2, V_3, \ldots, V_{M_2} \}$, *where M_2 is the size of the E^- set, and*

$E = \{ (V_i, V_j)$ *if and only if the i-th and the j-th examples in E^- are rejectable by a single clause (subject to the examples in E^+)* $\}$.

We call this graph the *rejectability graph* (or *R*-graph) of E^+ and E^-. Since the previous theorem indicates that it is computationally very easy to examine whether there is an edge between any two vertices of G, the rejectability graph G can be constructed by performing ${}_2(M_2-1)/$ simple rejectability examinations (where

M_2 is the size of the E^- set).

4.4 Properties of the Rejectability Graph

The rejectability graph G of a set of positive and a set of negative examples has a number of interesting properties. To illustrate these properties it is first necessary to consider the following theorem [Triantaphyllou and Soyster, 1996a]:

Theorem 3: *Suppose that the two sets E^+ and E^- are given and F is a family of K negative examples from E^- ($K \leq$ size of set E^-) which are rejected by a single clause subject to the positive examples in E^+. Then, the vertices which correspond to the K negative examples in the rejectability graph G, form a clique in G of size K.*

Although the previous theorem states that any set of negative examples which is rejected by a single clause corresponds to a clique in the rejectability graph, the inverse is not always true. That is, not any clique in the rejectability graph corresponds to a set of negative examples which are rejected by the same clause. In [Triantaphyllou and Soyster, 1996a] the previous statement is demonstrated via a counter-example.

4.4.1 On The Connected Components of The Rejectability Graph.

Another useful application of the rejectability graph G is based on the *connected components* of the graph G. First, observe that any subset of negative examples in the E^- set which is rejected by a single clause, subject to the examples in E^+, corresponds to a subset of vertices of the rejectability graph G which belong to the same connected component of the graph G.

The previous property may become critical when the sets of positive and negative examples are very large. In this case it is useful to first form the rejectability graph G (which is computationally a simple task). Next, determine all the connected components of the rejectability graph by applying an algorithm for finding the connected components. Then, one may solve the smaller clause inference problems which are formed by considering all the positive examples and the negative examples which correspond to the vertices of the connected components in G.

In [Pardalos and Rentala, 1990] there is an excellent survey of algorithms which determine the connected components of a graph. Once the connected components of the rejectability graph G have been found, clauses which reject negative examples, while they also accept all the positive examples, can be determined by considering only one connected component at a time. This approach clearly takes advantage of the connectivity structure of the G graph and provides the means for solving very large clause inference problems.

4.4.2 On The Minimum Clique Cover of The Rejectability Graph.

Consider two sets of negative and positive examples, denoted as E^+ and E^-, respectively. Suppose that C is a *family* of clauses which satisfy the constraints of the set of positive examples E^+ and the set of negative examples E^-. The following two theorems were proved in [Triantaphyllou and Soyster, 1996a] and are very critical in the sense that they establish lower bounds on the minimum number of clauses which reject all the members in E^-, while they accept all the members in E^+. It is important to observe that these theorems are independent of the inference algorithm used to process the positive and negative examples.

Theorem 4: *Suppose that E^+ and E^- are the sets of the positive and negative examples, respectively. Then, C_{min}, the minimum number of clauses which reject all the examples in E^-, while they accept all the examples in E^+, has as a lower bound $\omega(\bar{G})$, the size of the maximum clique of the graph $\bar{G}$. Where $\bar{G}$ is the complement of the rejectability graph G of E^+ and E^-. Moreover, the following relation is always true: $C_{min} \geq \omega(\bar{G}) = \alpha(G)$, where $\alpha(G)$ is the* ***stability number*** *of the rejectability graph G.*

In Carraghan and Pardalos [1990] a survey of algorithms which find the maximum clique in any graph is presented. They also present a very efficient algorithm which uses a partial enumeration and outperforms any other known algorithm. In that treatment random problems with 3,000 vertices and over one million edges were solved in rather short times (less than one hour on an IBM ES/3090-900E computer). In any graph G the *chromatic number* $\chi(\bar{G})$ is always equal to the clique cover number of G, (i.e., $k(G)$). Furthermore, the following is always true:

$$C_{min} \geq k(G) = \chi(\bar{G}) \geq \omega(\bar{G}) = \alpha(G).$$

Therefore, the following theorem [Triantaphyllou and Soyster, 1996a] is true:

Theorem 5: *Suppose that E^+ and E^- are the sets of the positive and negative examples, respectively. Then, C_{min}, the minimum number of clauses which reject all the examples in E^-, while they accept all the examples in E^+, has as a low bound $\chi(\bar{G})$, the chromatic number of the graph $\bar{G}$. Where $\bar{G}$ is the complement of the rejectability graph G of E^+ and E^-. Furthermore, this new bound is tighter than the bound given in the previous theorem. That is, the following relation is always true:*

$$C_{min} \geq \chi(\bar{G}) = k(G) \geq \omega(\bar{G}) = \alpha(G).$$

The previous theoretical results can be utilized to decompose large scale problems as follows. First, the connected components of the rejectability graph are determined and the Boolean function construction problem is solved within each connected component. The second approach is also motivated by partitioning the vertices of the rejectability graph into mutually disjoint sets. However, in this second approach, vertices are subdivided via a sequential construction of cliques.

First, the maximum clique of the rejectability graph is determined. The negative examples which correspond to the vertices of the maximum clique, along with *all* the positive examples, form the first sub-problem of this decomposition. Next, the maximum clique of the remaining graph is derived. The second sub-problem is formed by the negative examples which correspond to the vertices of the second clique and all the positive examples. This process continues until all the negative examples (or, equivalently, all the vertices in the rejectability graph) are considered.

We note that this sequence of cliques does not necessarily correspond to a minimum clique cover of the rejectability graph. This procedure is simply a greedy approach which approximates a minimum clique cover. Furthermore, it is possible that a single sub-problem (in which all the vertices in the rejectability graph form a clique) may yield more than one clause.

It should be noted at this point that the clique cover derived by using the above greedy approach may not always correspond to a minimum clique cover. Therefore, the number of cliques derived in that way, cannot be used as a lower bound on the number of clauses derivable from positive and negative examples. Obviously, if the number of cliques is equal to $\omega(G)$, then the previous clique cover is minimal. However, even if the previous clique cover is not of minimum size, it can still be very useful as it can lead to a decomposition of the original problem into a sequence of smaller problems. Some computational results described in [Triantaphyllou and Soyster, 1996a] provide some insight into the effectiveness of such a decomposition approach.

The two problem decomposition approaches described in this section can be combined into one approach. One first decomposes the original problem in terms of its connected components. Next, a clique cover, as described above, is derived for the individual problems which correspond to the connected components of the rejectability graph.

4.5 Clause Inference with Guided Input

The problem of inferring general CNF/DNF clauses with guided input was examined in [Triantaphyllou and Soyster, 1996b]. Suppose that the user can supply the expert with additional examples for correct classification. Then, the *problem of inference with guided input* is how to generate the next example. One obvious approach is to generate the next example randomly. However, this may result in generating many examples and still not achieving a good approximation of the unknown system. It is obviously desirable to consider a sequence of new examples which can lead to a good approximation of the unknown system as quickly as possible.

When a new example is considered, it is given to the expert for the correct classification. Two situations can occur. First, the current system (which is attempting to represent the unknown *"hidden logic"*) classifies the new example in a manner *identical* with the expert (who always correctly classifies each example). In the second case, the new example is classified in the *opposite* way by the expert and the current system. If the current system is not yet a good approximation of the *"hidden logic"*, then the last case is the most desirable scenario. This is true, because in this case one can re-execute a clause inference algorithm (for instance, the OCAT approach) again and, hopefully, derive a closer approximation of the unknown system.

If the current version of the Boolean function is an inaccurate approximation of the *"hidden logic"* and one generates new examples which fail to reveal any contradictions, then additional costs are incurred in classifying new examples, but no improvement is gained. Clearly, it is desirable to use a strategy for determining the next example, such that any possible contradiction between the current version of the Boolean function and the *"hidden logic"* will surface early in the interviewing process.

Next, consider two sets of positive and negative examples, E^+ and E^-, defined on t atoms. Let S_{SAMPLE} denote a logic system (Boolean function) that correctly classifies the sample data, i.e. the examples in E^+ are classified as positive and the examples in E^- are classified as negative (one such function can be obtained via the methods described in [Kamath *et al.*, 1992], [Triantaphyllou *et al.*, 1994], and [Triantaphyllou, 1994]). Also, define S_{HIDDEN} as the *"hidden logic"* Boolean function and $\bar{S}_{HIDDEN}$ as the complement of S_{HIDDEN}. The strategy proposed in [Triantaphyllou and Soyster, 1996b] is based on the following theorem:

Theorem 5: *Suppose that there exists an example $v \in \{0,1\}^t$ such that:*

$$S_{SAMPLE}(v) + S_{R\text{-}SAMPLE}(v) = 0 \quad \text{or:} \tag{5.a}$$

$$S_{SAMPLE}(v) + S_{R\text{-}SAMPLE}(v) = 2. \tag{5.b}$$

Furthermore, suppose that the example v is classified by the expert as either positive or negative. Then, one and only one of the following situations is true:

a) If (5.a) holds and v is a positive example, then system S_{SAMPLE} is not valid.

b) If (5.a) holds and v is a negative example, then system $S_{R\text{-}SAMPLE}$ is not valid.

c) If (5.b) holds and v is a positive example, then system $S_{R\text{-}SAMPLE}$ is not valid.

d) If (5.b) holds and v is a negative example, then system S_{SAMPLE} is not valid.

Therefore, the overall strategy, starting with two Boolean functions, is to attempt to generate a sequence of new examples $v_{k+1}, v_{k+2}, v_{k+3}, \ldots, v_m$, where each example is appropriately classified, as positive or negative, by the expert. Each additional example should have the property that it invalidates either S_{SAMPLE} or $S_{R\text{-}SAMPLE}$, i.e. one of the two Boolean functions must be modified. In doing so, it is expected that S_{SAMPLE} and $S_{R\text{-}SAMPLE}$ become more closely aligned with S_{HIDDEN} and $\bar{S}_{HIDDEN}$, respectively.

How does one find an example that invalidates either S_{SAMPLE} or $S_{R\text{-}SAMPLE}$? Conceptually it is quite simple. One strategy is to formulate and solve at most two *clause satisfiability problems*. The clause satisfiability problem has been examined with considerable success [Hooker, 1988a and 1988b]. A recent development reported in [Kamath *et al.*, 1992] uses an interior point algorithm developed by Karmakar and his associates [Karmakar, Resende, and Ramakrishnan, 1991] with considerable success. Some problem preprocessing techniques can be found in [Cavalier, Pardalos, and Soyster, 1990]). The interested reader may want to consult with the recent developments reported in the Second *DIAMACS* Challenge on Cliques, Coloring and Satisfiability [Trick and Johnson, 1995].

The two satisfiability problems for this problem are as follows: Determine an example v which results in a truth value TRUE for

$$\bar{S}_{SAMPLE}(\bar{v}) \quad \wedge \quad \bar{S}_{R\text{-}SAMPLE}(\bar{v}) \, , \qquad \text{or:} \qquad (6.a)$$

$$S_{SAMPLE}(\bar{v}) \quad \wedge \quad S_{R\text{-}SAMPLE}(\bar{v}) \qquad (6.b)$$

If (6.a) is TRUE (i.e., satisfied), then v is evaluated as positive by both systems (Boolean functions), and if (6.b) is TRUE, v is evaluated as negative by both systems. Observe that relations (6.a) and (6.b) are equivalent to relations (5.a) and (5.b), respectively. If an example is found which satisfies either (6.a) or (6.b), then one of the two functions is modified and the same process is repeated. Next, suppose that no example can be found that satisfies (6.a) or (6.b). Does this mean that $S_{SAMPLE} \equiv S_{HIDDEN}$? Unfortunately, the answer is no. In this case we revert to a random search process.

A number of computer experiments was conducted in [Triantaphyllou and Soyster, 1996b] in order to investigate the effectiveness of the proposed strategy compared with random input learning (that is, when new examples are generated randomly). Those results are very encouraging because they suggest that a *"hidden system"* can be inferred, on the average, with at most 50% of the data required with random input. This guided learning approach was also applied on the Wisconsin breast cancer data base [Murphy and Aha, 1994] with remarkable success.

A recent development on inferring a small set of clauses from positive and negative examples is reported in [Deshpande and Triantaphyllou, 1996]. In that approach the main branch-and-bound algorithm of [Triantaphyllou, 1994] is combined with a very fast and simple heuristic. That heuristic uses a randomization approach for performing a local search while forming a single clause at time (according to the OCAT philosophy). Computational experiments suggest that the new development is even more promising. Some random problems with 7,000 and 18,000 examples which were defined on 15 Boolean attributes were solved in matter of a few hours on an IBM

3090-600S computer. Before that work, the largest problem solved was up to 1,000 examples defined on 32 atoms.

The above idea of using randomization in a search algorithm has been explored recently by other researchers as well. For instance, Feo and Resende in [1995] have successfully used randomization (the GRASP approach) to solve clause satisfiability (SAT) problems. Also, in a recent book Motwani and Raghavan [1995] provide a comprehensive presentation of the theory on randomized algorithms. Randomization also offers a natural and intuitive way for implementing parallelism in algorithms.

5 INFERENCE OF MONOTONE BOOLEAN FUNCTIONS

The previous sections dealt with the case of inferring general Boolean functions (either in CNF or DNF form). However, in many real life applications the behavior of a system is *monotone* [Boros, Hammer and Hooker, 1994]. In a continuous environment the previous statement means that the output of the system is consistent with the magnitude of the data.

In [Boros, Hammer and Hooker, 1994] the problem of inferring monotone Boolean functions is presented. In that treatment it is assumed that some observations are available which describe the behavior of a monotone Boolean function. Each observation is described by the values of some of the attributes (denoted by the vector *x*), while an unknown number of other attributes (denoted as the vector *y*) has undetermined values. The Boolean function may return one of a discrete set of values, often different values for the inputs which correspond to the same *x* part (but to different *y* parts). Next, some penalty coefficients are defined (in a geometric sense) and the goal then is to determine a function which minimizes the sum of all these penalties and also to be monotone.

Also note that in the previous treatment the problem of inquiring values of the function for certain new inputs was not considered. Neither was considered the problem of inferring a small size Boolean function. These are, however, the focus of the developments discussed in the following sections. In the next paragraphs we give the formal definitions for monotonicity and some key concepts.

Let $\alpha,\beta \in E_n$, E_n where denotes the set of all binary vectors of length n. Then, we say that the vector $\alpha = (\alpha_1,\alpha_2,\alpha_3,\ldots,\alpha_n)$ *precedes* the vector $\beta = (\beta_1,\beta_2,\beta_3,\ldots,\beta_n)$ (denoted as: $\alpha \preceq \beta$) if and only if the following is true: $\alpha_i \leq \beta_i$, for all $1 \leq i \leq n$. A Boolean function $f(x)$ is *monotone* if for any vectors $\alpha,\beta \in E_n$, the relation $f(\alpha) \leq f(\beta)$ follows from the fact that $\alpha \preceq \beta$. Let M_n be the set of all monotone Boolean functions defined on n variables. A binary vector α of length n is said to be the *upper zero* of a function $f(\alpha)\in M_n$, if $f(\alpha) = 0$ and, for any vector $\beta \prec \alpha$, we have $f(\beta) = 0$.

A binary vector α of length n is said to be the *lower unit* of a function $f(\alpha)\in M_n$, if $f(\alpha) = 1$ and, for any vector β from E_n such that $\beta \prec \alpha$, we get $f(\beta) = 1$. We will also call the above defined monotone Boolean function to be an *increasing (isotone) monotone* Boolean function in contrast with a decreasing monotone Boolean function. A Boolean function is *decreasing (antitone) monotone*, if for any vectors $\alpha,\beta\in E_n$, the relation $f(\alpha) \leq f(\beta)$ follows from the fact that $\alpha \geq \beta$ [Rudeanu, 1974].

In terms of monotone Boolean function terminology note that the set of all upper zeros represents the *border elements* of the negative pattern. In an analogous manner we can also define the border of a positive pattern defined by using the lower units of the function. In this manner any monotone Boolean function represents two *"compact patterns"*. Please note that inferring the border elements of the positive and negative patterns is equivalent to the problem of inferring the structure of the target monotone Boolean function f from collections of positive and negative examples.

Some illustrative examples of monotone Boolean functions are: the constants 0 and 1, the identity function $f(x) = x$, the disjunction $x_1 \vee x_2$, the conjunction $x_1 \wedge x_2$, etc. Any function obtained by a composition of monotone Boolean functions is also monotone. In other words, the class of all monotone Boolean functions is closed. Moreover, the class of all monotone Boolean functions is one of the five *maximal (pre-complete) classes* in the set of all Boolean functions. That is, there is no closed class of Boolean functions, containing all monotone Boolean functions and distinct from the class of monotone functions and the class of all Boolean functions. The reduced *disjunctive normal form* (DNF) of any monotone Boolean function, distinct of 0 and 1, does not contain negations of variables. The set of functions $\{0, 1, (x_1 \vee x_2), (x_1 \wedge x_2)\}$ is a *complete system* (and moreover, a *basis*) in the class of all monotone Boolean functions [Alekseev, 1988].

Let $\psi(n)$ be the number of monotone Boolean functions defined on n variables. Then, it is known that:

$$\psi(n) = 2^{\binom{n}{\lfloor n/2 \rfloor}(1+\varepsilon(n))},$$

where $0 < \varepsilon(n) < c(\log n)/n$ and c is a constant (see, for instance, [Kleitman, 1969], and [Alekseev, 1988]).

Let a monotone Boolean function $f\in M_n$ be defined with the help of a certain operator A_f (also called an *oracle*). That is, when f is fed with a vector $\alpha = (\alpha_1,\alpha_2,\alpha_3,\ldots,\alpha_n)$, the operator A_f returns the value of $f(\alpha)$. The problem posed here is that of finding all upper and lower zeros of an arbitrary function $f\in M_n$ with the help of a certain number of accesses to the operator A_f.

In the following paragraphs we state a very important theoretical result regarding the inference of monotone Boolean functions. Let $\mathcal{F} = \{F\}$ be the set of all algorithms which can solve the above monotone Boolean function inference problem and $\varphi(F, f)$ be the number of accesses to the operator A_f required to completely infer a monotone function $f \in M_n$.

Next, the Shannon function $\varphi(n)$ is introduced as follows [Hansel, 1966]:

$$\varphi(n) = \min_{F \in \mathcal{F}} \max_{f \in M_n} \varphi(F, f). \tag{7}$$

It was shown in [Hansel, 1966] that when inferring a monotone Boolean function the following relation is true (also known as *Hansel's lemma*):

$$\varphi(n) = \binom{n}{\lfloor n/2 \rfloor} + \binom{n}{\lfloor n/2 \rfloor + 1}, \tag{8}$$

where $\lfloor n/2 \rfloor$ is the *floor* value of $n/2$.

It was shown in [Hansel, 1966] that restoration algorithms for monotone Boolean functions, and which use Hansel's lemma, are optimal in terms of the Shannon function. That is, they minimize the maximum input size requirements of any possible restoration algorithm. Finally, we state another related theorem [Kovalerchuk *et al.*, 1995]:

Theorem 6: *Each general Boolean function can be described in terms of several monotone Boolean functions.*

6 SOME KEY PROBLEMS AND ALGORITHM OUTLINES

FOR MONOTONE FUNCTIONS

PROBLEM 1: *The Main Inference Problem*
The central problem is how to infer a monotone Boolean function by issuing a sequence of membership inquires to an operator or "oracle". The solution of this problem is presented in detail in [Kovalerchuk *et al.*, 1996]. That approach uses the concept of Hansel chains and tries to identify border points of the previous "positive" and "negative" patterns. These patterns, in turn, lead to the definition of the target monotone Boolean function in DNF form. That algorithm is optimal in the sense of conditions (7) and (8), as explained in the previous section. There could be two different scenarios with this problem: one with no initial examples, and another with some initial examples.

PROBLEM 2: *Inference When Examples Have Different Costs*

The problem is similar to the fundamental problem presented in the previous paragraph. However, now it is assumed that different examples may be associated with different costs when one seeks to classify them by issuing a membership inquiry to an "oracle". In a related context, the user may somehow know the probability that a given example will have true (or, equivalently, false) value.

The basic idea of the proposed algorithm is as follows. Suppose that the example α, which needs to be classified by the "oracle" as part of the function restoration process, is associated with a known cost of classification, denoted as C_α. Moreover, suppose that the user knows (or can easily determine) the classification values of two other examples, say α_* and α^*, with costs $C_{\alpha*}$, and $C_\alpha*$, respectively which satisfy the following three conditions:

(*i*) $\alpha_* \leq \alpha \leq \alpha^*$ (i.e., α is between them),

(*ii*) $C_\alpha > (C_{\alpha*} + C_\alpha*)$. That is, their combined costs are lower than the cost of the target example α,

and (*iii*) $f(\alpha_*) = f(\alpha^*)$.

In a case like the above, the value of $f(\alpha)$ can be inferred from the values of two other examples at a lower cost than directly inquiring about the cost of the original target example α.

7. CONCLUDING REMARKS

In this paper we presented a survey of some results in inferring a small set of CNF or DNF clauses from positive and negative examples. A logical analysis approach was proposed for this purpose. The advantage of using a logical analysis approach is that the end result can be easily translated into a set of logical rules which can be more understandable by domain experts.

The paper considered the case of dealing with general Boolean functions and also the special case of dealing with monotone Boolean functions. For the case of general Boolean functions, the paper discussed a number of heuristics for inferring a small set of clauses and also some theoretical results on lower bounds on the number of clauses and ways for problem decomposition. This was achieved by exploiting the properties of a special graph which can be easily constructed from the input data.

For the monotone Boolean function case the main issue is that when inference is based in Hansel chains, then the target function can be inferred with a small number of membership inquires. Moreover, algorithms which use Hansel chains are optimal (in the sense of the Shannon function).

ACKNOWLEDGEMENTS

The first two authors gratefully recognize the partial support from the Office of Naval Research (ONR) grant N00014-95-1-0639 and LaSPACE (a NASA program). We thank the anonymous referee for his/her many thoughtful comments and suggestions made on an earlier version of this paper.

REFERENCES

ANGLUIN, D., 1988, "Queries and Concept learning," *Machine Learning*, Vol. 2, pp. 319-342.

ANGLUIN, D., 1992, "Computational Learning Theory: Survey and Selected Bibliography," *Proceedings of the 24-th Annual ACM Symposium on the Theory of Computing*, Victoria, BC, Canada, May 4-6, pp. 351-369.

ALEKSEEV, V.B., 1988, "Monotone Boolean Functions". Encyclopedia of Mathematics, v. 6, Kluwer Academic Publishers, 306-307.

BOROS, E., V. GURVICH, P.L. HAMMER, T. IBARAKI, AND A. KOGAN, 1994, "Structural Analysis and Decomposition of Partially Defined Boolean Functions," RRR 13-94, 25 pages, RUTCOR, Rutgers University, NJ.

BOROS, E., P.L. HAMMER, AND J.N. HOOKER, 1994, "Predicting Cause-Effect Relationships From Incomplete Discrete Observations," *SIAM J. on Discrete Math.*, Vol. 7, No. 4, pp. 531-543.

BLAIR, C.E., R.G. JEROSLOW, AND J.K. LOWE, 1985, "Some Results and Experiments in Programming Techniques for Propositional Logic," *Computers and Operations Research*. No. 13, pp.633-645.

CARRAGHAN, R., AND P.M. PARDALOS, 1990, "An Exact Algorithm for the Maximum Clique Problem," Operations Research Letters, 9, November, pp. 375-382.

CAVALIER, T.M., P.M. PARDALOS, AND A.L. SOYSTER, 1990, "Modeling and Integer Programming Techniques Applied to Propositional Calculus," J.P. Ignizio (ed.). *Computers and Operations Research*. 17:6, 561-570.

DESHPANDE, A.S., and E. TRIANTAPHYLLOU, 1996, "A Randomized Polynomial Heuristic for Inferring A Small Number of Logical Clauses from Examples," *Mathematical and Computer Modelling*, in print.

FEO, T.A. AND M.G.C. RESENDE, 1995, "Greedy Randomized Adaptive Search Procedures," *Journal of Global Optimization*, Vol. 6, pp. 109-133.

FU, L.M., 1993, "Knowledge-based connectionism for revising domain theories," *IEEE Transactions on Systems, Man and Cybernetics*, Vol. 23, No. 1, pp. 173-182.

HAMMER, P.L., AND E. BOROS, 1994, "Logical Analysis: An Overview," RUTCOR Research Report, Rutgers University, NJ.

HAMMER, P.L., AND A. KOGAN, 1992, "Horn Function Minimization and Knowledge Compression in Production Rule Bases," RUTCOR Research Report # 8-92, Rutgers University, NJ.

HANSEL, G., 1966, "Sur le nombre des fonctions Boolenes monotones den variables". *C.R. Acad. Sci. Paris*, v. 262, n. 20, 1088-1090.

HAUSSLER, D. 1989, "Learning conjunctive concepts in structural domains," *Machine Learning*, Vol. 4, pp. 7-40.

HAUSSLER, D., AND M. WARMUTH, 1993, "The Probably Approximately Correct (PAC) and Other Learning Models," Chapter in: *Foundations of Knowledge Acquisition: Machine Learning*, A.L. Meyrowitz and S. Chipman (Eds), Kluwer Academic Publishers, Norwell, MA, pp. 291-312.

HOOKER, J.N., 1988a, "Generalized Resolution and Cutting Planes," R.G. Jeroslow (ed.), *Annals of Operations Research*, Vol. 12, No. 1-4, pp. 217-239.

HOOKER, J.N., 1988b, "A Quantitative Approach to Logical Inference," *Decision Support Systems*. North-Holland, No. 4, pp. 45-69.

GIMPEL, J., 1965, "A Method of Producing a Boolean Function Having an Arbitrarily Prescribed Prime Implicant Table," *IEEE Trans. on Computers*, Vol. 14, pp. 485-488.

GOLDMAN, S.A., 1990, "Learning Binary Relations, Total Orders, and Real-Once Formulas." Ph.D. Thesis, Massachusetts Institute of Technology, September 1990. Available as Technical Report MIT/LCS/TR-483, MIT Laboratory for Computer Science.

GORBUNOV, Y. and B. KOVALERSCHUK, 1982, "An Interactive Algorithm For Restoring of a Monotone Boolean Function," *Izvestia AN USSR* (Proceedings of the Acad. of Sci.), STN, 2, Tashkent, pp. 3-16, in Russian.

JEROSLOW, R.G., 1988, "Computation-Oriented Reductions of Predicate to Prepositional Logic," *Decision Support Systems*. North-Holland, No. 4, pp. 183-197.

JEROSLOW, R.G., 1989, *Logic-Based Decision Support*, North-Holland.

QUINLAN, J.R., "Discovering rules by induction from large collections of examples," in *Expert Systems in the Micro Electronic Age*, D. Michie, Ed., Edinburgh, Scotland: Edinburgh University Press, 1979.

KAMATH, A.P., N.K. KARMAKAR, K.G. RAMAKRISHNAN, AND M.G.C. RESENDE, 1990, "Computational Experience with an Interior Point Algorithm on the Satisfiability Problem," *Annals of Operations Research*. P.M. Pardalos and J.B. Rosen (eds.). Special issue on: Computational Methods in Global Optimization, Vol 25, pp. 43-58.

KAMATH, A.P., N.K. KARMAKAR, K.G. RAMAKRISHNAN, AND M.G.C. RESENDE, 1992, "A Continuous Approach to Inductive Inference," *Math. Progr.*, Vol. 57, pp. 215-238.

KAMATH, A.P., N.K. KARMAKAR, K.G. RAMAKRISHNAN, AND M.G.C. RESENDE, 1994, "An Interior Point Approach to Boolean Vector Synthesis," *Proceedings of the 36th MSCAS*, pp. 1-5.

KARMAKAR, N.K., M.G.C. RESENDE, AND K.G. RAMAKRISHNAN, 1991, "An Interior Point Algorithm to Solve Computationally Difficult Set Covering Problems," *Math. Progr.*, Vol. 52, pp. 597-618.

KLEITMAN, D., 1969, "On Dedekind's problem: the number of monotone Boolean functions". *Proc. Amer. Math. Soc.* 21, 677-682.

KLEITMAN, D., 1969, "On Dedekind's problem: the number of monotone Boolean functions". Proc. Amer. Math. Soc. 21, 677-682.

KEARNS, M., MING LI, L. PITT, AND L.G. VALIANT, 1987, "On the Learnability of Boolean Formulae," *Journal of the Association for Computing Machinery*, No. 9, pp. 285-295.

KOVALERCHUK, B, E. TRIANTAPHYLLOU, and E. VITYAEV, 1995, "Monotone Boolean Function Learning Techniques Integrated with User Interaction," *Proceedings of the 12-th Inter'l Conference in Machine Learning*, Lake Tahoe, CA, U.S.A., July 9-12, pp. 41-48.

KOVALERCHUK, B., E. TRIANTAPHYLLOU, A.S. DESHPANDE, AND E. VITYAEV, 1996, "Interactive Learning of Monotone Boolean Functions," *Information Sciences*, 20 pages, (in print).

McCLUSKEY, E., 1956, "Minimization of Boolean Functions," *Bell Syst. Tech. J.*, Vol. 35, pp. 1417-1444.

MANGASARIAN, O.L., R. SETIONO, AND W.H. WOLBERG, 1991, "Pattern Recognition Via Linear Programming: Theory and Application to Medical Diagnosis," in: *Large-Scale Numerical Optimization*, Eds. T.F. Coleman, and Y. Li, SIAM, pp. 22-30.

MANSOUR, Y., 1992, "Learning of DNF Formulas", *Proceedings of the Fifth Annual Workshop on Computational Learning Theory*, pp. 53-59.

MOTWANI, R. AND P. RAGHAVAN, 1995, *Randomized Algorithms*, Cambridge University Press, New York, NY, USA.

MURPHY AND D.W. AHA, 1994, UCI Repository of Machine Learning Databases, Machine readable data repository, Irvine, CA, University of California, Dept. of Information and Computer Science.

PARDALOS, P.M., AND C.S. RENTALA, 1990, "Computational Aspects of a Parallel Algorithm to Find the Connected Components of a Graph," Technical Report, Dept. of Computer Science, Pennsylvania State University.

PEYSAKH, J., 1987, "A Fast Algorithm to Convert Boolean Expressions into CNF," IBM Computer Science RC 12913 (#57971), Watson, NY.

PITT, L. AND L.G. VALIANT, 1988, "Computational Limitations on Learning from Examples," *Journal of the Association for Computing Machinery*. Vol. 35, No. 4, pp. 965-984.

RUDEANU, S., 1974, *Boolean Functions and Equations*, North-Holland, NY.

SHAVLIK, J.W., 1994, "Combining Symbolic and Neural Learning," *Machine Learning*, Vol. 14, pp. 321-331.

TRIANTAPHYLLOU, E, A.L. SOYSTER, AND S.R.T. KUMARA, 1994, "Generating Logical Expressions from Positive and Negative Examples via a Branch-And-Bound Approach," *Computers and Operations Research*, Vol. 21, No. 2, pp. 185-197.

TRIANTAPHYLLOU, E., 1994, "Inference of A Minimum Size Boolean Function From Examples by Using A New Efficient Branch-And-Bound Approach," *Journal of Global Optimization*, Vol. 5, No. 1, pp. 69-94.

TRIANTAPHYLLOU, E. AND A.L. SOYSTER, 1995, "An Important Relationship Between CNF and DNF Systems Which are Derived from Examples,"

ORSA Journal on Computing, Vol. 7, No. 3, pp. 283-285.

TRIANTAPHYLLOU, E. AND A.L. SOYSTER, 1996a, "On the Minimum Number of Logical Clauses Which Can be Inferred From Positive and Negative Examples," *Computers and Operations Research*, Vol. 23, No. 8, pp. 783-799.

TRIANTAPHYLLOU, E. AND A.L. SOYSTER, 1996b, "An Approach to Guided Learning of Boolean Functions," *Computers and Mathematical Modelling*, Vol. 23, No. 3, pp. 69-86.

TRICK, M. AND D. JOHNSON, 1995, "Second DIAMACS Challenge on Cliques, Coloring and Satisfiability," American Mathematical Society, Rutgers University, Summer.

TURKSEN, I.B., AND H. ZHAO, "An Equivalence Between Inductive Learning and Pseudo-Boolean Logic Simplification: A Rule Generation and Reduction Scheme," *IEEE Transactions on Systems*, Man and Cybernetics, Vol. 23, No. 3, pp. 907-917, 1993.

VALIANT, L.G., 1984, "A Theory of the Learnable," *Comm. of ACM*, Vol. 27, No. 11, pp. 1134-1142.

WILLIAMS, H.P., 1986, "Linear and Integer Programming Applied to Artificial Intelligence," Preprint series, University of Southampton, Faculty of Mathematical Studies, pp. 1-33.

YABLONSKII, S., 1986, *Introduction to Discrete Mathematics*, Moscow, Nauka Publ. (in Russian).

10

COMPUTATIONAL ANALYSIS OF A $G/G/1$ QUEUE WITH VACATIONS AND EXHAUSTIVE SERVICE

Huan Li and Yixin Zhu

Wireless Networks, NORTEL,
MS-D207,
2201 Lakeside Blvd.,
Richardson, TX 75082, USA

ABSTRACT

A class of G/G/1 queueing systems with exhaustive service and Erlang vacations is studied. We develop a computationally tractable algorithm to calculate the moments of the stationary waiting time and sojourn time of the system. Numerical examples show that the algorithm is easy to implement and very powerful.

1 INTRODUCTION

In this paper, we consider a $G/G/1$ queue with exhaustive service and Erlang vacations. It is assumed that the interarrival times, service times and vacation times are three independent renewal sequences. The vacation policy can be described as follows: When all customers in the system are cleared, the server takes a vacation with probability p_0 or it waits for a customer to arrive with probability $1 - p_0$. Upon returning from the jth vacation ($j = 1, 2, \ldots$), the server resumes the service if at least one customer is present in the system. Otherwise it takes another vacation with probability p_j. With probability $1-p_j$, it waits for a customer to start a new busy period without taking additional vacations. Once the service to customer is resumed, it is continued until the system becomes empty. The server can take at most N vacations in a row. If there are no backlogged customers in the system when the server returns from the Nth vacation, it stays in the system to wait for customers. The vacation times are assumed to have an Erlang distribution. Observe that the multiple vacation model is not a special case of this model. However, we can either modify the approach described in this paper to analyze a G/G/1 queue with multiple vacations or approximate it by using our model with a relatively large N. We notice that Kella[1990] studied a model with similar dynamics in the context of M/G/1 queues.

The literature on single server queueing systems with server vacations is extensive. Many variations have been investigated in detail and expressions for important performance measures have been obtained. Recently, the interest in such systems has been further enhanced by its applicability in communication, computer, and production systems. Most of the earlier work, as well as a complete set of reference, on analysis of vacation models can be found in Doshi[1986],[1990], Shanthikumar and Sumita[1989], Takagi[1991] and Teghem[1986]. The major general result for vacation models is the stochastic decomposition property, which allows the system in the steady state to be analyzed by considering separately the system behavior without vacations and the additional delay due to the vacations. Shanthikumar and Sumita[1989] studied vacation models in its full generality by defining the modified Lindley process with replacement (MLPR). Under some mild conditions, they established the stochastic decomposition property. Their results lead to the stochastic decomposition property of our model immediately. However, the analysis of the additional delay due to the vacations are extremely difficult, particularly for models with complicated vacation policies. Moreover, very few computational tractable results are available for the performance measures of a general G/G/1 queue with vacations (e.g. single vacation model, multiple vacation model, etc.).

Recently, Gong and Hu[1992] have provided a simple recursive procedure calculating the MacLaurin series of the moments of the stationary waiting time and the system time in terms of a parameter in the service time distribution for G/G/1 queueing systems. This method is applied to G/G/1 queues with setup time and some other models successfully in Li and Zhu[1994], Hu et al.[1993] and Zhu and Li[1993]. The numerical experiments have shown that the MacLaurin expansion method is fast and quite accurate. In this paper, We seek to extend the MacLaurin expansion method to develop an algorithm to compute the moments of the stationary waiting time and system time in the G/G/1 queues with exhaustive service and Erlang vacations. Many authors have studied vacation models with general vacations in the literature. However, they either focus on Poisson arrival case or obtain computationally intractable results with general arrival case. We assume that the vacation times have an Erlang distribution in this paper. Further study for modifying the analysis to cover vacations having a phase type distribution is being undertaken. It will certainly give a good understanding of G/G/1 queues with general vacations since any distribution can be approximated by a phase type distribution.

The rest of the paper is organized as follows. In section 2, based on the fact that the waiting time process at the arrival epochs in a G/G/1 queue with exhaustive service and vacations is a modified Lindley process with replacement (MLPR) defined by Shanthikumar and Sumita[1989], we obtain infinite linear equations for the moments of stationary waiting time in terms of a parameter in the service times and vacation times. In section 3, we prove that if the service times and vacations are scaled by a deterministic parameter θ, then the kth moments of the stationary waiting time and the system time in our model are dominated by θ^k $(k > 0)$. Based on the results in section 3, a simple recursive algorithm for computing the MacLaurin series of the moments is developed in section 4. Some numerical examples are given in section 5.

2 BASIC SYSTEMS

Observe that the waiting time sequence $\{W_n, n \geq 1\}$ at the arrival epochs in the queueing vacation model described above satisfies the so-called Modified Lindley Process with Replacement (MLPR) (see Shanthikumar and Sumita[1989]):

$$W_{n+1} = \begin{cases} W_n + S_{n+1} - A_{n+1} & \text{if } W_n + S_{n+1} - A_{n+1} \geq 0; \\ R_{n+1} & \text{if } W_n + S_{n+1} - A_{n+1} < 0, \end{cases} \tag{10.1}$$

Where $W_1 = 0$.

S_{n+1} is the service time for the nth customer.

A_{n+1} is the nth interarrival time.

$\{A_n, n \geq 1\}$ and $\{S_n, n \geq 1\}$ are two independent renewal sequences.

V_j is the jth vacation time.

The server's replacement R_{n+1} is a nonnegative random variable with conditional distribution:

$$\begin{aligned} & P\{R_{n+1} \leq x | W_n + S_{n+1} - A_{n+1} = -t, t \geq 0\} \\ = \; & 1 - \sum_{m=1}^{N} \Big(\prod_{l=0}^{m-1} p_l\Big) P\{t + x - V_m < \sum_{j=1}^{m-1} V_j < t\} \quad \text{for } x > 0. \end{aligned} \tag{10.2}$$

The system time for the nth customer is given by

$$T_n = W_n + S_{n+1}. \tag{10.3}$$

Let A be a generic interarrival time with a bounded probability density function (p.d.f.) $f(\cdot)$. Let S be a generic service time, V be a generic vacation time and R be a generic replacement.

We assume that the process $\{W_n, n \geq 1\}$ is stable and ergodic, i.e., $W_n \xrightarrow{d} W$, $T_n \xrightarrow{d} T$, as $n \to \infty$. However, this can be shown under some mild conditions by using the Markov property of the process $\{W_n, n \geq 1\}$ (see Shanthikumar and Sumita[1989]). Letting $n \to \infty$ in equations (10.1) and (10.3), we have

$$\begin{aligned} W &= (T-A)1_{(T\geq A)} + R1_{(T<A)} & (10.4) \\ T &= W + S. & (10.5) \end{aligned}$$

Where $1_{(E)}$ is the indicator function of the event E.

We shall assume that $f(x) = \sum_{j=0}^{\infty} \frac{f^{(j)}(0)}{j!} x^j$, where $f^{(j)}(0)$ is the jth derivative of $f(x)$ at $x = 0$, for all $x \in [0, \infty)$. It is true for the phase type distribution. Suppose the service times and vacation times are scaled by a parameter θ, i.e., $S_n = \theta X_n$, $V_n = \theta Y_n$, $S = \theta X$, $V = \theta Y$, where the deterministic parameter θ is independent of $\{X_n, n \geq 1\}$ and $\{Y_n, n \geq 1\}$, and $Y_1, Y_2, \cdots$, are i.i.d. random variables with p.d.f. $h(x) = \frac{\nu^\gamma x^{\gamma-1} e^{-\nu x}}{(\gamma-1)!}$.

Define the p.d.f. and c.d.f. of $\sum_{j=1}^{m} Y_j$ as $\psi_m(\cdot)$ and $\Psi_m(\cdot)$. Then, from (10.2), the conditional distribution of R is given by

$$\begin{aligned} & G(x,t) \\ = \; & P\{R \leq x | T - A = -t, t \geq 0\} \\ = \; & 1 - p_0[1 - \Psi_1(\frac{x+t}{\theta})] - \sum_{m=2}^{N}(\prod_{l=0}^{m-1} p_l) \int_0^{\frac{t}{\theta}} [1 - \Psi_1(\frac{t+x}{\theta} - y)]\psi_{m-1}(y)dy \end{aligned}$$

$$\begin{aligned} & g(x,t) \\ = \; & \frac{\partial G(x,t)}{\partial x} \\ = \; & p_0 h(\frac{x+t}{\theta})\frac{1}{\theta} + \sum_{m=2}^{N}(\prod_{l=0}^{m-1} p_l) \int_0^{\frac{t}{\theta}} h(\frac{x+t}{\theta} - y)\frac{1}{\theta}\psi_{m-1}(y)dy \\ = \; & p_0 \sum_{i=0}^{\gamma-1} \binom{\gamma-1}{i} (\frac{x}{\theta})^i (\frac{t}{\theta})^{\gamma-1-i} \frac{1}{\theta} \frac{\nu^\gamma e^{-\nu \frac{x+t}{\theta}}}{(\gamma-1)!} + \end{aligned} \quad (10.6)$$

$$\sum_{m=2}^{N}(\prod_{l=0}^{m-1} p_l)\int_0^{\frac{t}{\theta}}\sum_{i=0}^{\gamma-1}\binom{\gamma-1}{i}(\frac{x}{\theta})^i(\frac{t}{\theta}-y)^{\gamma-1-i}\frac{1}{\theta}\frac{\nu^\gamma e^{-\nu(\frac{x+t}{\theta}-y)}}{(\gamma-1)!}\psi_{m-1}(y)dy.$$

Then , for $k = 1, 2, \ldots$, we have

$$\begin{aligned}
& E[\frac{R^k}{k!}1_{(T<A)}|T-A=-t, t>0] \\
= & \int_0^\infty \frac{x^k}{k!}g(x,t)dx \\
= & p_0\sum_{i=0}^{\gamma-1}\binom{\gamma-1}{i}\int_0^\infty \frac{x^k}{k!}(\frac{x}{\theta})^i\frac{1}{\theta}\nu e^{-\nu\frac{x}{\theta}}dx\frac{\nu^{\gamma-1}(\frac{t}{\theta})^{\gamma-1-i}}{(\gamma-1)!}e^{-\nu\frac{t}{\theta}} + \\
& \sum_{m=2}^{N}(\prod_{l=0}^{m-1}p_l)\sum_{i=0}^{\gamma-1}\binom{\gamma-1}{i}\int_0^\infty \frac{x^k}{k!}(\frac{x}{\theta})^i\frac{1}{\theta}\nu e^{-\nu\frac{x}{\theta}}dx \\
& \int_0^{\frac{t}{\theta}}\frac{\nu^{\gamma-1}(\frac{t}{\theta}-y)^{\gamma-1-i}}{(\gamma-1)!}e^{-\nu(\frac{t}{\theta}-y)}\psi_{m-1}(y)dy \\
= & p_0\sum_{i=0}^{\gamma-1}\binom{i+k}{i}\frac{\theta^k}{\nu^{k+1}}\frac{\nu^{\gamma-i}(\frac{t}{\theta})^{\gamma-1-i}}{(\gamma-i-1)!}e^{-\nu\frac{t}{\theta}} + \sum_{m=2}^{N}(\prod_{l=0}^{m-1}p_l)\sum_{i=0}^{\gamma-1} \\
& \binom{i+k}{i}\frac{\theta^k}{\nu^{k+1}}\int_0^{\frac{t}{\theta}}\frac{\nu^{\gamma-i}(\frac{t}{\theta}-y)^{\gamma-i-1}}{(\gamma-i-1)!}e^{-\nu(\frac{t}{\theta}-y)}\psi_{m-1}(y)dy \\
= & \sum_{m=1}^{N}(\prod_{l=0}^{m-1}p_l)\sum_{i=0}^{\gamma-1}\binom{i+k}{i}\frac{\theta^k}{\nu^{k+1}}\frac{\nu^{m\gamma-i}(\frac{t}{\theta})^{m\gamma-i-1}}{(m\gamma-i-1)!}e^{-\nu\frac{t}{\theta}}. \qquad (10.7)
\end{aligned}$$

The last equality follows from the fact that the convolution of Erlang distributions is still an Erlang distribution. Hence, by using (10.7), we have the following equation after some algebraic manipulations:

$$\begin{aligned}
E[\frac{R^k}{k!}1_{(T<A)}] = & \frac{\theta^k}{\nu^k}\sum_{m=1}^{N}(\prod_{l=0}^{m-1}p_l)\sum_{i=0}^{\gamma-1}\binom{i+k}{i}\sum_{j=0}^{\infty}f^{(j)}(0) \\
& \sum_{r=0}^{j}\binom{m\gamma-i-1+r}{r}\frac{\theta^{r+1}}{\nu^{r+1}}E[\frac{T^{j-r}}{(j-r)!}]. \qquad (10.8)
\end{aligned}$$

Moreover, for k=1,2,..., we have

$$\begin{aligned}
& E[\frac{(T-A)^k}{k!}1_{(T\geq A)}] \\
= & E[\frac{1}{k!}\int_0^T (T-x)^k f(x)dx]
\end{aligned}$$

$$
\begin{aligned}
&= E[\frac{1}{k!}\int_0^T (T-x)^k \sum_{j=0}^{\infty} \frac{f^{(j)}(0)}{j!} x^j \, dx] \\
&= \sum_{j=0}^{\infty} f^{(j)}(0) E[\frac{T^{j+k+1}}{(j+k+1)!}]. \qquad (10.9)
\end{aligned}
$$

From (10.4), for $k = 1, 2, \ldots$, we then obtain:

$$
\begin{aligned}
& E[\frac{W^k}{k!}] \\
&= E[\frac{(T-A)^k}{k!} 1_{(T \geq A)}] + E[\frac{R^k}{k!} 1_{(T<A)}] \\
&= \sum_{j=0}^{\infty} f^{(j)}(0) E[\frac{T^{j+k+1}}{(j+k+1)!}] + \frac{\theta^k}{\nu^k} \sum_{m=1}^{N} (\prod_{l=0}^{m-1} p_l) \sum_{i=0}^{\gamma-1} \binom{i+k}{i} \\
& \sum_{j=0}^{\infty} f^{(j)}(0) \sum_{r=0}^{j} \binom{m\gamma - i - 1 + r}{r} \frac{\theta^{r+1}}{\nu^{r+1}} E[\frac{T^{j-r}}{(j-r)!}]. \qquad (10.10)
\end{aligned}
$$

From equation (10.5),

$$
\begin{aligned}
E[\frac{T^k}{k!}] &= E[\frac{(W+S)^k}{k!}] \\
&= \sum_{j=0}^{k} E[\frac{S^{k-j}}{(k-j)!}] E[\frac{W^j}{j!}] \\
&= \sum_{j=0}^{k} \theta^{k-j} E[\frac{X^{k-j}}{(k-j)!}] E[\frac{W^j}{j!}].
\end{aligned}
$$

Thus,

$$
E[\frac{T^k}{k!}] = \sum_{j=0}^{k} \theta^{k-j} \beta_{k-j} E[\frac{W^j}{j!}], \qquad (10.11)
$$

where $\beta_i = E[\frac{X^i}{i!}]$.

3 STOCHASTIC BOUND

In this section, an upper bound is obtained for our model. However, the results are valid for the model where the vacation times have a general distribution.

Then, we show that $E[\frac{W^k}{k!}] = o(\theta^{k-1})$ and $E[\frac{T^k}{k!}] = o(\theta^{k-1})$. These results will be used in the next section to develop an algorithm for our model.

We begin with introducing the definition of stochastic ordering $\leq_{st}$ which is frequently used in the literature.

Definition 1 *A random variable X is stochastically less than a random variable Y, written $X \leq_{st} Y$, if $P(X > a) \leq P(Y > a)$ for all a.*

Some important results which we need in this paper for the ordering $\leq_{st}$ are listed as follows: If $X \leq_{st} Y$, then (i) $E[\frac{X^k}{k!}] \leq E[\frac{Y^k}{k!}]$ for $k \geq 1$; (ii) $X^+ \leq_{st} Y^+$; (iii) If Z_1 and Z_2 are independent of X and Y respectively and $Z_1 \leq_{st} Z_2$, $X + Z_1 \leq_{st} Y + Z_2$. The proofs are available in Ross[1983].

Lemma 1 *There is a random variable Z which is independent of the arrival process and the service times such that $(R_{n+1}|W_n + S_{n+1} - A_{n+1} < 0) \leq_{st} Z$ for all $n \geq 1$.*

Proof: Let V be the generic vacation. We define

$$G(x) = 1 - \sup\{\frac{P(V > x + t)}{P(V > t)}, t > 0\} \quad \text{for } x \in (0, \infty).$$

One can verify that G is a cumulative distribution function (c.d.f.). Therefore, there is a nonnegative random variable Z whose c.d.f. is $G(\cdot)$. Obviously, Z is independent of the arrival process and the service times. By the definition of G, we also have

$$P(V > x + t) \leq P(Z > x)P(V > t), \quad \text{for } x \geq 0, t > 0. \tag{10.12}$$

Let $\phi_m(\cdot)$ be the p.d.f. of $\sum_{j=1}^m V_j$ for $m \geq 1$. Then, from (10.2), for any $x \geq 0$ and $t > 0$,

$$\begin{aligned} & P(R_{n+1} > x | W_n + S_{n+1} - A_{n+1} = -t, t > 0) \\ = & \sum_{m=1}^{N} (\prod_{l=0}^{m-1} p_l) P(x + t < \sum_{j=1}^{m} V_j, \sum_{j=1}^{m-1} V_j \leq t) \\ = & \sum_{m=1}^{N} (\prod_{l=0}^{m-1} p_l) \int_0^t P(V_m > x + t - y) \phi_{m-1}(y) dy \end{aligned}$$

$$
\begin{aligned}
&= \sum_{m=1}^{N}(\prod_{l=0}^{m-1} p_l)\int_0^t P(V > x+t-y)\phi_{m-1}(y)dy \\
&\leq \sum_{m=1}^{N}(\prod_{l=0}^{m-1} p_l)\int_0^t P(Z > x)P(V > t-y)\phi_{m-1}(y)dy \\
&= P(Z > x)P(R_{n+1} > 0|W_n + S_{n+1} - A_{n+1} = -t, t > 0) \\
&\leq P(Z > x).
\end{aligned}
$$

Let $f_n(\cdot)$ be the p.d.f. of $W_n + S_{n+1} - A_{n+1}$. Then

$$
\begin{aligned}
& P(R_{n+1} > x|W_n + S_{n+1} - A_{n+1} < 0) \\
&= \int_0^\infty P(R_{n+1} > x|W_n + S_{n+1} - A_{n+1} = -t, t > 0)f_n(t)dt \\
&\leq P(Z > x)\int_0^\infty f_n(t)dt \\
&= P(Z > x)P(W_n + S_{n+1} - A_{n+1} < 0) \\
&\leq P(Z > x).
\end{aligned}
$$

We arrive at the desired result. *Q.E.D.*

Now, we construct a new G/G/1 queueing system based on the original system as follows: Let $\{Z_n, n \geq 1\}$ be a sequence of independent random variables which have an identical distribution as Z does. The service time for the nth customer is modified as $S_{n+1} + Z_{n+1}$ while the arrival process is the same as that of the original system and the server doesn't take vacation. Then the sequence of the waiting times $\{\widetilde{W}_n, n \geq 1\}$ in this system satisfies the so-called Lindley equation:

$$\widetilde{W}_{n+1} = (\widetilde{W}_n + (S_{n+1} + Z_{n+1}) - A_{n+1})^+, \tag{10.13}$$

where $\widetilde{W}_1 = 0$ and $x^+ = \max\{x, 0\}$. $\widetilde{T}_n = \widetilde{W}_n + S_{n+1} + Z_{n+1}$ is the system time for the nth customer in the modified system.

Theorem 1 $T_n \leq_{st} \widetilde{T}_n$ *for* $n \geq 1$.

Proof: We prove it by mathematical induction. Since Z_2 is independent of S_2,

$$T_1 = S_2 \leq_{st} S_2 + Z_2 = \widetilde{T}_1 .$$

Assume it is true for $n = k$, i.e. $T_k \leq_{st} \widetilde{T}_k$. We want to show it is also true for $n = k+1$. From (10.1) and (10.3),

$$T_{k+1} = (T_k - A_{k+1})^+ + R_{k+1}1_{(T_k - A_{k+1} < 0)} + S_{k+2}.$$

From Lemma 1, we have,

$$\begin{aligned}
& P(R_{k+1}1_{(T_k - A_{k+1}<0)} + (T_k - A_{k+1})^+ > x) \\
= \ & P(R_{k+1} > x | T_k - A_{k+1} < 0) P(T_k - A_{k+1} < 0) + P((T_k - A_{k+1})^+ > x) \\
\leq \ & P(Z > x) P(T_k - A_{k+1} < 0) + P((T_k - A_{k+1})^+ > x) \\
\leq \ & P(Z1_{(T_k - A_{k+1}<0)} + (T_k - A_{k+1})^+ > x) \\
\leq \ & P(Z + (T_k - A_{k+1})^+ > x) \\
= \ & P(Z_{k+2} + (T_k - A_{k+1})^+ > x).
\end{aligned}$$

The last equality holds since Z, Z_{k+2} are identically distributed and independent of $(T_k - A_{k+1})^+$.
Hence,

$$R_{k+1}1_{(T_k - A_{k+1}<0)} + (T_k - A_{k+1})^+ \leq_{st} Z_{k+2} + (T_k - A_{k+1})^+$$

By the induction hypothesis, we have

$$T_k - A_{k+1} \leq_{st} \widetilde{T}_k - A_{k+1},$$

because of the independence of A_{k+1} and T_k, $\widetilde{T}_k$. Thus,

$$(T_k - A_{k+1})^+ \leq_{st} (\widetilde{T}_k - A_{k+1})^+.$$

Therefore, we have

$$\begin{aligned}
T_{k+1} &= (T_k - A_{k+1})^+ + R_{k+1}1_{(T_k - A_{k+1}<0)} + S_{k+2} \\
&\leq_{st} (T_k - A_{k+1})^+ + Z_{k+2} + S_{k+2} \\
&\leq_{st} (\widetilde{T}_k - A_{k+1})^+ + Z_{k+2} + S_{k+2} \\
&= \widetilde{T}_{k+1}\,.
\end{aligned}$$

We arrive at the desired result. *Q.E.D.*

Lemma 2 *There is a random variable $\widetilde{Z}$ which is independent of θ such that $Z \stackrel{d}{=} \theta\,\widetilde{Z}$.*

Proof: Let $\tilde{Z}$ be a random variable with c.d.f. $\tilde{G}(x) = 1-\sup\{\frac{P(Y>x+t)}{P(Y>t)}, t>0\}$ for $x \geq 0$. By using the fact that $V = \theta Y$, one can verify that $Z \stackrel{d}{=} \theta \tilde{Z}$ and θ, $\tilde{Z}$ are independent. *Q.E.D.*

Theorem 2

$$E[\frac{W^k}{k!}] = o(\theta^{k-1}), \quad k \geq 1, \tag{10.14}$$
$$E[\frac{T^k}{k!}] = o(\theta^{k-1}), \quad k \geq 1. \tag{10.15}$$

Proof: By Lemma 2, there is a sequence of i.i.d. random variables $\{\tilde{Z}_n, n \geq 1\}$ such that $Z_n \stackrel{d}{=} \theta \tilde{Z}_n$ for $n \geq 1$. ¿From (10.13),

$$\tilde{W}_{n+1}(\theta) = (\tilde{W}_n(\theta) + S_{n+1} + Z_{n+1} - A_{n+1})^+,$$

where $\tilde{W}_1(\theta) = 0$. Thus,

$$\frac{\tilde{W}_{n+1}(\theta)}{\theta} = (\frac{\tilde{W}_n(\theta)}{\theta} + X_{n+1} + \tilde{Z}_{n+1} - \frac{A_{n+1}}{\theta})^+.$$

One can verify that $\frac{\tilde{W}_n(\theta)}{\theta}$ is stochastically increasing in $\theta \in (0, \epsilon]$ by mathematical induction. Thus,

$$E[(\frac{\tilde{W}_n(\theta)}{\theta})^k] \leq E[(\frac{\tilde{W}_n(\epsilon)}{\epsilon})^k].$$

That is, $E[\frac{\tilde{W}_n^k(\theta)}{k!}] = o(\theta^{k-1})$. Since $\tilde{T}_n(\theta) = \tilde{W}_n(\theta) + \theta X_{n+1} + \theta Y_{n+1}$, we also have $E[\frac{\tilde{T}_n^k(\theta)}{k!}] = o(\theta^{k-1})$. From Theorem 1, we immediately have

$$E[\frac{T_n^k}{k!}] \leq E[\frac{\tilde{T}_n^k(\theta)}{k!}], \quad n, k \geq 1.$$

That is, $E[\frac{T_n^k}{k!}] = o(\theta^{k-1})$. $E[\frac{W_n^k}{k!}] = o(\theta^{k-1})$ follows from $W_n = T_n - \theta X_{n+1}$. By letting $n \to \infty$, we arrive at the desired result. *Q.E.D.*

4 THE ALGORITHM

In this section, we give a computational scheme for the coefficients of the MacLaurin series of $E[\frac{W^k}{k!}]$ and $E[\frac{T^k}{k!}]$ $(k = 1, 2, \ldots)$ with respect to the parameter θ based on equations (10.10) and (10.11).

Assuming that

$$E[\frac{W^k}{k!}] = \sum_{n=1}^{\infty} b_{kn}\theta^n, \tag{10.16}$$

$$E[\frac{T^k}{k!}] = \sum_{n=1}^{\infty} a_{kn}\theta^n, \tag{10.17}$$

We shall find the coefficients a_{kn}'s and b_{kn}'s of the MacLaurin series of $E[\frac{T^k}{k!}]$ and $E[\frac{W^k}{k!}]$. Note that the MacLaurin series in (10.16) and (10.17) might not converge for all θ. However, it is possible for the series to converge in a neighborhood for $\theta = 0$, for example, if the cumulative distribution function of the interarrival time can be expanded as a MacLaurin series over $[0, \infty)$ (see Gong and Hu[1992]).

Theorem 3 *The coefficients a_{kn}'s and b_{kn}'s in equations (10.16) and (10.17) are given by*

$$a_{kn} = \begin{cases} 0 & \text{if } n < k \\ \beta_k + \sum_{j=1}^{k} \beta_{k-j} b_{jj} & \text{if } n = k \\ \sum_{j=1}^{k} \beta_{k-j} b_{j,n-k+j} & \text{if } n > k \end{cases}$$

$$b_{kn} = \begin{cases} 0 & \text{if } n \le k \\ f(0)a_{k+1,k+1} + \frac{f(0)}{\nu^{k+1}} \sum_{m=1}^{N} (\prod_{l=0}^{m-1} p_l) \sum_{i=0}^{\gamma-1} \binom{i+k}{i} & \text{if } n = k+1 \\ \sum_{j=0}^{n-k-1} f^{(j)}(0) a_{j+k+1,n} + \frac{f^{(n-k-1)}(0)}{\nu^n} \sum_{m=1}^{N} (\prod_{l=0}^{m-1} p_l) & \\ \sum_{i=0}^{\gamma-1} \binom{i+k}{i} \binom{m\gamma-i-1+n-k-1}{n-k-1} + \sum_{j=1}^{n-k-1} f^{(j)}(0) \sum_{r=0}^{j-1} & \\ \frac{a_{j-r,n-k-1-r}}{\nu^{k+r+1}} \sum_{m=1}^{N} (\prod_{l=0}^{m-1} p_l) \sum_{i=0}^{\gamma-1} \binom{i+k}{i} \binom{m\gamma-i-1+r}{r} & \text{if } n > k+1 \end{cases}$$

$$\textit{for } k = 1, 2, \ldots$$

Proof: From Theorem 2,

$$E[\frac{W^k}{k!}] = o(\theta^{k-1}) \tag{10.18}$$

$$E[\frac{T^k}{k!}] = o(\theta^{k-1}) \tag{10.19}$$

It is easy to see that (10.18) and (10.19) imply that $a_{kn} = 0$ and $b_{kn} = 0$ for $n < k$. Thus, we may rewrite (10.16) and (10.17) as

$$E[\frac{W^k}{k!}] = \sum_{n=k}^{\infty} b_{kn}\theta^n, \tag{10.20}$$

$$E[\frac{T^k}{k!}] = \sum_{n=k}^{\infty} a_{kn}\theta^n. \tag{10.21}$$

Substituting equations (10.20) and (10.21) into equation (10.11), we have

$$\begin{aligned} \sum_{n=k}^{\infty} a_{kn}\theta^n &= \beta_k\theta^k + \sum_{j=1}^{k} \beta_{k-j}\theta^{k-j} \sum_{n=j}^{\infty} b_{jn}\theta^n \\ &= \beta_k\theta^k + \sum_{n=k}^{\infty}\sum_{j=1}^{k} \beta_{k-j} b_{j,n-k+j}\theta^n. \end{aligned} \tag{10.22}$$

Therefore

$$a_{kn} = \begin{cases} 0 & \text{if } n < k \\ \beta_k + \sum_{j=1}^{k} \beta_{k-j} b_{jj} & \text{if } n = k \\ \sum_{j=1}^{k} \beta_{k-j} b_{j,n-k+j} & \text{if } n > k \end{cases}$$

Similarly, by substituting (10.20) and (10.21) into (10.10) and after some algebraic manipulations, we have

$$\begin{aligned} &\sum_{n=k}^{\infty} b_{kn}\theta^n \\ = &\sum_{n=k+1}^{\infty} \theta^n \sum_{j=0}^{n-k-1} a_{j+k+1,n} f^{(j)}(0) + \sum_{n=k+1}^{\infty} \theta^n \frac{f^{(n-k-1)}(0)}{\nu^n} \sum_{m=1}^{N} (\prod_{l=0}^{m-1} p_l) \\ &\sum_{i=0}^{\gamma-1} \binom{i+k}{i} \binom{m\gamma - i - 1 + n - k - 1}{n-k-1} + \sum_{n=k+2}^{\infty} \theta^n \sum_{j=1}^{n-k-1} f^{(j)}(0) \sum_{r=0}^{j-1} \\ &\frac{a_{j-r,n-k-1-r}}{\nu^{k+r+1}} \sum_{m=1}^{N} (\prod_{l=0}^{m-1} p_l) \sum_{i=0}^{\gamma-1} \binom{i+k}{i} \binom{m\gamma - i - 1 + r}{r}. \end{aligned} \tag{10.23}$$

Thus, the desired result for b_{kn}'s follows from (10.23). *Q.E.D.*

Therefore, we have following algorithm for calculating a_{kn}'s and b_{kn}'s:

Step 1. Let $n = 1$.

Step 2. Let $k = n$. Evaluate a_{kk}.

step 3. Let $k - 1 \Rightarrow k$. If $k > 0$, go to next step. Otherwise, go to step 5.

Step 4. Evaluate b_{kn} and then a_{kn}. Return to step 3.

Step 5. Let $n + 1 \Rightarrow n$ and return to step 2 if necessary.

5 NUMERICAL RESULTS

In this section, we provide numerical examples to demonstrate the validity and the power of our algorithm. In all examples, the vacation time distribution is assumed to be Erlang of order γ with p.d.f. $\frac{\nu^{\gamma}x^{\gamma-1}e^{-\nu x}}{(\gamma-1)!}$. The numerical results, along with the simulation results or exact results, for the first two moments of the system time are tabulated against the parameter θ in Tables 1, 2 and 3. M in these tables is the least number of terms in the MacLaurin series needed to ensure the convergence of numerical values in the calculations.

The first example refers to $E_{k_1}/E_{k_2}/1$ queues with N vacations where the p.d.f. of the interarrival time and the service time are given by $\frac{\lambda^{k_1}x^{k_1-1}e^{-\lambda x}}{(k_1-1)!}$ and $\frac{\mu^{k_2}x^{k_2-1}e^{\mu x}}{(k_2-1)!}$ respectively, while the second example refers to $H_{k_1}/H_{k_2}/1$ queues with N vacations where the p.d.f's of the hyperexponential interarrival times and the hyperexponential service times are given by $\sum_{i=1}^{k_1} q_i\lambda_i e^{-\lambda_i x}$ and $\sum_{i=1}^{k_2} r_i\mu_i e^{-\mu_i x}$ respectively. We compare the first two moments of the system time calculated by our algorithm with those obtained from simulation. As we can see in Tables 1 and 2, the numerical results approach to the simulation results vary quickly.

Finally, in Table 3, we consider $M/E_{k_2}/1$ queues with N vacations. The numerical results for the first two moments of the system time are compared with the exact values calculated based on the results in Kella [1990]. It is easy to see that the numerical results converge rapidly to the exact values. The approximations with a relatively small M are usually accurate enough for practical purpose. The computation time to obtain the numerical results for all cases in these examples is less than one second CPU time on an VAX 8086.

REFERENCES

[1] Doshi, B., "A Note on Stochastic Decomposition in a $GI/G/1$ Queue with Vacation or Set-up times", *Journal of Applied Probability* **23**, 1985, pp. 419–428.

[2] Doshi, B., "Queueing Systems with Vacations – A Survey" , *Queueing Systems* **1**, 1986, pp. 29–66.

[3] Doshi, B., "Single Server Queues with Vacations",*Stochastic Analysis of Computer and Communication Systems*, **H. Takagi ed.**, Elsevier Science Publishers B.V. (North-Holland), 1990, pp. 217–266.

[4] Gong, W.B. and Hu, J.Q., "The Maclaurin Expansion for the $GI/G/1$ Queue", *Journal of Applied Probability* **29**, 1992, pp. 176–184.

[5] Hu, J.Q., Nananukul, S. and Gong, W.B., "A New Approach to (s, S) Inventory Systems", *Journal of Applied Probability* **30**, 1993, pp. 898–912.

[6] Kella, O., "Optimal Control of the Vacation Scheme in an M/G/1 Queue", *Operations Research* **38**, 1990, pp. 725–728.

[7] Li, H. and Zhu, Y., "A New Approach to $G/G/1$ Queues with Generalized Setup Time and Exhaustive Service ", *Journal of Applied Probability* **31**, 1994, pp. 1083–1097.

[8] Ross, S.M., " Stochastic Processes", *John Wiley & Sons*, 1983.

[9] Shanthikumar, J.G. and Sumita, U., "Modified Lindley Process with Replacement : Dynamic Behavior, Asymptotic Decomposition and Application", *Journal of Applied Probability* **26**, 1989, pp. 552–565.

[10] Takagi, H., "Queueing Analysis, Foundation of Performance Evaluation, Vol 1: Vacation and Priority Systems, Part 1", *North-Holland, Amsterdam*, 1991.

[11] Teghem, J.,Jr., "Control of the Service Process in Queueing System", *European J. of Operational Research* **23**, 1986, pp. 141–158.

[12] Zhu, Y. and Li, H., "The MacLaurin Expansion for a G/G/1 Queue with Markov-Modulated Arrivals and Services", *Queueing Systems* **14**, 1993, pp. 125-134.

Inputs	N	θ	M	$E[T]$		$E[\frac{T^2}{2!}]$	
				algo.	simul.	algo.	simul.
$k_1 = 2$	2	0.1	10	0.03343	0.03345	0.00084	0.00084
$\lambda = 1$		0.2	10	0.06740	0.06741	0.00342	0.00341
$k_2 = 2$	$p_0 = 1$	0.4	20	0.13849	0.13810	0.01452	0.01434
$\mu = 6$	$p_1 = 1$	0.5	20	0.17616	0.17625	0.02358	0.02362
$\gamma = 1$		0.6	20	0.21550	0.21590	0.03540	0.03545
$\nu = 3$		0.7	20	0.25654	0.25425	0.05043	0.05046
		0.8	30	0.29904	0.30100	0.07011	0.06957
		0.9	30	0.31718	0.31770	0.10653	0.09158
$k_1 = 2$	4	0.1	10	0.03340	0.03341	0.00084	0.00084
$\lambda = 1$		0.2	10	0.06716	0.06716	0.00338	0.00339
$k_2 = 2$	$p_0 = 1$	0.4	10	0.13681	0.13685	0.01405	0.01400
$\mu = 6$	$p_1 = \frac{1}{2}$	0.5	20	0.17311	0.17370	0.02250	0.02271
$\gamma = 2$	$p_2 = \frac{1}{3}$	0.6	20	0.21060	0.21070	0.03329	0.03336
$\nu = 8$	$p_3 = \frac{1}{4}$	0.7	30	0.24940	0.24985	0.04668	0.04677
		0.8	30	0.28960	0.28940	0.06293	0.06262
		0.9	40	0.33125	0.33223	0.08236	0.08220

Table 1 Numerical results for $E_{k_1}/E_{k_2}/1$ queues with Erlang vacations (E_γ).

Inputs	N	θ	M	$E[T]$ algo.	$E[T]$ simul.	$E[\frac{T^2}{2!}]$ algo.	$E[\frac{T^2}{2!}]$ simul.
$k_1 = 2$	3	0.1	10	0.01871	0.01876	0.00040	0.00040
$\lambda = (1.0, 0.9)$		0.2	10	0.04358	0.04360	0.00219	0.00220
$q = (0.5, 0.5)$	$p_0 = 1$	0.4	10	0.10906	0.10910	0.01302	0.01276
$k_2 = 2$	$p_1 = 1$	0.5	10	0.14837	0.14845	0.02331	0.02327
$\mu = (6.0, 8.0)$	$p_3 = 1$	0.6	20	0.19129	0.19180	0.03753	0.03756
$r = (0.7, 0.3)$		0.7	20	0.23732	0.23680	0.05607	0.05582
$\gamma = 1$		0.8	20	0.28603	0.28523	0.07928	0.07924
$\nu = 3$		0.9	20	0.33707	0.33700	0.10743	0.10677
		1.0	30	0.39014	0.39380	0.14081	0.14588
$k_1 = 2$	5	0.1	10	0.01819	0.01818	0.00038	0.00037
$\lambda = (1.0, 0.9)$		0.2	10	0.04240	0.04239	0.00209	0.00205
$q = (0.5, 0.5)$	$p_0 = 1$	0.4	10	0.10655	0.10700	0.01258	0.01251
$k_2 = 2$	$p_1 = \frac{1}{2}$	0.6	20	0.18797	0.18775	0.03659	0.03625
$\mu = 6.0, 8.0)$	$p_2 = \frac{1}{3}$	0.7	20	0.23391	0.23410	0.05490	0.05506
$r = (0.6, 0.4)$	$p_3 = \frac{1}{4}$	0.8	20	0.28278	0.28340	0.07793	0.07756
$\gamma = 2$	$p_4 = \frac{1}{5}$	0.9	20	0.33424	0.33560	0.10602	0.10649
$\nu = 4$		1.0	20	0.38800	0.38855	0.13947	0.14060

Table 2 Numerical results for $H_{k_1}/H_{k_2}/1$ queues with Erlang vacations (E_γ).

				$E[T]$		$E[\frac{T^2}{2!}]$	
Inputs	N	θ	M	algo.	formula	algo.	formula
$k_1 = 1$	2	0.1	10	0.034858	0.034832	0.000931	0.000929
$\lambda = 0.5$		0.2	10	0.072754	0.072557	0.004117	0.004091
$k_2 = 2$	$p_0 = 1$	0.4	10	0.157560	0.156162	0.019649	0.019267
$\mu = 6$	$p_1 = 1$	0.5	10	0.204414	0.201840	0.033214	0.032326
$\gamma = 1$		0.6	20	0.254197	0.250000	0.051471	0.049722
$\nu = 3$		0.7	20	0.306882	0.300591	0.075058	0.071977
		0.8	20	0.362444	0.353576	0.104627	0.099624
		0.9	20	0.420868	0.408937	0.140849	0.133217
		1.0	20	0.482139	0.466667	0.184422	0.173333
$k_1 = 1$	4	0.1	10	0.034155	0.034149	0.000877	0.000877
$\lambda = 0.5$		0.2	10	0.069969	0.069927	0.003688	0.003684
$k_2 = 2$	$p_0 = 1$	0.4	10	0.146693	0.146378	0.016259	0.016194
$\mu = 6$	$p_1 = \frac{1}{2}$	0.5	10	0.187666	0.187074	0.026637	0.026483
$\gamma = 2$	$p_2 = \frac{1}{3}$	0.6	20	0.230428	0.229444	0.040191	0.039879
$\nu = 8$	$p_3 = \frac{1}{4}$	0.7	20	0.230428	0.229444	0.040191	0.039879
		0.8	20	0.321488	0.319322	0.078316	0.077383
		0.9	20	0.369881	0.366909	0.103706	0.102251
		1.0	20	0.420255	0.416327	0.133918	0.131758

Table 3 Numerical results for $M/E_{k_2}/1$ queues with Erlang vacations (E_γ).

11

STABILITY AND QUEUEING-TIME ANALYSIS OF A READER-WRITER QUEUE WITH WRITER PREFERENCE

L. C. Puryear and V. G. Kulkarni*

Software Solutions, T40/062, IBM Corp., Research Triangle Park, NC 27709

** Dept. of Oper. Res., Univ. of N. Carolina, Chapel Hill, NC 27599-3180*

ABSTRACT

A reader-writer queue manages two classes of customers: readers and writers. An unlimited number of readers can be processed in parallel; whereas writers are processed serially in First-In, First-Out (FIFO) order. Both classes arrive according to Poisson processes. Reader and writer service times are general iid random variables. There is infinite room in the queue for waiting customers.

The writer-preference scheduling discipline gives writers non-preemptive prior- ity over readers. A customer arriving to an empty system begins service immediately. Once writers begin service, they are served exhaustively; i.e. until none remain in the system. The first writer to arrive to a system serving readers must wait until those readers complete service. Additional reader arrivals queue up.

This system is analyzed to produce a stability condition and Laplace-Stieltjes Transforms (LSTs) for the steady state queueing times of readers and writers. An example and simulation results are also given.

Keywords: reader-writer queue, writer preference, M/G/1 queue, M/G/∞ queue, semi-Markov process, concurrency control, locking.

1 INTRODUCTION

A reader-writer queue manages two classes of customers: readers and writers. No other customer can receive service while a writer is being processed, but multiple readers can be processed simultaneously.

Applications of the reader-writer queueing model are found in computer systems, in which multiple users require read and write access to a given resource. To maintain resource integrity, i.e. preserve all updates to the resource, the following policy is adopted. Readers, or customers that access the resource without changing its state, can concurrently use the resource. A writer, or customer that can change the state of the resource, must be accorded exclusive use of the resource. Consequently, a writer cannot be processed with another writer or reader. These scheduling constraints are enforced by using exclusive and non-exclusive locks. (Non-exclusive locks are also known as shared locks.)

A replicated database can be thought of as one or more shared resources. Database replication enables more than one read query, for the same resource, to be processed at once, with the level of parallelism limited by the number of replications. This is accomplished by placing copies of the database on independent Direct Access Storage Devices (DASDs). To maintain synchronization of the copies, any given write query must be applied to each database image. Since this processing proceeds in parallel, there is negligible overhead due to replication.

In order to further increase parallelism, the database can be partitioned into N resources or items, with each customer request (or transaction) requiring one or more of these items. The lock granularity in this case is N. The minimum granularity is attained when $N = 1$ and the entire database is the single item. (The minimum granularity is assumed in this paper.) While higher granularity can result in greater concurrency, it can also induce more overhead, in the form of more processing and larger lock tables. See Ullman [24] for further details.

Shared computer memory can also be viewed as one or more resources. The case in which this memory is accessed by multiple processors is logically equivalent to that of the replicated database. Here the servers are processors, rather than DASDs.

In this paper, a simplified reader-writer queueing system, operating under the writer-preference scheduling discipline, is considered. Some characteristics of the above applications are altered for the sake of tractability. Specifically, readers and writers arrive according to Poisson processes, with infinite waiting room. An unlimited number of readers can be processed simultaneously. However, only one writer can be processed at a time. (No readers can receive service while a writer is in service.)

The writer-preference scheduling discipline is such that writers are accorded non-preemptive priority over readers. Thus an arriving writer is served immediately if the system is idle at the time of arrival. Otherwise, writers are served in FIFO order when all of the readers in service at the time of the first writer arrival have finished. Readers can begin service only when no writers are in the system.

Mitra and Weinberger [15] and Mitra [16] consider stochastic models of database systems. They assume that the database is divided into N items. They make the important assumption that if the requested items are not available immediately, the query terminates. Thus there is no queue of waiting queries.

Several papers analyze reader-writer queues in which writers have priority over readers: Coffman et al. [4] and [3], Tay et al. [22], Nelson and Iyer [17], Baccelli et al. [2] and [1], Johnson [10], and Reiman and Wright [19]. Coffman et al. [4] assume writers have preemptive priority over an infinite supply of readers. Also, a finite degree of parallelism, N, is assumed. The Laplace transform for the effective reader service-time distribution is obtained. See also Coffman et al. [3] in which the number of servers (or replications) is optimized. In another paper, Baccelli and Coffman [1], replace the saturated reader queue with Poisson reader arrivals. The steady-state expected reader waiting time is derived and used to optimize N. Nelson and Iyer [17] analyze a replicated database system in which both readers and writers arrive according to Poisson processes and the writer priority is non-preemptive. Processing times are exponential for readers and phase-type for writers. Average delay is used to compute the optimal number of copies. Tay et al. [22] approximate a reader-writer system by assuming that the probability of more than one waiting customer is negligible.

The stability of reader-writer queueing systems is examined in several papers. Courcoubetis et al. [5] assume N heterogeneous servers. That is, requests require a specific server and each server can have a distinct service rate. The case of N homogeneous servers is analyzed by Baccelli et al. [2], who develop a stability condition for a more general setting: the arrival and service processes are only assumed to be ergodic and stationary with no independence stipulation. Reiman and Wright [19] assume an infinite number of homogeneous servers, Poisson reader arrivals, renewal process writer arrivals, and general service-time distributions. Priority scheduling disciplines are addressed by none of these papers. Courcoubetis and Reiman [6] assume the reader-writer queue is saturated with readers and derive the form of optimal scheduling disciplines.

Thomasian and Nicola [23] compare the stability and performance measures of various scheduling disciplines for reader-writer queues with N homogeneous servers. Expressions for the overall throughput are obtained for the First-Come, First-Served (FCFS), alternating-exhaustive, and threshold-fastest-emptying disciplines. With the alternating-exhaustive scheduling discipline, a given class of customer is served exhaustively (i.e. until none remain in the system) before beginning service for the other class. The threshold-fastest-emptying discipline gives non-preemptive priority to writers when K of them have accumulated.

Omahen and Schrage [18] examine a system similar to a reader-writer queue with degree of parallelism two. Five scheduling disciplines are addressed: FCFS, preemptive reader priority, non-preemptive reader priority, dynamic (i.e. priority is granted under certain conditions), and preemptive writer priority.

Kulkarni and Puryear [12] and [13] analyze the reader-writer queue under the reader-preference and alternating-exhaustive scheduling disciplines, respectively. Under the reader-preference discipline, readers are given non-preemptive priority over writers.

In Section 2, the stochastic model is described and a Semi-Markov Process (SMP) is constructed to establish a sufficient stability condition. In Section 3, the steady-state queueing-time results are developed. The final section includes an example and simulation results.

2 MODEL AND STABILITY ANALYSIS

Readers arrive according to a Poisson process of rate λ. Their processing times are iid with distribution $G(\cdot)$ and mean τ. Writers arrive according to a Poisson process with rate θ and have iid service-time distribution $H(\cdot)$ with mean γ. Writers have a non-preemptive priority over readers. The system is said to be in 'read mode' when serving readers and in 'write mode' when serving writers.

Let $X(t)$ and $Y(t)$ be the number of readers and writers, respectively, in the system at time t. The process $\{(X(t), Y(t)), t \geq 0\}$ is regenerative and regenerates whenever the system enters state $(0,0)$. Figure 1 illustrates typical sample paths of the $X(t)$ and $Y(t)$ processes during the regenerative cycle. A stability condition for $\{(X(t), Y(t)),\ t \geq 0\}$ will be derived by constructing an SMP, $\{Z(t),\ t \geq 0\}$, from the original system, such that the stability of the SMP implies the stability of the regenerative process. The stability of the SMP will be established by showing that the embedded Discrete-Time Markov Process (DTMC), $\{Z_n,\ n \geq 0\}$, is positive recurrent and has finite limiting expectation. First, a preliminary result is developed.

Suppose that a write mode has ended at time zero with maximum pending reader service requirement z. Also assume the first writer arrival occurs at time t. Let $\hat{T}(t)$ be the time this first writer has to wait to enter service, and let

$$\begin{aligned} F(x \mid z, t) &= P\{\hat{T}(t) \leq x \mid \\ &\quad \text{Max. initial pending reader service requirement } z\}. \end{aligned} \tag{2.1}$$

The next lemma gives an expression for $F(x \mid z, t)$.

Lemma 2.1

$$F(x \mid z, t) = I_z(x + t)Q(x, t), \tag{2.2}$$

where

$$Q(x, t) = \exp(-\lambda \int_x^{x+t} (1 - G(u))\, du) \tag{2.3}$$

and for $z \in (-\infty, \infty)$,

$$I_z(t) = \begin{cases} 0 & : \quad z > t \\ 1 & : \quad z \leq t. \end{cases} \tag{2.4}$$

Proof: As above, assume that a write mode has ended at time zero with z the maximum pending reader service requirement, and t the time of the first writer arrival. Let $Q(\cdot, t)$ denote the distribution of the maximum residual service requirement of the readers arriving in $[0, t]$ and present at time t. In other words, $Q(\cdot, t)$ is the distribution of the time from t until all readers arriving in $[0, t]$ and present at time t have left the system. Define $R(t)$ to be the number of reader arrivals up to time t. Given that $R(t) = n$, let $U_1, U_2, \ldots, U_n$ be iid arrival times from distribution $U_t(\cdot)$, of the n readers, where $U_t(\cdot)$ is the uniform distribution over $[0, t]$ (See Scheaffer and Mendenhall [21]). Also let $S_1, S_2, \ldots, S_n$ be the iid service times from distribution $G(\cdot)$, of the n readers.

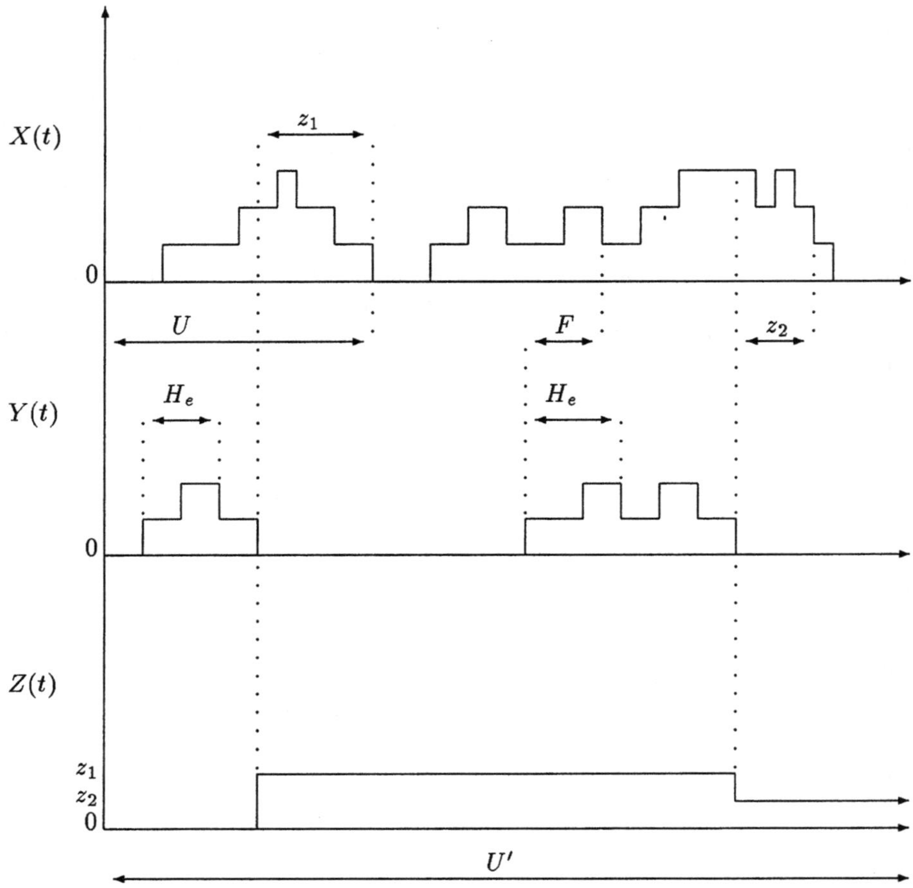

Figure 1 Regenerative Cycles Initiated by Reader and Writer, Respectively

Let $*$ be the convolution operator. Then

$$F(x \mid z, t) \;\; = \;\; \sum_{n=0}^{\infty} P\{\hat{T}(t) \leq x \mid \text{Maximum initial pending reader}$$

$$
\begin{aligned}
& \text{service requirement } z, R(t) = n\}e^{-\lambda t}\frac{(\lambda t)^n}{n!} \\
= & \sum_{n=0}^{\infty} P\{z - t \le x, U_1 + S_1 - t \le x, U_2 + S_2 - t \le x, ..., \\
& U_n + S_n - t \le x\}e^{-\lambda t}\frac{(\lambda t)^n}{n!} \\
= & \sum_{n=0}^{\infty} I_z(x+t)(U_t * G(t+x))^n e^{-\lambda t}\frac{(\lambda t)^n}{n!} \\
= & I_z(x+t)\exp(-\lambda t(1 - U_t * G(t+x))) \\
= & I_z(x+t)\exp(-\lambda t(1 - \frac{1}{t}\int_0^t G(t+x-u)\,du)) \\
= & I_z(x+t)\exp(-\lambda \int_0^t (1 - G(t+x-u))\,du) \\
= & I_z(x+t)\exp(-\lambda \int_x^{x+t} (1 - G(u))\,du) \\
= & I_z(x+t)Q(x,t).
\end{aligned}
$$

The above yields the result. □

Note that $F(0 \mid z,t) > 0$, i.e. there is a positive probability that the first writer arrival does not wait at all. Now define a process $\{Z(t),\, t \ge 0\}$ with piecewise constant sample paths, on state space $[0,\infty)$. Let $T_1, T_2, ...$ be the epochs at which the write modes end and let z_n be the maximum service requirement of readers in the system at time T_n. Define $\{Z(t),\, t \ge 0\}$ as

$$Z(t) = z_n, \quad T_n \le t < T_{n+1}. \tag{2.5}$$

The next theorem shows that $\{Z(t),\, t \ge 0\}$ is an SMP. First, $\{Z(t),\, t \ge 0\}$ has been defined to have piecewise constant paths. Suppose $Z(t)$ has undergone a transition at time zero. Let T be the time of the next transition. Define

$$D(x,y \mid z) = P\{T \le x, Z(T+) \le y \mid Z(0+) = z\}, x, y, z \ge 0. \tag{2.6}$$

Let $H_e(\cdot \mid z,t)$ be the convolution of $F(\cdot \mid z,t)$ with $H(\cdot)$. Also let $C_e(\cdot \mid z,t)$ be the distribution of the busy period, initiated by a single customer, in an M/G/1 queue with Poisson arrivals of rate θ, exceptional first-service-time distribution $H_e(\cdot \mid z,t)$, and regular service-time distribution $H(\cdot)$.

Theorem 2.1 Process $\{Z(t),\ t \geq 0\}$ is an SMP with state space $[0, \infty)$ and with kernel given by

$$D(x, y \mid z) \;=\; \int_0^x \int_0^{x-t} e^{-\lambda u(1-G(y))}\, dC_e(u \mid z, t)\theta e^{-\theta t}\, dt, x, y, z \geq 0.$$

Proof: The semi-Markov nature of $\{Z(t), t \geq 0\}$ follows from its construction. Assume process $\{Z(t), t \geq 0\}$ enters state z at time zero and consider $D(x, y \mid z)$ with $x, y, z \geq 0$. Suppose the first writer arrives at time t. This writer enters service at time $t + \hat{T}(t)$, given state z was entered at time zero. The write mode ends when an M/G/1 queue busy period (with Poisson arrivals of rate θ, exceptional first-service-time distribution $H_e(\cdot \mid z, t)$, and regular service-time distribution $H(\cdot)$), starting at time t, ends. Conditioning on the time of the first writer arrival, and applying the above, yields the following.

$$\begin{aligned}
D(x, y \mid z) &= \int_0^x P\{T \leq x, Z(T+) \leq y \mid Z(0+) = z, \text{First writer arrives} \\
&\quad \text{at time } t\}\theta e^{-\theta t}\, dt \\
&= \int_0^x \int_0^{x-t} P\{Z(T+) \leq y \mid \text{Next transition is at time } t + u\} \times \\
&\quad dC_e(u \mid z, t)\theta e^{-\theta t}\, dt \\
&= \int_0^x \int_0^{x-t} P\{Z(T+) \leq y \mid n \text{ readers waiting at the end} \\
&\quad \text{of the sojourn}\} \left[\sum_{n=0}^{\infty} e^{-\lambda u} \frac{(\lambda u)^n}{n!}\right] dC_e(u \mid z, t)\theta e^{-\theta t}\, dt \\
&= \int_0^x \int_0^{x-t} \left[\sum_{n=0}^{\infty} e^{-\lambda u} \frac{(\lambda u G(y))^n}{n!}\right] dC_e(u \mid z, t)\theta e^{-\theta t}\, dt \\
&= \int_0^x \int_0^{x-t} e^{-\lambda u(1-G(y))}\, dC_e(u \mid z, t)\theta e^{-\theta t}\, dt.
\end{aligned}$$

□

Next a stability condition for $\{Z(t),\ t \geq 0\}$ is derived. For this, the stability of the embedded DTMC, $\{Z_n,\ n \geq 0\}$, is studied, where Z_n is the state of SMP $\{Z(t),\ t \geq 0\}$ after the nth transition. The limiting 'expected value' of Z_n is also shown to be finite.

Define

$$C_e(\cdot \mid z) = \int_0^{\infty} C_e(\cdot \mid z, t)\theta e^{-\theta t}\, dt, \tag{2.7}$$

and $\tilde{C}_e(s \mid z)$ to be the LST of $C_e(\cdot \mid z)$. Using a first-step analysis (see Kulkarni [11]),

$$\tilde{C}_e(s \mid z) = \tilde{H}_e(s + \theta(1 - \tilde{C}(s)) \mid z), \tag{2.8}$$

where $\tilde{C}(s)$ is the LST of the distribution for the length of a busy period, initiated by a single customer, in an M/G/1 queue with rate θ Poisson arrivals and common service-time distribution $H(\cdot)$. The transition probabilities for $\{Z_n, n \geq 0\}$ are given by

$$P\{Z_{n+1} \leq y \mid Z_n = z\} = \tilde{C}_e(\lambda(1 - G(y)) \mid z), y, z \geq 0. \tag{2.9}$$

Let σ^2 be the variance of $G(\cdot)$.

Theorem 2.2 Assume $\theta\gamma < 1$ and $\sigma^2 < \infty$. Then Z_n converges in distribution to Z as n tends to infinity and $E(Z) < \infty$.

Proof: The proof is similar to the corresponding proof in Kulkarni and Puryear [13]. □

Theorem 2.2 is next used to infer positive recurrence of the SMP.

Theorem 2.3 SMP $\{Z(t), t \geq 0\}$ is positive recurrent if $\theta\gamma < 1$ and $\sigma^2 < \infty$.

Proof: The proof is similar to the corresponding proof in Kulkarni and Puryear [13]. □

The above theorem can now be used to prove the following.

Theorem 2.4 The bivariate process $\{(X(t), Y(t)), t \geq 0\}$ is stable if $\theta\gamma < 1$ and $\sigma^2 < \infty$.

Proof: It will now be shown that the bivariate process $\{(X(t), Y(t)), t \geq 0\}$ is stable if SMP $\{Z(t), t \geq 0\}$ is stable. Assume the SMP has entered state zero at time zero, which is then a regeneration time for $\{(X(t), Y(t)), t \geq 0\}$. Let U be the time until the next regeneration, and let U' be the time until the SMP returns to state zero. Since $\{(X(t), Y(t)), t \geq 0\}$ may regenerate during a nonzero state of the SMP, $U \leq U'$. If the conditions for positive recurrence in the SMP are satisfied, $E(U')$, and hence $E(U)$, is finite. □

3 QUEUEING TIMES

In this section, the LSTs of the steady-state queueing times for readers and writers are computed. The stability conditions from Theorem 2.4 are assumed to be satisfied; i.e. $\theta\gamma < 1$ and $\sigma^2 < \infty$.

First the embedded DTMC $\{Z_n,\ n \geq 0\}$ is used in the following theorem to derive the limiting probabilities, $\Phi(z) = \lim_{t\to\infty} P\{Z(t) \leq z\}, z \geq 0$. Let $\mu(z)$ be the expected sojourn time of the SMP in state $z \geq 0$, and $\pi(z) = \lim_{n\to\infty} P\{Z_n \leq z\}, z \geq 0$. Also let $F(\cdot \mid z, t)$ be as in (2.0). Define

$$F(\cdot \mid z) = \int_0^\infty F(\cdot \mid z, t)\theta e^{-\theta t}\, dt \tag{3.1}$$

and let γ_z be the first moment of $F(\cdot \mid z)$.

Theorem 3.1 Assume $\theta\gamma < 1$ and $\sigma^2 < \infty$. Then for $z \geq 0$,

$$\Phi(z) = \frac{\int_0^z \mu(x)\, d\pi(x)}{\int_0^\infty \mu(x)\, d\pi(x)} \tag{3.2}$$

where

$$\mu(z) = \frac{1 + \theta\gamma_z}{\theta(1 - \theta\gamma)}. \tag{3.3}$$

Proof: The result follows from SMP theory (Ross [20]). □

Let W_r and W_w be the steady-state queueing times (i.e. service time is excluded) of the readers and writers, respectively.

3.1 Reader Queueing Times

Theorem 3.2 Assume $\theta\gamma < 1$ and $\sigma^2 < \infty$. Then the LST of the steady-state queueing time for readers is given by

$$E(e^{-sW_r}) = \int_0^\infty \left[\frac{(s + \theta(1 - \tilde{C}_e(s \mid z)))(1 - \theta\gamma)}{s(1 + \theta\gamma_z)}\right] d\Phi(z). \tag{3.4}$$

Proof: Consider an arbitrary reader arriving to the system. Let Z be the state of SMP $\{Z(t),\ t \geq 0\}$ as seen by this arriving customer. Conditioning upon Z yields

$$E(e^{-sW_r}) = \int_0^\infty E(e^{-sW_r} \mid Z = z)\, d\Phi(z). \tag{3.5}$$

If there are no writers in the system when the reader arrives, then $W_r = 0$ and $E(e^{-sW_r} \mid Z = z) = 1$. On the other hand, if there are writers in the system when this reader arrives, it must wait until no writers remain in the system. This queueing time is the remaining lifetime associated with an M/G/1 queue busy period with Poisson arrivals of rate θ, exceptional first-service-time distribution $H_e(\cdot \mid z)$, and regular service-time distribution $H(\cdot)$. $H_e(\cdot \mid z)$ is given by

$$H_e(x \mid z) = \int_0^\infty H_e(x \mid z, t)\theta e^{-\theta t}\, dt. \tag{3.6}$$

Combining these two cases,

$$\begin{aligned} E(e^{-sW_r} \mid Z = z) \quad &= \quad P\{\text{No writers in the system} \mid Z = z\} + \\ &\qquad \{\text{LST for remaining lifetime of M/G/1 busy period} \\ &\qquad \text{with exceptional first service} \mid Z = z\} \times \\ &\qquad P\{\text{Writers present in the system} \mid Z = z\} \\ &= \quad \frac{\frac{1}{\theta}}{\mu(z)} + \frac{(1 - \tilde{C}_e(s \mid z))(1 - \theta\gamma)}{s(\gamma + \gamma_z)}(1 - \frac{\frac{1}{\theta}}{\mu(z)}) \\ &= \quad \frac{1 - \theta\gamma}{1 + \theta\gamma_z} + \frac{(1 - \tilde{C}_e(s \mid z))(1 - \theta\gamma)}{s(\gamma + \gamma_z)} \frac{\theta(\gamma + \gamma_z)}{1 + \theta\gamma_z}. \end{aligned} \tag{3.7}$$

The result follows. □

Now let ${\alpha_z}^2$ be the second moment of $F(\cdot \mid z)$ and define α^2 to be the second moment of $H(\cdot)$.

Corollary 3.1 The expected value of W_r is given by

$$E(W_r) = \int_0^\infty \frac{\theta}{2(1 + \theta\gamma_z)(1 - \theta\gamma)} \left[2\gamma\gamma_z + {\alpha_z}^2 + \alpha^2 + \frac{\theta\alpha^2(\gamma + \gamma_z)}{1 - \theta\gamma}\right] d\Phi(z). \tag{3.8}$$

Proof: Taking the derivative of the formula for $E(e^{-sW_r} \mid Z = z)$ given in (3.7) yields

$$E(W_r \mid Z = z) = \frac{\tilde{C}_e''(0 \mid z)}{-2\tilde{C}_e'(0 \mid z)} \frac{\theta(\gamma + \gamma_z)}{1 + \theta\gamma_z}. \tag{3.9}$$

Now taking the first and second derivatives of (2.8), the first and second moments of $C_e(\cdot \mid z)$ can be obtained. These are given by the following.

$$\begin{aligned} -\tilde{C}_e'(0 \mid z) &= \frac{-\tilde{H}_e'(0 \mid z)}{1-\theta\gamma}, \\ \tilde{C}_e''(0 \mid z) &= \frac{\tilde{H}_e''(0 \mid z)}{(1-\theta\gamma)^2} + \frac{-\theta\tilde{H}_e'(0 \mid z)\alpha^2}{(1-\theta\gamma)^3} \end{aligned}$$

where

$$\begin{aligned} -\tilde{H}_e'(0 \mid z) &= \gamma + \gamma_z, \\ \tilde{H}_e''(0 \mid z) &= 2\gamma\gamma_z + \alpha^2 + {\alpha_z}^2. \end{aligned}$$

Combining the above yields the result. □

3.2 Writer Queueing Times

The corresponding results for the queueing times of writers are now presented. Let $\tilde{\Omega}(s \mid z)$ be the LST of the steady-state queueing time distribution for the M/G/1 queue with exceptional first-service-time distribution $H_e(\cdot \mid z)$ (See (3.6) for the definition of $H_e(\cdot \mid z)$).

Theorem 3.3 Assume $\theta\gamma < 1$ and $\sigma^2 < \infty$. Then the LST of the steady-state queueing time for writers is given by

$$E(e^{-sW_w}) = \int_0^\infty \left[\tilde{\Omega}(s \mid z) + (\tilde{F}(s \mid z) - 1)\left(\frac{1-\theta\gamma}{1+\theta\gamma_z}\right)\right] d\Phi(z), \tag{3.10}$$

where

$$\tilde{\Omega}(s \mid z) = \frac{(1-\theta\gamma)(\theta\tilde{H}(s)[\tilde{F}(s \mid z) - 1] - s)}{(1+\theta\gamma_z)(\theta[1-\tilde{H}(s)] - s)}. \tag{3.11}$$

Proof: Consider an arbitrary writer arriving to the system. Let Z be the state of the SMP as seen by this arrival. Condition upon Z to get

$$E(e^{-sW_w}) = \int_0^\infty E(e^{-sW_w} \mid Z = z)\, d\Phi(z). \tag{3.12}$$

If the writer arrives to a system with no other writers present, it must wait for any readers currently in service to complete. Otherwise, if the writer arrives to a system with writers already present, it must wait until these predecessor writers have left the system. This queueing time is the same as that for a customer arriving to a busy M/G/1 queue with exceptional first-service-time distribution $H_e(\cdot\,|\,z)$. Thus

$$\begin{aligned}
E(e^{-sW_w}\,|\,Z=z) &= \tilde{F}(s\,|\,z)P\{\text{No writers in the system}\,|\,Z=z\}+ \\
&\quad \tilde{\Omega}(s\,|\,z,\text{Arrival must wait})\times \\
&\quad P\{\text{Writers present in the system}\,|\,Z=z\} \\
&= \tilde{F}(s\,|\,z)\frac{1-\theta\gamma}{1+\theta\gamma_z}+\tilde{\Omega}(s\,|\,z)- \\
&\quad \tilde{\Omega}(s\,|\,z,\text{No wait})P\{\text{No writers in the system}\,|\,Z=z\} \\
&= (\tilde{F}(s\,|\,z)-1)\left(\frac{1-\theta\gamma}{1+\theta\gamma_z}\right)+\tilde{\Omega}(s\,|\,z). \qquad (3.13)
\end{aligned}$$

The formula for $\tilde{\Omega}(s\,|\,z)$ is given in Welch [25]. The result follows. □

Corollary 3.2 The expected value of W_w is given by

$$E(W_w)=\int_0^\infty\left[\frac{\theta\alpha^2}{2(1-\theta\gamma)}+\frac{2\gamma_z+\theta{\alpha_z}^2}{2(1+\theta\gamma_z)}\right]d\Phi(z). \qquad (3.14)$$

Proof: Taking the derivative of the formula for $E(e^{-sW_w}\,|\,Z=z)$ given in (3.13) gives

$$E(W_w\,|\,Z=z)=-\tilde{F}'(0\,|\,z)\frac{1-\theta\gamma}{1+\theta\gamma_z}-\tilde{\Omega}'(0\,|\,z), \qquad (3.15)$$

where

$$-\tilde{\Omega}'(0\,|\,z)=\frac{\theta(2\gamma\gamma_z+{\alpha_z}^2)}{2(1+\theta\gamma_z)}+\frac{\theta\alpha^2}{2(1-\theta\gamma)}. \qquad (3.16)$$

The result follows. □

The LST of the queueing time for an arbitrary arrival is given by

$$\frac{\lambda}{\lambda+\theta}E(e^{-sW_r})+\frac{\theta}{\lambda+\theta}E(e^{-sW_w}) \qquad (3.17)$$

and the average queueing time is given by

$$\frac{\lambda}{\lambda+\theta}E(W_r)+\frac{\theta}{\lambda+\theta}E(W_w). \qquad (3.18)$$

4 EXAMPLE AND SIMULATION RESULTS

In this section, a specific example is considered, with numerical results presented graphically. The section is concluded with more general simulation results.

For the example, assume that reader service times are a constant τ, and writer service times are exponentially distributed with mean γ. The reader arrival rate λ and writer arrival rate θ are left as parameters.

SMP $\{Z(t),\, t \geq 0\}$ in this case has state space $\{0, \tau\}$. It is assumed here that $\tau = \gamma = 1$. The LSTs $\tilde{F}(s \,|\, z)$ and $\tilde{C}(s)$ are as follows:

$$\begin{aligned} \tilde{F}(s \,|\, \tau) &= \frac{s e^{-(\lambda+\theta)\tau} - (\lambda+\theta) e^{-s\tau}}{s - \lambda - \theta}, \\ \tilde{F}(s \,|\, 0) &= \frac{\theta}{\lambda+\theta} + \frac{\lambda}{\lambda+\theta} \tilde{F}(s \,|\, \tau), \\ \tilde{C}(s) &= \frac{1 + \gamma(s+\theta) - \sqrt{(1+\gamma(s+\theta))^2 - 4\theta\gamma}}{2\theta\gamma}. \end{aligned}$$

Letting $s(\lambda, \theta) = \lambda + \theta(1 - \tilde{C}(\lambda))$, and using (2.9) and (2.8), the transition probability matrix for embedded DTMC $\{Z_n,\, n \geq 0\}$ is

$$\begin{bmatrix} \frac{\tilde{F}(s(\lambda,\theta) \,|\, 0)}{1+\gamma s(\lambda,\theta)} & 1 - \frac{\tilde{F}(s(\lambda,\theta) \,|\, 0)}{1+\gamma s(\lambda,\theta)} \\ \frac{\tilde{F}(s(\lambda,\theta) \,|\, \tau)}{1+\gamma s(\lambda,\theta)} & 1 - \frac{\tilde{F}(s(\lambda,\theta) \,|\, \tau)}{1+\gamma s(\lambda,\theta)} \end{bmatrix}.$$

The unnormalized steady-state solution is given by

$$\pi(0) = 1, \pi(\tau) - \pi(0) = \frac{1 + \gamma s(\lambda,\theta) - \tilde{F}(s(\lambda,\theta) \,|\, 0)}{\tilde{F}(s(\lambda,\theta) \,|\, \tau)}. \tag{4.1}$$

The formulas for γ_z can be obtained by taking the first derivative of $\tilde{F}(s \,|\, z)$ to yield

$$\begin{aligned} \gamma_\tau = -\tilde{F}'(0 \,|\, \tau) &= \tau + \frac{e^{-(\lambda+\theta)\tau} - 1}{\lambda + \theta}, \\ \gamma_0 = -\tilde{F}'(0 \,|\, 0) &= \frac{\lambda}{\lambda+\theta} \gamma_\tau. \end{aligned}$$

The mean sojourn times $\mu(z)$ of SMP $\{Z(t),\, t \geq 0\}$ are given in (3.3). Now (3.2) can be used to obtain $\Phi(z)$. Then $\alpha_z{}^2$ is derived by taking the second derivative of $\tilde{F}(s \mid z)$ which yields

$$\alpha_\tau{}^2 = \tilde{F}''(0 \mid \tau) = \tau^2 + \frac{2(1-(\lambda+\theta)\tau - e^{-(\lambda+\theta)\tau})}{(\lambda+\theta)^2},$$

$$\alpha_0{}^2 = \tilde{F}''(0 \mid 0) = \frac{\lambda}{\lambda+\theta}\alpha_\tau{}^2.$$

Finally, using $\alpha^2 = 2\gamma^2$, one can apply Corollaries 3.1 and 3.2 to obtain $E(W_r)$ and $E(W_w)$.

Figures 2 and 3 show the reader, writer, and general steady-state queueing times, as a function of the mean writer processing times. Mean reader processing times of one-fourth, one, and four, and unit reader and writer arrival rates are assumed. Figures 3 and 4 show the same queueing times as a function of the writer arrival rate, assuming reader arrival rates of one-fourth, one, and four, and unit mean reader and writer processing times.

Table 1 lists the ninety-five percent confidence intervals for simulated steady-state reader and writer queueing times. Reader and writer arrival rates, λ and θ, are assumed equal to one. Mean reader and writer service times, τ and γ, are assumed equal to one half, with service time distributions variable.

A batching routine, ABATCH, was employed to compute the statistics (See Fishman [7], Fishman and Yarberry [9], Yarberry [26]). A significance level of .10 was used for testing batch independence. Random uniform numbers were obtained from the following multiplicative congruential generator

$$Z_i \equiv AZ_{i-1}(\text{mod}M), i \geq 1, \tag{4.2}$$

where modulus $2^{31} - 1$ and multiplier $A = 950706376$ are from Fishman and Moore [8]. A warmup interval of $100,000$ samples was used, followed by 9 million recorded samples. Each of ten runs took approximately thirty minutes on a 90MHz Intel Pentium processor. The programming languages used were Microsoft Fortran and Borland C++.

The 'Type' column of Table 1 uses abbreviations that are explained below. The character 'd' denotes a deterministic distribution (yielding a constant value of .5), 'x' an exponential distribution (parameter 2), 'm' a mixture of exponential distributions (parameter 3 or 1.5 with equal probability), and 'c' a convolution of two exponential distributions (parameters 6 and 3). The first character of the abbreviation applies to the reader service-time distribution and the second to the writer service-time distribution. Thus 'dx' signifies constant reader service times and exponential writer service times. The first row for each table, denoted

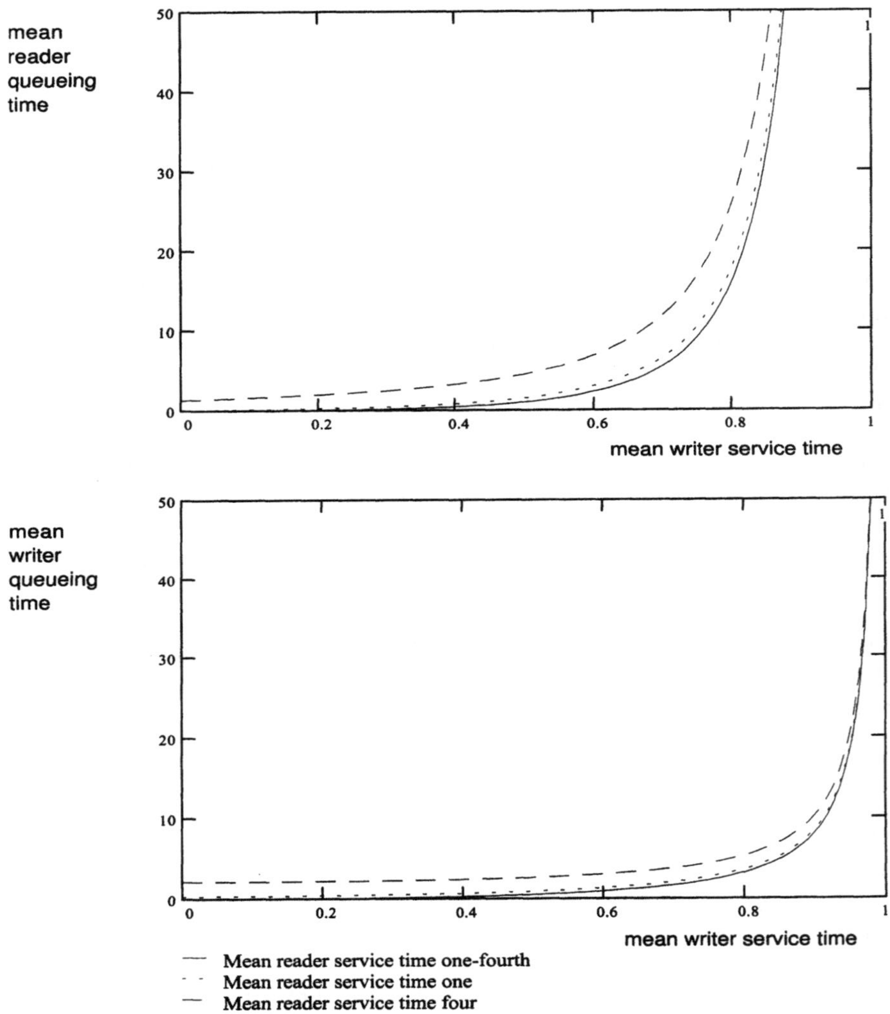

Figure 2. Mean steady state queueing times.

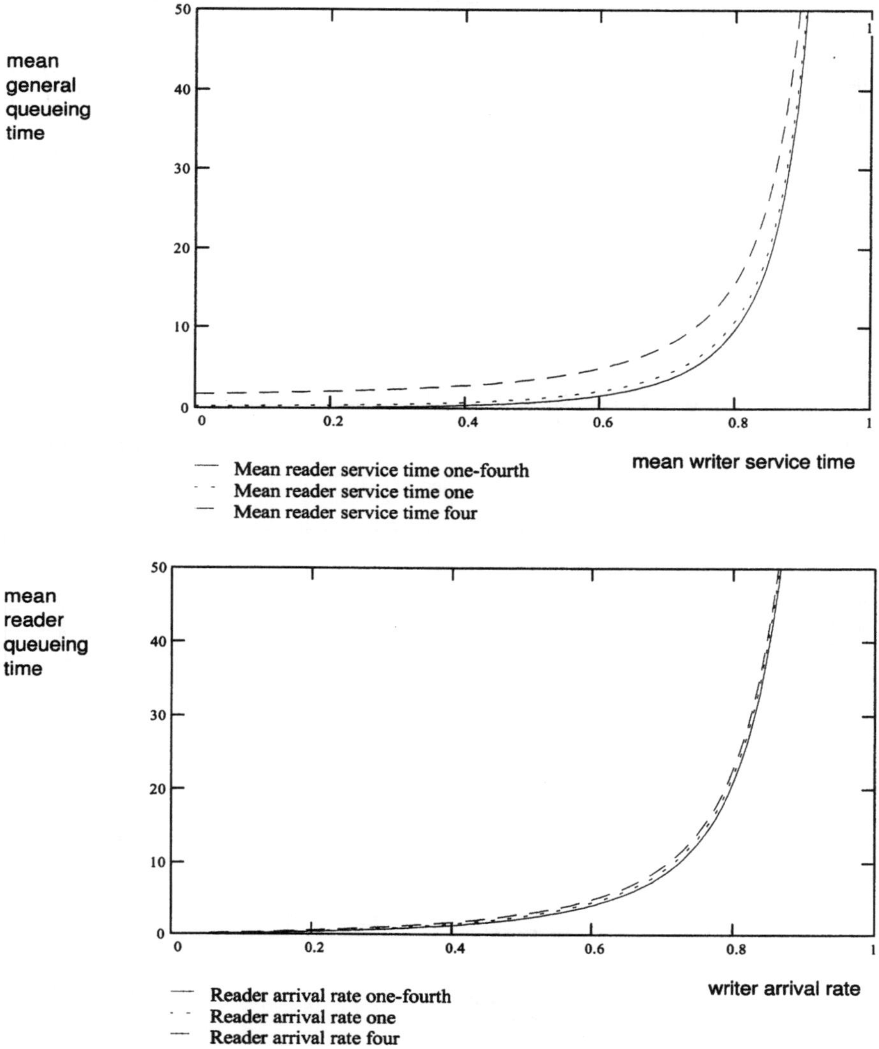

Figure 3. Mean steady state queueing times.

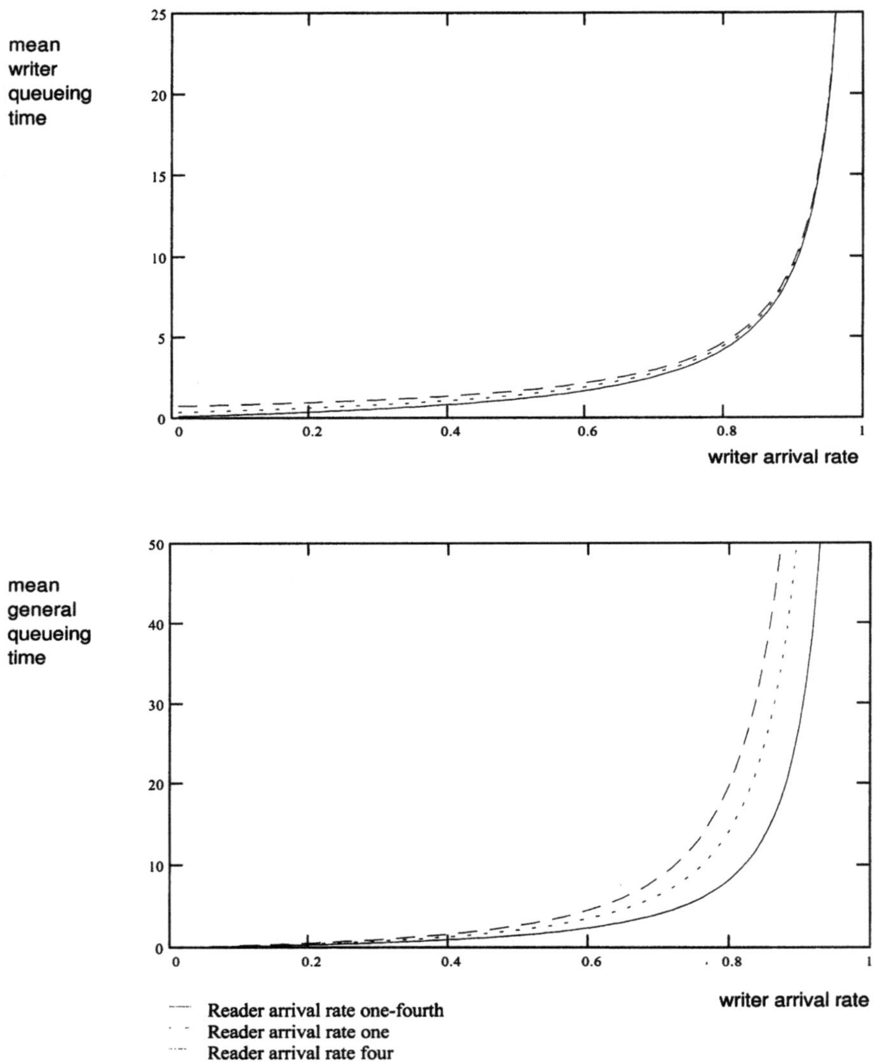

Figure 4. Mean steady state queueing times.

Service-Time Distribution	Reader	Writer
DX	1.165	.6434
dx	(1.163, 1.178)	(.6421, .6471)
cc	(1.140, 1.152)	(.6723, .6764)
cm	(1.458, 1.477)	(.8289, .8357)
cx	(1.355, 1.371)	(.7779, .7833)
mc	(1.382, 1.396)	(.8234, .8287)
mm	(1.698, 1.719)	(.9780, .9856)
mx	(1.594, 1.613)	(.9269, .9333)
xc	(1.294, 1.307)	(.7686, .7734)
xm	(1.612, 1.632)	(.9247, .9319)
xx	(1.501, 1.519)	(.8739, .8801)

Table 1 Simulated Queueing Times

by type 'DX', lists analytical results for constant reader service times ($\tau = .5$) and exponentially distributed writer service times ($\gamma = .5$). Note that the analytical values are covered by the corresponding confidence intervals. Also note that the queueing times are significantly affected by varying the service-time distributions.

REFERENCES

[1] Baccelli, F., Coffman Jr., E. G., "A Data Base Replication Analysis Using an M/M/m Queue with Service Interruptions," Performance Evaluation Review, 1983, **11**(4), pp. 102-107.

[2] Baccelli, F., Courcoubetis, C. A., Reiman, M. I., "Construction of the Stationary Regime of Queues with Locking," Stochastic Processes and their Applications, 1987, **26**, pp. 257-265.

[3] Coffman Jr., E., Gelenbe, E., Platteau, B., "Optimization of the Number of Copies in a Distributed Database," IEEE Trans. Software Eng., 1981, **7**(1), pp. 78-84.

[4] Coffman, E. G., Pollak, H. O., Gelenbe, E., Wood, R. C., "An Analysis of Parallel-read Sequential-write Systems," Performance Evaluation, 1981, **1**, pp. 62-69.

[5] Courcoubetis, C. A., Reiman, M. I., Simon, B., "Stability of a Queueing System with Concurrent Service and Locking," SIAM J. Computing, 1987, **16**(1), pp. 169-178.

[6] Courcoubetis, C. A., Reiman, M. I., "Optimal Control of a Queueing System with Simultaneous Service Requirements," IEEE Transactions on Automatic Control, 1987, **32**(8).

[7] Fishman, G. S., "Principles of the Monte Carlo Method," 1993, Dept. of Oper. Res., Univ. of N. Carolina, Chapel Hill, NC.

[8] Fishman, G. S., Moore, L. R., "An Exhaustive Analysis of Multiplicative Congruential Random Number Generators with Modulus $2^{31} - 1$," SIAM Journal on Scientific and Statistical Computation, 1986, **7**, pp. 24-45.

[9] Fishman, G. S., Yarberry, L. S., "An Implementation of the Batch Means Method," 1993, UNC/OR/TR 93-1, Dept. of Oper. Res., Univ. of N. Carolina, Chapel Hill, NC.

[10] Johnson, T., "Approximate Analysis of Reader and Writer Access to a Shared Resource," Conference proceedings, 1990 ACM Sigmetrics Conference on Measurement and Modelling of Computer Systems, Boulder, CO, 1990, pp. 106-114, ACM Press, New York.

[11] Kulkarni, V. G., "Modeling and Analysis of Stochastic Systems," 1995, Chapman and Hall.

[12] Kulkarni, V. G., Puryear, L. C., "A Reader-Writer Queue with Reader Preference," Queueing Systems Theory and Applications, 1994, **15**, pp. 81-97.

[13] Kulkarni, V. G., Puryear, L. C., "A Reader-Writer Queue with Alternating Exhaustive Priorities," Queueing Systems Theory and Applications, 1995, **19**, pp. 81-103.

[14] Meyn, S. P., Tweedie, R. L., "Markov Chains and Stochastic Stability," 1993, Springer-Verlag.

[15] Mitra, Debasis, Weinberger, P. J., "Probabilistic Models of Database Locking: Solutions, Computational Algorithms, and Asymptotics," Journal of the Association for Computing Machinery, 1984, **31**(4), pp. 855-878.

[16] Mitra, Debasis, "Probabilistic Models and Asymptotic Results for Concurrent Processing with Exclusive and Non-exclusive Locks," Society for Industrial and Applied Mathematics Journal of Computing, 1985, **14**(4), pp. 1030-1051.

[17] Nelson, R., Iyer, B., "Analysis of a Replicated Database," Performance Evaluation, 1985, **5**(3), pp. 133-148.

[18] Omahen, K., Schrage, L., "A Queueing Analysis of a Multiprocessor System with Shared Memory," Conference proceedings, Symposium on Computer-Communications Networks and Teletraffic, Polytechnic Institute of Brooklyn, 1972.

[19] Reiman, M. I., Wright, P. E., "Performance Analysis of Concurrent-Read Exclusive-Write," Performance Evaluation Review, 1991, **19**(1), pp. 168-177.

[20] Ross, S. M., "Stochastic Processes," 1982, Wiley.

[21] Scheaffer, R. L., Mendenhall, W., "Introduction to Probability Theory and Applications," 1975, North Scituate, Massachusetts.

[22] Tay, Y. C., Suri, R., Goodman, N., "A Mean Value Performance Model for Locking in Databases: The No Waiting Case," Journal of the Association for Computing Machinery, 1985, **32**(3), pp. 618-651.

[23] Thomasian, A., Nicola, V. F., "Performance Evaluation of a Threshold Policy for Scheduling Readers and Writers," IEEE Transactions on Computers, 1993, **42**(2).

[24] Ullman, J. D., "Principles of Database Systems," 1982, Second edition Computer Science Press, Rockville, Maryland.

[25] Welch, Peter D., "On a Generalized M/G/1 Queueing Process in which the First Customer of Each Busy Period Receives Exceptional Service," Operations Research, 1964, **12**, pp. 736-752.

[26] Yarberry, L. S., "Incorporating a Dynamic Batch Size "Selection Mechanism in a Fixed-Sample-Size Batch Means Procedure," Ph.D. dissertation, 1993, Dept. of Oper. Res., Univ. of N. Carolina, Chapel Hill, NC.

12

IMPORTANCE SAMPLING IN LATTICE PRICING MODELS

Soren S. Nielsen

Management Science and Information Systems
University of Texas at Austin
Austin, TX 78712
nielsen@guldfaxe.bus.utexas.edu

ABSTRACT

Binomial lattice models find widespread uses in the valuation of derivative financial instruments. When these instruments are path-dependent (non-Markovian), it is usually necessary to resort to Monte Carlo simulation. However, a crude Monte Carlo sampling of the lattice can be very inefficient since a large number of sample paths may need to be calculated.

We develop procedures for sampling the paths of a binomial lattice based on importance sampling and related methods. Such procedures can lead to a significant reduction in sample variance, but due to the very large number of sample paths, it is problematic to specify the "importance" of each path. A two-stage sampling procedure for doing this is presented, and also a method based on varying the lattice branching probabilities. The methods can be fitted to instrument characteristics, either based on prior knowledge of the instruments' payoff functions, or adaptively.

1 INTRODUCTION

Binomial lattice models represent a very general and popular class of models for the pricing and valuation of derivative instruments. When these instruments are path-independent (Markovian), each lattice node corresponds to a unique price, which can be calculated efficiently, but for path-dependent instruments, one must usually resort to Monte Carlo simulation. A set of random paths through the lattice is generated, and the price of the instrument under consideration is calculated along each path. The resulting prices are then averaged to yield an unbiased price estimate.

However, for instruments that have high variance of returns, crude Monte Carlo can be very inefficient since a large number of samples may be needed to reduce the variance of the price estimate. *Importance sampling* (see, e.g., Kleijnen and Rubinstein [1994]) is a technique to overcome this problem. The lattice is sampled non-uniformly (but still randomly), in a way that tends to generate more "important" samples, i.e., samples which result in a low variance price estimator. One problem in importance sampling is to specify a good sampling distribution *a priori*, i.e., before sampling, and to represent this distribution conveniently for a possibly very large sample space.

We exploit in this paper the structure of the recombining, binomial lattice to design such distributions, although our methods are easily applicable to the cases of higher-order or non-recombining trees. Two ideas are explored. One is a two-stage sampling procedure based on conditioning on the final states of the sample paths, and which is closely related to stratified sampling. For many financial instruments, even path-dependent ones, the final state of the underlying factor carries high "importance", or impact, on the price, and it makes sense to sample accordingly. The other idea is to specify "branching probabilities", i.e., a probability, associated with each lattice node, that the process passing through that node will branch up rather than down. Each of these possibilities leads to specific measures of "importance" which are manageable despite the potentially huge number of sample paths. A general way of automatically estimating the attributes of an instrument's payoff function which are needed to construct importance measures is presented, but we also show a way to incorporate analysts' knowledge of the payoff function into the procedure to possibly further reduce the sample variance. This involves the use of a non-linear optimization program to generate branching probabilities such that the random paths approximate "desired" sample paths. Finally, we propose a method for adaptive sampling, where branching probabilities are adjusted dynamically as more information about the underlying instrument's payoff function is learned. These sampling procedures can all be used in conjunction with standard variance-reduction techniques such as antithetic sampling and the use of a control variate.

The remainder of this paper is organized as follows. Section 2 gives an introduction to importance sampling and the use of binomial lattices, and shows how one can integrate the two. In Section 3 we present the various strategies for specifying or calculating branching probabilities leading to efficient sam-

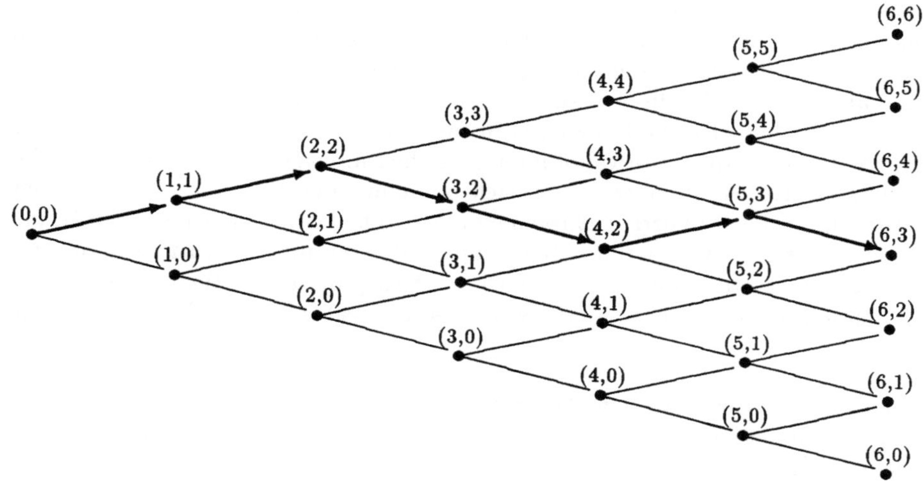

Figure 1 Binomial tree with $N = 6$ time periods. Coordinates (i, j) and the sample path $\omega = (1, 1, 0, 0, 1, 0)$ are shown.

pling. Finally, Section 4 concludes the paper. A list of symbols is given in Appendix A.

2 LATTICES AND IMPORTANCE SAMPLING

In this section we describe the use of binomial lattices for options pricing, followed by an overview of variance reduction techniques for use in sampling and simulation, with special focus on importance sampling. It is then shown how to integrate the two in a framework for importance sampling on lattices.

2.1 Lattice Pricing Models

Lattice models are used for pricing and hedging derivative instruments, such as stock options. It is usually true (or assumed) that the prices of such instruments derive from one or a few factors. For instance, the value of a stock option depends only on the price of the underlying stock through time. The underlying factors are in principle time- and space-continuous stochastic processes, but the resulting stochastic differential equations cannot always be solved analytically, and are then usually discretized in both time and state. One method of doing this for single-factor models results in the binomial lattice, Figure 1. A set of N points in the time interval of interest is chosen, usually equally spaced. The process starts out in state 0. At each point in time, the process can branch either up or down to arrive at the next state. The lattice shown is *recombining* in the sense that an equal number of up- and down-moves returns the process to the same state, no matter in which order they occurred. Hence, numbering the time points $i = 0, ..., N$, there are $i + 1$ possible states of the process at time i.

Knowledge of the state j in the grid of a process at any time i is not enough to price the derived instrument. What is missing is knowledge of the values of the underlying process that corresponds to grid nodes (i, j). In most cases it is desirable to have an arbitrage-free model, so the lattice must be *calibrated* to price observable market instruments correctly. This generally involves assuring that the lattice accurately represents (at least) the two first moments of the underlying, continuous-time process, i.e., expected value and volatility. Examples of binomial lattice models are the model by Cox, Ross and Rubinstein [1979], used for stock option pricing, and the interest rate model by Black, Derman and Toy [1988]. An overview of such models and their calibration, both for equity and fixed-income derivatives, is given in Hull [1993].

The calibration of the lattice means that certain parameters associated with each node (i, j) of the lattice must be determined:

- Branching probability: p_{ij}, the probability that a process passing through the node will branch up. Of course, in a binomial lattice, $1 - p_{ij}$ is the probability the process will branch down.
- Factor levels: F_{ij}, the level of the underlying factor associated with the node, e.g., stock prices or interest rates.

A *path* through the lattice is denoted by $\omega \in \{0, 1\}^N$, i.e., by an N-dimensional vector of 0's and 1's, where $\omega = 0$ indicates that the process branches down at time i (for $i = 0, ..., N - 1$), and $\omega = 1$ indicates that the process branches up. To price a derivative we further need a function from sample paths (or factor levels) to prices, $f : \{0, 1\}^N \mapsto \Re$, which will allow the calculation of the price,

$f(\omega)$, of the derivative instrument under path ω. Sample paths are also called *scenarios*. A list of symbols used is given in Appendix A.

The case where a derivative price P_{ij} can be determined for each lattice node (i, j) as a function of only the factor level f_{ij} at that node deserves special mention. Such instruments are called ***path-independent***, because the path taken by the process to the node (i, j) is not relevant to the price. If the price of a path-independent instruments is known at some time i for all states $j = 0, ..., i$ (for instance an European option at maturity), then risk-neutral valuation, discounting at the risk-neutral rate r,[1] dictates that, for a time discretization Δt,

$$P_{ij} = (p_{ij} P_{i+1,j+1} + (1 - p_{ij}) P_{i+1,j}) \exp(-r\Delta t),$$

so we can solve for P_{ij} backwards in time for the desired price, P_{00}. However, our focus in this paper is on ***path-dependent*** instruments, where knowledge of the whole process ***up to*** time i, not only its state ***at*** time i, is needed to determine the price at (i, j). In general, simulation is needed to price such instruments, and that is the topic of the next section.

2.2 Importance Sampling

Consider the problem of computing, or estimating, the value $\bar{f} = E\{f(\mathbf{x})\}$, i.e., the expectation of a function f with respect to a random variable $\mathbf{x}$ defined on some probability space $(\Omega, \mathcal{S}, p)$, where f is $\mathcal{S}$-measurable. For the purposes of this paper we assume that f is a positive, real-valued function and that $\mathbf{x}$ is finite and discrete (and can hence take $\mathcal{S} = 2^{\Omega}$). Then we can write the expectation as

$$\bar{f} = \sum_{x \in \Omega} p(x) f(x),$$

where $p(x)$ is the probability of x. In cases where the size of the sample space is very large, this expression cannot be calculated directly, but an unbiased estimate of $\bar{f}$ can be obtained by generating S independent, random samples, $x^1, x^2, ..., x^S$, from Ω, by

$$\hat{f} = \frac{1}{S} \sum_{s=1}^{S} f(x^s).$$

This estimator has variance

$$\mathrm{var}\{\hat{f}\} = \frac{1}{S} \mathrm{var}\{f(x)\},$$

[1] **If the factor modeled is short rates, the appropriate discount factor is the rate associated with each lattice node.**

where we can of course estimate $\text{var}\{f(x)\}$ by its sample variance,

$$\frac{1}{S-1}\sum_{s=1}^{S}(f(x^s)-\hat{f})^2$$

(by "random samples" we mean that the probability of obtaining a sample x is $p(x)$).

The above method is known as *crude Monte Carlo* sampling. It is usually very robust and easy to implement, but since the standard deviation of $\hat{f}$ is of the order $S^{-1/2}$, the desire for a small confidence interval for $\hat{f}$ could necessitate a large sample size, which could be computationally expensive. One standard method for reducing the variance in pricing models is the use of a *control variate*, which is another function $g(x)$ with known expectation $\bar{g}$, and which is positively correlated with f. By sampling instead of $f(x)$ the difference $f(x) - g(x)$, we obtain (by adding back $\bar{g}$) an estimate for f with (possibly) reduced variance. For pricing, this means finding another instrument which behaves similarly to the instrument under consideration, but for which exact pricing formulas are known (note that the "instruments" are represented by f and g). Another standard technique is *antithetic* sampling. Here, one looks for a bijection b on the sample space such that whenever $f(x)$ is "high", $f(b(x))$ is "low", and extend the sampling procedure to use $\frac{1}{2}(f(x) + f(b(x)))$ for each sample value. Similarly to the control variate technique, the argument is that the two values are negatively correlated, so the resulting sample variance is reduced. Boyle [1977] demonstrated that these techniques can lead to a significant reduction in variance.

Assume now that we sample the space Ω according to a sample distribution q instead of p (technically, $q(x)$ must be positive whenever $p(x)$ is). By this we mean that the probability of generating a sample x is $q(x)$ rather than $p(x)$. Then an unbiased estimator for $\bar{f}$ is

$$\tilde{f} = \frac{1}{S}\sum_{s=1}^{S}\frac{p(x^s)f(x^s)}{q(x^s)}. \tag{12.1}$$

The question here is how to chose q. If the true mean $\bar{f}$ were known, then chosing

$$q(x) = \frac{p(x)f(x)}{\bar{f}} \tag{12.2}$$

would result in an exact "estimate" of $\bar{f}$ for any sample size, even just $S = 1$, i.e., reducing the sample variance to zero! But, alas, not only is $\bar{f}$ what we wish to estimate, it would also be just as difficult to obtain the values $f(x)$ needed as to calculate $\bar{f}$ directly.

The significance of (12.2) lies in the fact that using a sampling distribution which is even roughly proportional to $p(x)f(x)$ (which can be thought of as the *importance* of x) can lead to a significant reduction in sample variance. When valuing a financial instrument, it is often known quite precisely under which circumstances x it will have a high or low value $f(x)$, and also which circumstances are more or less likely to occur, i.e., $p(x)$. Hence, good sampling distributions are often readily available, but it can still be a problem to provide a convenient representation of these for use in sampling, due the the (usually) huge size of Ω. The structure of lattice pricing models, however, offers some possible solutions to this problem.

For a more formal overview of variance reduction, including importance sampling, see Kleijnen and Rubinstein [1994]. Computational techniques are presented in Press et al. [1992]. An early empirical study of control variate and antithetic sampling for options pricing was given by Boyle [1977]. Very recently, Joy, Boyle and Tan [1994] showed the application of quasi-random lattice sampling with very encouraging results. Importance sampling was used in the context of stochastic programming by Dantzig and Glynn [1990] and by Infanger [1990], and in stochastic simulation by Glynn and Iglehart [1989].

2.3 Importance Sampling in Lattices

The set of possible paths through a binomial lattice grows exponentially in the number of time periods, $|\ \Omega\ | = 2^N$, and can be very large for a realistically sized lattice. It does not seem practicable to assign specific importance measures to each of these paths without some simplifying system. We explore two such simplifying ideas: Sampling by *conditioning* on final states and the assignment of *importance branching probabilities* to lattice nodes.

Sampling by conditioning on final states, ω_N, of the sample paths is motivated by the observation that the final state is often a major determinant of a financial instrument's price. For a simple European option, for instance, it is the only determinant, but even for path-dependent instruments such as barrier options, mortgage derivatives etc., the possible levels of interest rates at maturity obviously influence the price. The procedure is as follows. Let V^j be an estimate of the average price of the security on paths which terminate in state j. Such an estimate can be obtained simply by pricing a few such paths, selected randomly or judicially (See Section 3.1). In a binomial lattice with N periods there are $N+1$ final states, with $C_{Nj} = \frac{N!}{(N-j)!j!}$ paths ending in state j, $j = 0, ..., N$. The following two-stage procedure is now suggested for sampling:

1. Select a (final) state j from $\{0, ..., N\}$ according to the distribution

$$d^j = \frac{V^j C_{Nj}}{\sum_{j=0}^{N} V^j C_{Nj}}$$

   This can be done easily by obtaining a random sample from the uniform distribution $U(0,1)$, then evaluate the inverse of the cumulative distribution function of d at that point. In the discrete case, this simplifies to, e.g., a binary search in an ordered array.

2. Select a random path ω among the C_{Nj} paths which terminates in state j. This can be done by traversing the lattice from time $i = 0$ to $N-1$, setting $\omega_i = 1$ (i.e., branching up) with probability $(j - \sum_{k=0}^{i-1} \omega_k)/(N-i)$.

The probability that a specific path is selected by this procedure, i.e., the importance, is

$$q^j = \frac{V^j}{\sum_{j=0}^{N} V^j C_{Nj}} \tag{12.3}$$

where j is the final state of the path. The closer the payoff function is to being path-independent (depending only on j), the better this procedure works in reducing sample variance. For instance, for a European option that expires at N, this leads to a zero-variance "estimate" in (12.1).

The two-stage approach given above ignores the branching probabilities p_{ij} of the tree, and will work well only if the probability of branching up (or down) is close to $\frac{1}{2}$. When this is not the case, the definition of d^j can be amended to take into account that some final states have higher probability. For instance, if the up-probabilities are constant, $p_{ij} \equiv p$, it is correct to use

$$d^j = \frac{V^j b(j; N, p)}{\sum_{j=0}^{N} V^j b(j; N, p)}$$

where $b(j; N, p) = C_{Nj} p^j (1-p)^{N-j}$ is the binomial probability mass function.[2]

[2] Elsewhere we use superscript j to represent different final states, but in the formula for $b(j; N, p)$ the superscripts represent powers.

Our second approach to importance sampling in lattices is based on assigning *importance branching probabilities* q_{ij} to each node in the lattice, such that the probability that a sampled process branches up at node (i, j) is q_{ij}. The sampling probability of a path $\omega \in \{0,1\}^N$ is then given by

$$q(\omega) = \prod_{i=0}^{N-1} \left(q_{ij(i)}\omega_i + (1 - q_{ij(i)})(1 - \omega_i)\right), \tag{12.4}$$

where the notation $q_{ij(i)}$ refers to the up-probability at the node where the process ω finds itself at time i, i.e., $j(i) = \sum_{k=0}^{i-1} \omega_k$.

The procedure for sampling given the underlying process probabilities p_{ij} and the sampling probabilities q_{ij}, is to generate a set of sample paths, $\omega^s, s = 1, ..., S$. This is done by drawing, for each process and each time point, from the Bernoulli($q_{ij(i)}$)-distribution. Along the way, the two probabilities $p(\omega^s)$ and $q(\omega^s)$ are calculated. Then the derivative instrument's value under each scenario, $f(\omega^s)$, is calculated, and the estimate

$$\tilde{f} = \frac{1}{S} \sum_{s=1}^{S} \frac{p(\omega^s) f(\omega^s)}{q(\omega^s)} \tag{12.5}$$

for $\bar{f}$ is obtained. This procedure is completely general for any choice of $q(\omega)$, as long as $q(\omega) > 0$ for any path ω with $p(\omega) > 0$. Expressed in terms of path probabilities, this means that the condition

$$p_{ij} > 0 \Rightarrow q_{ij} > 0 \text{ and } p_{ij} < 1 \Rightarrow q_{ij} < 1 \tag{12.6}$$

must be satisfied. Violating this condition could result in areas of the grid with positive sample probability not being sampled.

We illustrate the use of sample branching probabilities with a simple example. Sampling randomly (Crude Monte Carlo, corresponding to chosing $q_{ij} = \frac{1}{2}$) results in sample paths which have the probability $2^{-i} C_{ij}$ of going through state j at time i. Such sampling tends to undersample extreme states, as illustrated in Figure 2. However, if we use the probabilities $q_{ij} = (j+1)/(i+2)$ of branching up, the probability of a process being in state j at time i is $1/(i+1)$, so the processes "fill" the lattice in a uniform manner, as illustrated in Figure 3. This clearly illustrates how non-uniform branching probabilities can be used to guide the sample path behavior. We point out that, although the sampling probabilities used in this example seem to lead to "uniform" coverage of the lattice, it corresponds to optimal importance sampling of an instrument which has payoff proportional to $1/C_{Nj}$ at time N, which would be a quite peculiar "straddle". We investigate in the next section some more natural possibilities for selecting sampling probabilities.

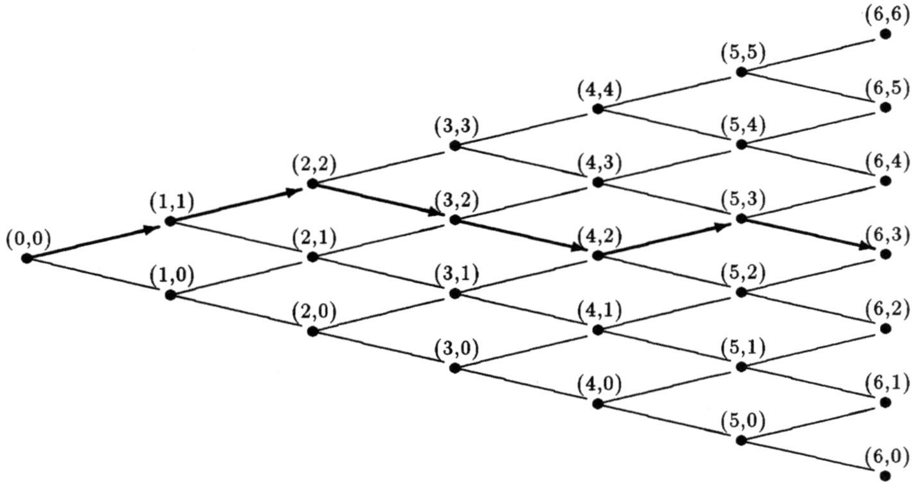

Figure 2 100 uniformly sampled paths through a lattice with $N = 36$ time steps. At each grid node is indicated the number of samples passing through the node. There is a strong centralizing tendency, and the extreme states of the lattice are undersampled, especially towards the end of the time horizon.

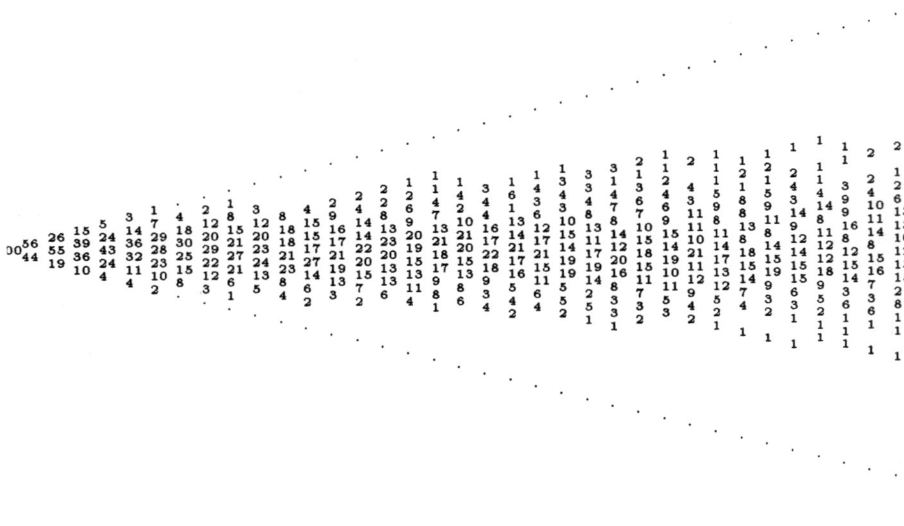

Figure 3 100 paths sampled according to the distribution $q_{ij} = \frac{j+1}{i+2}$, where q_{ij} is the probability of branching up at node (i, j).

3 SPECIFYING BRANCHING PROBABILITIES

We now present two methods of selecting importance branching probabilities associated with lattice nodes. One is based on the notion of "tracking" desirable (or important) paths, the other is an adaptive procedure which updates the branching probabilities dynamically during the simulation.

3.1 Tracking Desirable Sample Paths

Branching probabilities can be assigned independently of the instrument's payoff function, but in many cases we have some knowledge about sample paths which are particularly "important" for the instrument's value. We now give a procedure for incorporating such knowledge into the sampling procedure in a flexible way. This could be used to construct importance sampling probabilities which lead to even lower sample variances than the methods previously described.

Knowledge of the instrument's payoff function is represented by a set of probabilities, d_{ij}, which represents the proportion of sample paths that we would like to go through node (i, j). Such a set can be obtained, for example, by having an analyst draw a few "desirable" paths through a lattice, then assigning non-zero values d_{ij} to the lattice nodes where these paths pass through. We assume that $\sum_{j=0}^{i} d_{ij} = 1$ for every i, which is just a matter of scaling.

Not every possible set of d_{ij}'s is possible, in that there may not be sampling probabilities q_{ij} between 0 and 1 that lead exactly to the desirable sampling pattern. To overcome this problem, we suggest the use of the closest set of state probabilities, u_{ij}, which leads to feasible sampling probabilities. This can be found using the following non-linear tracking program:

$$\text{Minimize} \quad \sum_{i=0}^{N-1} \sum_{j=0}^{i} (d_{ij} - u_{ij})^2 \tag{12.7}$$

$$\text{Subject to} \quad u_{i,j-1}q_{i,j-1} + u_{ij}(1 - q_{ij}) = u_{i+1,j} \quad \text{for } j = 1, ..., i-1 \tag{12.8}$$

$$u_{ij}(1 - q_{ij}) = u_{i+1,j} \quad \text{for } j = 0 \tag{12.9}$$

$$u_{i,j-1}q_{i,j-1} = u_{i+1,j} \quad \text{for } j = i \tag{12.10}$$

$$\sum_{j=0}^{i} u_{ij} = 1 \quad \text{for } i = 0, ..., N-1$$

$$\epsilon \leq q_{ij} \leq 1 - \epsilon,$$

where the variables are the state probabilities u_{ij} and the sampling probabilities q_{ij}. The constraints of this program just express that the probability, u_{ij}, of being in state j at time i, which, in the objective, we are trying to get to match the desired probabilities, can be calculated by conditioning on the previous state, using the transition probabilities q_{ij}. The probabilities q_{ij} are constrained from assuming the values 0 or 1 in order to ensure that (12.6) is satisfied. Although this program is non-linear in both the objective and constraints, it appears to be easily solved using optimization software such as GAMS/Minos (See GAMS, [1988]). Naturally, one can also attempt direct solution by setting $u_{ij} = d_{ij}$ and then use the transition relations (12.8)–(12.10) to solve for q_{ij}. If this leads to a feasible solution (with respect to the bounds on q_{ij}), then that solution is also optimal.

3.2 Adaptive Sampling

As an alternative (or supplement) to specifying sampling probabilities q_{ij} before sampling, we consider now the possibility of gathering information during the sampling process which could allow us to improve these probabilities dynamically, i.e., adapt to the instrument as we learn more about its payoff function.

This is approached as an optimization problem. The objective is to find branching probabilities which would result in exact (or nearly exact) price estimates for each sample:

$$q(\omega) = \frac{p(\omega)f(\omega)}{\bar{f}}$$

for each $\omega \in \{0,1\}^N$, where $q(\omega)$ is defined in (12.4). The following non-linear program addresses this problem:

$$\text{Minimize} \quad \frac{1}{2}\sum_{\omega}\left(q(\omega) - \frac{p(\omega)f(\omega)}{\bar{f}}\right)^2 \qquad (12.11)$$

$$\text{Subject to} \quad \epsilon \leq q_{ij} \leq 1 - \epsilon, \qquad (12.12)$$

where, again, we bound the branching probabilities away from 0 and 1 by some small amount. It is easily shown that if $\tilde{f}$ is used as an estimate of $\bar{f}$, this objective is equivalent to

$$\text{Minimize } 2^{-N}\sum_{\omega}\left(\frac{p(\omega)f(\omega)}{q(\omega)} - \tilde{f}\right)^2,$$

i.e., finding branching probabilities which minimize the sample variance, which is precisely what importance sampling seeks to do.

This objective cannot be specified, because we do not wish to specify data for the complete set of possible paths, and we do not know the instrument's true price, $\bar{f}$. Still, each sample gives os pieces of this information, which can be used to improve the sample probabilities. For a specific sample, ω, the quantity

$$q(\omega) - p(\omega)f(\omega)/\bar{f} \tag{12.13}$$

is a ***stochastic gradient*** which indicates the amount by which the current $q(\omega)$ fails to be optimal with respect to (12.11), i.e., the *dual error*. Using instead of the unknown $\bar{f}$ the current average price, $\tilde{f}$, we can update $q(\omega)$ so that (12.13) becomes zero.

Recall that $q(\omega)$ is not itself a decision variable, but is defined by (12.4), so the variables which need to be updated are the q_{ij}. The gradient of some q_{ij} for a node (i,j) that ω passes through and where it branches up, is $q(\omega)/q_{ij}$ (refer to (12.4), and substitute into the objective function), and where it branches down, $q(\omega)/(1-q_{ij})$. Hence, we can update any q_{ij} along the path in this direction until (12.13) is zero, or we can update *all* the probabilities along the path, in the direction given by their gradients. It is appealing to do the latter, and to update the probabities in proportion to their gradients, but this leads to an unmanageable update. Instead, we propose to update the probabilities along the path by the same share of the error in (12.13), as follows:

$$q_{ij} \leftarrow \begin{cases} \alpha \cdot q_{ij} & \text{if } \omega_i = 1, \\ 1 - \alpha \cdot (1 - q_{ij}) & \text{if } \omega_i = 0, \end{cases}$$

where

$$\alpha = \left(\frac{q(\omega)\tilde{f}}{p(\omega)f(\omega)} \right)^{-1/N}.$$

It is easy to verify that this update would result in a zero gradients, and this procedure converges under liberal conditions (Bertsekas and Tsitsiklis [1989]). Of course, the update needs to be truncated at the bounds away from 0 and 1, which amounts to an underrelaxed update, and does not destroy convergence.

4 CONCLUSION

We have in this paper introduced the notion of applying importance sampling to financial pricing using binomial lattices. Several methodologies for chosing importance sampling distributions, both based on conditioning on the final state reached by a path, and by prescriptively or adaptively assigning importance branching probabilities to lattice nodes, were given.

Further research currently being pursued centers on three issues: (1) An implementation of the methods presented with the aim of demonstrating their effectiveness over crude Monte Carlo sampling, (2) a comparison of these methods to other variance-reduction techniques, such as stratified sampling, (3) combining these methods with standard techniques, such as quasi-Monte Carlo sampling and the use of control variates.

REFERENCES

[1] D.P. Bertsekas and J.N. Tsitsiklis. *Parallel and Distributed Computation: Numerical Methods.* Prentice Hall, Englewood Cliffs, New Jersey, 1989.

[2] F. Black, E. Derman, and W. Toy. A one-factor model of interest rates and its application to treasury bond options. Discussion paper no. 1, Goldman Sachs, June 1988.

[3] P.P. Boyle. Options: A Monte Carlo approach. *Journal of Financial Economics*, 42:323–338, 1977.

[4] A. Brooke, D. Kendrick, and A. Meeraus. *GAMS: A User's Guide*, 1988.

[5] J.C. Cox, S.A. Ross, and M. Rubinstein. Option pricing: A simplified approach. *Journal of Financial Economics*, 7:229–263, October 1979.

[6] G.B. Dantzig and P.W. Glynn. Parallel processors for planning under uncertainty. *Annals of Operations Research*, 22:1–21, 1990.

[7] P.W. Glynn and Iglehart. Importance sampling for stochastic simulation. *Management Science*, 35:1367–1392, 1989.

[8] J.C. Hull. *Options, Futures, and other Derivative Securities.* Prentice Hall, Englewood Cliffs, New Jersey 07632, 1993. Second Edition.

[9] G. Infanger. Monte Carlo (importance) sampling within a Benders decomposition algorithm for stochastic linear programs. Technical report SOL 89-13R, Systems Optimization Laboratory, Department of Operations Research, Stanford University, Stanford, California 94305-4022, August 1990.

[10] C. Joy, P.P. Boyle, and K.S. Tan. Quasi-Monte Carlo methods in numerical finance. Technical report, Enron Corporation, 1400 Smith Street, Houston, TX 77002–7361, October 1994.

[11] J.P.C. Kleijnen and R.Y. Rubinstein. Monte Carlo sampling and variance reduction techniques. In S. Gass and C. Harris, editors, *Encyclopedia of Operations Research and Management Science*, October 1993.

[12] W.H. Press, B.P. Flannery, S.A. Teukolsky, and W.T. Vetterling. *Numerical Recipes in C.* Cambridge University Press, Cambridge, 1992. Second Edition.

Appendix A: List of Symbols

$\omega \in \{0,1\}$: A sample path. $\omega_i = 0$ designates a downward branch in the binomial tree, $\omega_i = 1$ an upwards branch.

p_{ij}: The probability that a process passing through lattice node (i,j) branches up.

$p(x)$: Probability of x.

F_{ij}: Levels of the underlying factor (e.g., stock price or interest rate) at lattice node (i,j).

Δt: The time discretization step.

$\bar{f}$: Mean of the function f.

$\hat{f}$: Monte Carlo estimate of f.

$\tilde{f}$: Estimate for f obtained by importance sampling.

Ω: The set of 2^N sample paths through an N-step lattice.

q_{ij} : "Importance" probability of branching up at lattice node (i,j).

$q(x)$ (or $q(\omega)$): The importance measure of x (or ω).

d^j: The probability of final state j in two-stage sampling.

V^j: The price estimate for paths ending in final state j.

q^j: The importance measure of paths ending in final state j.

d_{ij}, u_{ij}: u_{ij} represents the set of lattice node passing proportions which most closely tracks a set of "desirable" proportions, d_{ij} (refer to Section 3.1).

13

DATA AND OPTIMISATION MODELLING: A TOOL FOR ELICITATION AND BROWSING (DOME)

Hossein Mousavi

Gautam Mitra

Cormac Lucas

ABSTRACT

The prototype design of a data and optimisation modelling and elicitation tool (DOME) is presented in this paper. The design objectives of DOME are to support structuring of the data and structuring of the model. A suitable graphical user interface (GUI) has been implemented to elicitate the problem owner's LP application. In order to aid the model development and analysis process a mechanism to browse through the data model, algebraic model, the constraint matrix and the solution representation has been introduced. Multi-dimensional tables and the "slice and dice" facilities to present different views of data are also incorporated within this tool. An example of developing and investigating applications using this tool is presented.

1. INTRODUCTION

Mathematical Programming (MP) optimisation models belong to an important class of decision tools; LP/IP models and their extensions such as stochastic optimisation (*Kall & Wallace 1994*), robust optimisation (*Mulvey et al 1994*) and goal programming have many industrial applications (*Mitra 1988*).

There are other decision support tools which are used in industry. Often these tools are data intensive and use data modelling techniques as supported by hierarchical, network and relational database systems. These tools are generally called Decision Support Systems (DSS) or Executive Information Systems (EIS); they have modelling capabilities such as simulation, decision analysis and often sit on top of a database system (*ButlerBloor 1994, Holtham 1992, P. Carnelley 1994, Business Intelligence 1992*). In the mean time, LP/IP modelling systems have evolved at an accelerating pace during the last 15 years, for example, (*Ellison & Mitra 1982, Fourer 1983, Ashford & Daniel 1987, Bisschop & Entriken 1993, Fourer et al 1993*). Only recently has there been further developments which have brought data modelling and database tools closer to LP/IP modelling (*Mitra et al 1995, Choobineh 1991*).

Most of the current MP modelling systems are text based (*Murphy & Stohr 1986, Jones 1990, Maturana 1994*). Recently, however some progress has been made in introducing graphical modelling system (*Senne & Lucas 1994, Collaud & Pasquier-Boltuck 1994*). These systems are, of course, easier to use but the implementation is difficult and considerable effort is directed to this area. We are, however, more interested in providing suitable data modelling support in these modelling tools.

The world of EIS has also moved forward and new analytical tools have been progressively introduced. Codd (*Codd et al 1993*) has put forward new concepts in terms of On-Line Analytical Processing (OLAP) and multidimensional databases which are of considerable interest to the EIS and database communities. Savage (*Savage 1992*), well known for his work in incorporating spreadsheets into management modelling, has recently introduced the idea of connecting multidimensional spreadsheets to LP/IP modelling tools (*Lotus Improv for Windows Handbook 1993*). We have also found that for many real life LP/IP applications involving model conceptualisation, model development and model investigation the following techniques play a progressive role:

- model structuring and data categorisation,
- specification of the algebraic model in a symbolic form,
- instantiation using data held in a database,
- displaying data, model and results for the purpose of model investigation.

In this report we summarize our prototype system, DOME, which is based on the concepts mentioned above. In particular, data categorisation and structured LP models, multidimensional databases, hypertext systems are all design features of our prototype. The rest of this paper is organised as follows. In section 2 the above three concepts are discussed. The design objectives of DOME are explained in section 3. The prototype is illustrated by an example in section 4. Section 5 gives an overview of the system and its architecture. Finally, section 6 provides a brief discussion.

2. BACKGROUND AND CONCEPTS

The DOME system is based on three major concepts, namely categorisation of data and structuring of LP model components, multidimensional database (MDDB) tables, and the use of hypertext to navigate through the integrated system. These concepts are discussed in subsections 2.1, 2.2 and 2.3 respectively.

2.1 Structured LP Models

Creating LP models in a structured fashion is an important design issue to enhance model building. All LP modelling systems cater for this by defining sets and indices (subscripts in algebraic notation) which are used in the symbolic definition of the model at the time of conceptualisation. These sets and indices are elicited for data modelling and subsequently used for LP modelling. A comprehensive analysis of LP modelling systems is presented by Sharda (*Sharda 1993*). Two aspects of structure, important in modelling support are that:

- the modeller follows the structure to define the model in a progressive and systematic way,
- the structure is treated as dynamic whereby different models can be instantiated by varying parameter and table sizes. In IT terminology, this is stated as data independence.
- Geoffrion (*Geoffrion 1987*), makes the following assertions concerning structured LP models:
- they have a clear general structure quite distinct from whatever particulars are needed to specify a specific model instance,
- they are abstract and hence general,
- they are likely to be vertically integrated by incorporating aspects often left outside the scope of the model as coming before it (data development) or after it (e.g. solution analysis), i.e. they are within an integrated environment,
- they are easy to understand,
- they are flexible enough to evolve in response to changing conditions and requirements.

2.2 Multidimensional Databases

The efficient operation and success of organisations is very much dependent on the sophistication of their information systems, and their ability to process and analyse information. Therefore, MDDB and its related concept, OLAP, are subjects of considerable interest in the EIS community (*Codd et al 1993*). An up to date review of the progressive developments in the field of MIS, DSS, EIS and OLAP can be found in (*Koutsoukis & Mitra 1995*).

OLAP involves consolidating, viewing and analysing data according to multiple business dimensions. This is not easily achievable with a relational database system (RDBS) in which flat two dimensional tables store data. Multiple dimensions are currently mapped into two dimensional tables and composed (joined) together as and when required to represent multiple dimensions. This becomes cumbersome for large data due to the multiple joins while it is also necessary to know the inter-relationship between tables. This has led to the development of MDDB which facilitates ready and flexible access as well as analysis of a large volume of complex and inter-related data (*Kenan Technologies 1993, DBMS 1994*).

The fundamental difference between RDBS and MDDB is that the former is value-based whereas the latter is reference-based where direct access is accomplished by specifying the appropriate dimensions. This leads to a highly structured organisation of the data with easy access and retrieval. Moreover, it makes data browsing and manipulation intuitive. (*Arbor Software 1993, Codd et al 1993*)

MDDB constitutes the core of OLAP tools (*Codd et al 1993, Kenan Technologies 1993, Arbor Software 1993, Finklestein 1994*) and possesses the following features:

- **Slicing**: Allowing rotation of dimensions so that data can be viewed from different perspectives,
- **Dicing**: Filtering the dimensions based on user specification so that a new multidimensional view is defined. All rotation and computational operations can be performed on any filter of the original data structure,
- **Hierarchies, Roll-up and Drill-down**: The ability to define hierarchies which enable very quick data manipulation and detailed analysis along different levels within all the dimensions of the MDDB. Moving up and down levels in a hierarchy are referred to as roll-up and drill-down respectively. Each view as defined by the chosen level along each dimension can then be sliced or diced.

However, MDDB is analysis oriented; it has a complementary role to RDBS and provides:

- data integration from existing systems,
- data structure that maps to end-users' understanding of business issues,

- an analysis environment for complex applications by offering data manipulation and presentation capabilities.

For a discussion of these issues, the readers are referred to (*Codd et al 1993, Finklestein 1994*). In diagram 2.1, the positioning of MDDB in relation to the existing systems is illustrated:

Diagram 2.1 Role of OLAP Servers

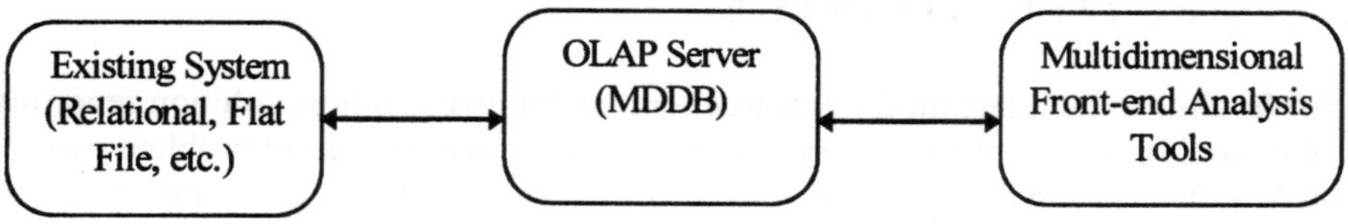

From the LP modelling perspective, we find MDDB support extremely valuable as it can be used to hold instantiated data. Additionally, MDDB representation and analysis tools can be used to manipulate LP/IP data. Sets and indices in LP structured models, discussed above, are directly translated into the dimensions (categories) of the MDDB on which slicing, dicing and other OLAP operations can be performed. Bisschop (*Bisschop 1995*) has also discussed linking graphical objects to sets and indices so that data can be viewed from different perspectives as specified by the user.

2.3 Hypertext Systems and Browsing

Shneiderman (*Shneiderman & Kearsley 1989*) defines hypertext as a database that has active cross-references and allows the reader to traverse through any part of the database when desired. It can be conceptualised as a network of nodes and links where documents are the nodes and the links are the cross-references. Features of a hypertext system can be summarized as follows:

- **Structure**: It has a web structure so that the user has freedom to explore. A web is a set of documents with little or no overall structure. However, some systems follow a hierarchical design which limits the user navigation process. To avoid limitations of the hierarchical systems, an open ended navigation is supported by a map or a schematic which shows the present location and past links selected.

- **Direct manipulation**: It provides easy-to-use interaction such that the desktop is transparent and the user concentrates on the task at hand. It avoids typing errors. Menus complement the capabilities of the system.
- **Filtering**: Filtering information by specifying attributes of links and nodes is used to arrive at the most accurate location in a large hypertext system.

- **Reversing:** Users can backtrack their interaction with the system

- **Integration:** A versatile and usable hypertext system can load information to and from many existing systems and integrates all the programs running on a machine. This provides a consistent user interface to the other programs.

3. OBJECTIVES OF DOME

DOME aims to facilitate model construction, model investigation, solution reporting and overall browsing of an LP application. It is designed to support problem owners and MS/OR professionals who wish to construct and investigate LP/IP models. The objectives of DOME are summarized as follows:

- it is an integrated modelling tool which supports the entire modelling life cycle,
- it provides a consistent environment within which data models, algebraic (symbolic) models, solution reports and analysis can be specified,
- it can communicate with the solver via machine readable models,
- it is hospitable to the modeller, that is, MS/OR professionals.

Menus and screens are the means of progressively defining an LP model which can later be revised. During construction, models are checked for integrity of structural components, syntactical correctness, and consistency of units. For a discussion of unit consistency we refer the readers to (*Bradley et al, 1987*). An important feature of DOME is that it supports stepwise modelling such that the system takes the modeller through all steps of constructing LP models, from parameter definition (global entities) to individual constraint terms. However, an experienced modeller can construct the model in a non-sequential fashion and the embedded checks in the system notifies the modeller of any missing components. For example, if variables are defined before sets, then the modeller is prompted to define the sets and their corresponding indices.

DOME aids the user by automating syntax checking during the model development process. Furthermore, menus, screens and pick-up lists minimise key strokes. The modeller does not need to know about any particular programming language structure as the menus and screens are designed such that the model components can be defined intuitively. Model documentation, a listing of all the major model components, and the generated LP model (algebraic statements) provide alternative hardcopy representations (views) of the model; these alternative representations help communication between multiple constituents (see Appendix A). Exploiting the features of MDDB in defining model instances and solution reporting make it a friendly environment for the end user as the data items are presented in terms of categories and units which are meaningful to the problem owner.

Data and model definitions are utilised by the system to enable the modeller to browse an LP model at all levels of abstraction, that is from model definition to the generated matrix coefficients.

4. PROTOTYPE DESIGN ILLUSTRATED BY AN EXAMPLE

In this section a small example, Nuts, Bolts and Washers (*Ellison & Mitra 1982*) is used to illustrate the process of construction and navigation of an LP model by browsing through different parts of the results and other components of the model using DOME. Once the model is specified, generated and solved, the model and the results can be investigated through the model browser.

4.1 Model Statement

A company manufactures three products P1, P2, P3 (Nuts, Bolts and Washers) and has at its disposal three machines M1, M2, M3. The company can undertake normal and overtime production and needs to plan for two time periods, say WINTER and SUMMER. Any product left after the second time period has very little resale value. The necessary information concerning the operation of the company is set out in tables 1-3. It is necessary to find an LP formulation that maximises the profit of the company's operation over the two periods.

4.2 Mathematical Formulation

Let the four indices i,j,k,l be defined as

$i = 1..I$ Index for the two time periods, Summer and Winter (I=2)
$j = 1..J$ Index for the two production modes, Normal and Overtime (J=2)
$k = 1..K$ The index for the three product types, P1, P2, P3 (K=3)
$l = 1..L$ The index for the three machines, M1, M2, M3. (L=3)

The following information relating to the problem are available in table 1.

t_{ijkl} the number of hours required to produce one unit of product k on machine l, in time period i, using normal or overtime production j

a_{ijl} the machine availability in hours for machine l in period i and mode j

In table 2

c_{ijkl} the production cost in the category i,j,k,l

In table 3

p_{ik}	the selling price for product k in time period i
d_{ik}	the demand for product k in time period i
s_k	the storage cost for product k in one time period
h_k	the corresponding storage capacity for product k
r_k	the final resale value at the final time period for product k

The decision variables of the problem are as follows:

x_{ijkl}	denotes the quantity to be produced in the category i,j,k,l
y_{ik}	denotes the quantity of product k stored in time period i
z_{ik}	denotes the quantity of product k sold in time period i

Table 4.1 Table of Machine-Hour

	Summer								*Winter*							
	Normal			*Overtime*			*Available Hours*		*Normal*			*Overtime*			*Available Hours*	
	P1	P2	P3	P1	P2	P3	N	O	P1	P2	P3	P1	P2	P3	N	O
M1	4	5	6	3	4	5	100	80	5	6	7	4	5	5	110	90
M2	7	6	6	6	5	5	100	90	8	7	7	7	6	6	110	100
M3	3			2			40	30	4			3			50	40

Note: P1=Nuts, P2=Bolts, P3=Washers

Table 4.2 Table of Production Cost

	Summer						*Winter*					
	Normal			*Overtime*			*Normal*			*Overtime*		
	P1	P2	P3	P1	P2	P3	P1	P2	P3	P1	P2	P3
M1	2	3	4	3	4	5	3	4	5	4	5	6
M2	4	3	2	5	4	3	5	4	3	6	5	4
M3	1			2			2			3		

Note: P1=Nuts, P2=Bolts, P3=Washers

Table 4.3 Table of Additional Data

	Summer			*Winter*		
	Nuts	*Bolts*	*Washers*	*Nuts*	*Bolts*	*Washers*
Sale Price	10	10	9	11	11	10
Min. Demand	25	30	30	30	25	25
Storage Data						
Capacity	20	20	-			
Cost	1	1	1			
Resale Value				2	2	1

The profit function of the problem may be expressed as

$$profit = \sum_{i=1}^{I}\sum_{j=1}^{J} P_{ik}\, z_{ik} - \sum_{k=1}^{K} S_k\, y_{ik} + \sum_{k=1}^{K} r_k\, y_{2k} - \sum_{i=1}^{I}\sum_{j=1}^{J}\sum_{k=1}^{K}\sum_{l=1}^{L} C_{ijkl}\, x_{ijkl}$$

In an optimal plan profit must be maximised subject to the constraints:

machine availability $\sum_{k=1}^{K} t_{ijkl}\, x_{ijkl} \leq a_{ijl}$ for all i,j,l

stock balance in the two periods

$$\sum_{j=1}^{J}\sum_{l=1}^{L} x_{1jkl} - y_{1k} - z_{1k} = 0 \quad \text{for period 1, and all } k$$

and

$$\sum_{j=1}^{J}\sum_{l=1}^{L} x_{2jkl} + y_{1k} - y_{2k} - z_{2k} = 0 \quad \text{for period 2, and all } k$$

minimum demand to be satisfied $z_{ik} \geq d_{ik}$ for all i, and k

upper bound on storage $y_{1k} \leq h_k$ for all k

non-negativity of the variables

$y_{ik} \geq 0$ for all i,k $z_{ik} \geq 0$ for all i,k and $x_{ijkl} \geq 0$ for all i,j,k,l

4.3 Model Elicitation

The example is illustrated by displaying various screens followed by a brief explanation of each. It is presented in the logical sequence followed when building structured models. The following screen defines the model to be worked with:

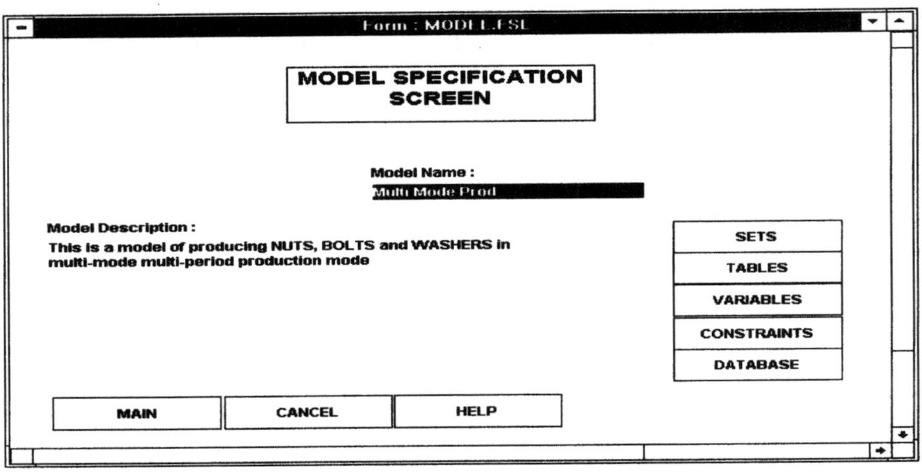

Here a model name and a brief description of the application are entered. The next step is either to define the sets or capture from a database the data which instantiates the sets and the model. If the 'DATABASE' option is chosen all the model data is imported into set specification and table definition screens. The modeller can define the model in the subsequent screens. However, on pressing 'NEXT' button the 'Set Specification Screen' appears.

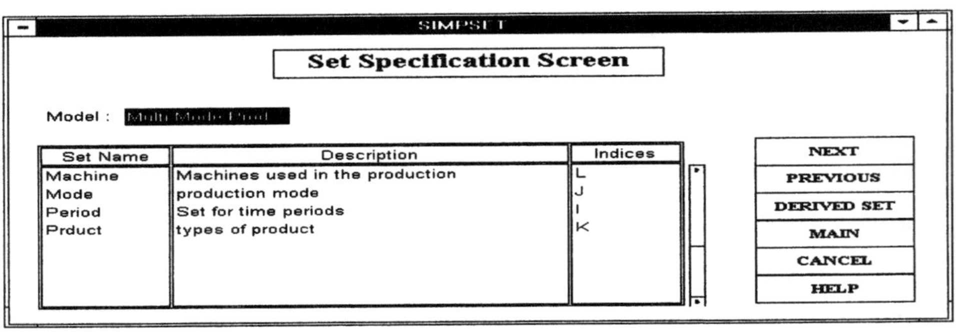

Here sets and their corresponding indices are defined. Derived sets can also be defined based on existing sets. The derived sets are created by manipulating other sets following the criteria discussed by Geoffrion (*Geoffrion 1987*). This example does not involve derived sets, hence the corresponding screen is not detailed here. Sets (indices) defined here, translate to categories (dimensions) by which the data is structured in the MDDB. On pressing the "NEXT" button, the 'Table/Scalar Definition Screen' is displayed.

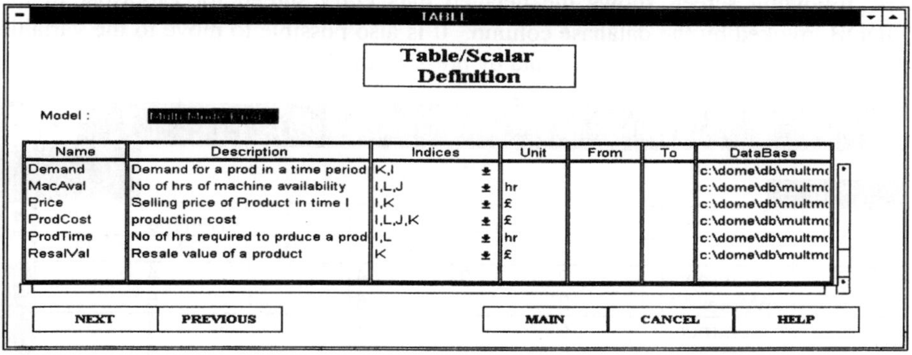

Data tables are defined based on the sets (indices) already defined by the modeller; the indices and units of measurement are chosen from pick-up lists. If a table is defined using a set which does not exist, the modeller is prompted to define the new set or choose an existing set. The unit field enables the modeller to provide the unit for a given data table. It also enables the modeller to choose from a pre-defined list which already exists within the system; the list can of course be extended. If a new unit is defined (basic or derived), the modeller is prompted to define its relation to existing ones. DOME uses an underlying database for unit conversion purposes.

The database column provides access to the MDDB where detailed data of a model instance are specified. This can be achieved in two ways:

- **Model to MDDB:** In this case, the detailed data tables have not already been defined and the modeller specifies members (items within a category) of each set, on which the data table is defined. The following example illustrates such table definition: By defining the Demand table a two dimensional table is displayed with categories Period (I) and Product (k). The user enters Summer and Winter for Period, and 'P1','P2','P3' for Product. Later, when defining MacAval a three dimensional table is displayed with categories Period (I), Machine(l) and Mode(j) where the set Period is already instantiated with Summer and Winter.
- **MDDB to Model:** The detailed data have already been defined within the MDDB and accessing it provides a data instance by which the current model is defined.

However, sometimes the data definition can be based on a mixture of the above scenarios where parts of the data is captured from a database and parts of it is defined by the modeller and exported back to the database.

The following screen shows the detailed data entry screen for **ProdCost** in the MDDB invoked by the database column. It is also possible to move to the variable definition screen before entering data by pressing the 'NEXT' button.

Worksheet1 · View1 · PRODCOST

		Normal			Overtime		
		P1	P2	P3	P1	P2	P3
M1	Summer	2	3	4	3	4	
	Winter	3	4	5	4	5	
M2	Summer	4	3	3	5	4	
	Winter	5	4	3	6	5	
M3	Summer	1			2		
	Winter	2			3		

Machine (L)

As items for all the categories (**Period(i), Machine(l), Product(k), Mode(j)**) have been defined in the previous tables, the template for **ProdCost** tables is filled in with those item names. The analysis and presentation tools of the MDDB can be used in manipulating the data. For example, the following picture shows a different view of the data by means of slicing:

Worksheet1 · View2 · PRODCOST

			P1	P2	P3
M1	Normal	Summer	2	3	4
		Winter	3	4	5
	Overtime	Summer	3	4	
		Winter	4	5	
M2	Normal	Summer	4	3	3
		Winter	5	4	3
	Overtime	Summer	5	4	
		Winter	6	5	
M3	Normal	Summer	1		
		Winter	2		
	Overtime	Summer	2		
		Winter	3		

Mode (J)

Other views can be obtained by slicing different dimensions; it is also possible to define 'page' dimensions based on existing dimensions; this is a very useful facility for viewing large quantities of data in large models. On completion of the data section the decision variables are specified using the following screen

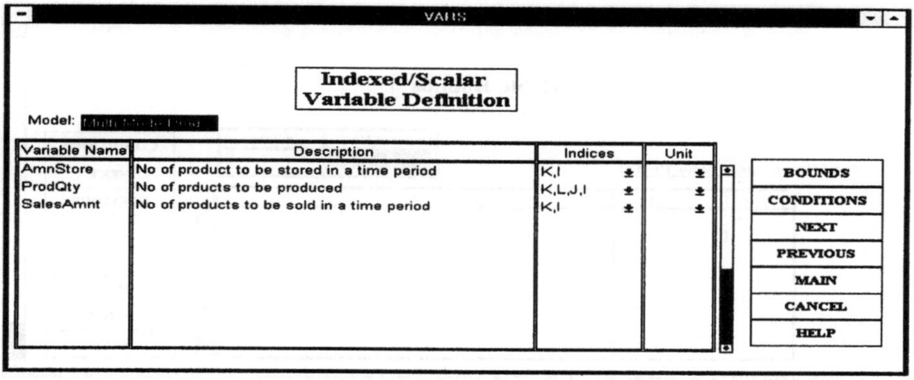

Variables are defined in a similar fashion to the data tables where their dimension and units are specified. The variable structure is translated into the MDDB format. After optimisation the solution values of the variables are returned to the MDDB to produce solution reports. Often, it is the case that a variable is not defined for all of the range of its dimensions. This conditional generation of variables is specified using the following screen.

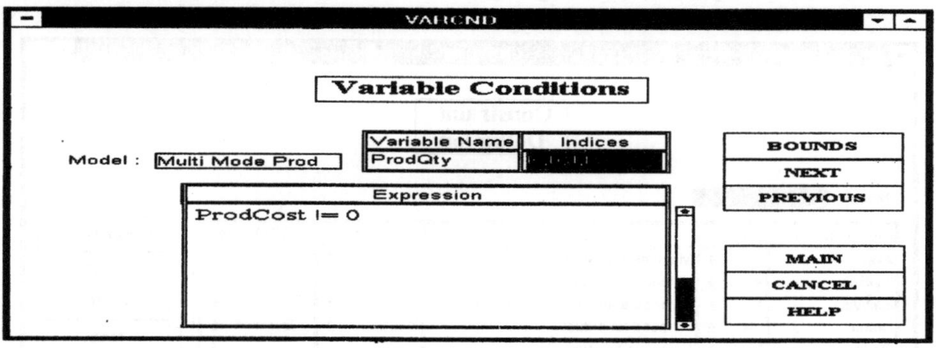

In our example, the variable ProdQty is defined only when ProdCost exists. In this model, the production cost is zero for any disallowed combination of product, production mode, time period and machine. By including this restriction in the variable definition, the disallowed combinations do not need to be explicitly specified in the algebraic modelling statements, without which the model is unbounded. Once the model is specified, generated and solved, the model and the results can be investigated through the model browser. The "!=" operator means 'not equal to'. Bounds on variables are specified as set out below.

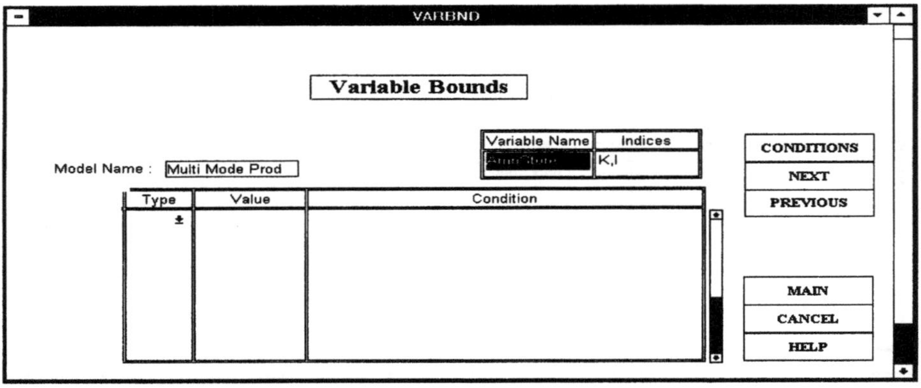

One or more bounds can be defined for a specific variable by specifying its type (integer, >=, <=, =, free, binary, SOS1, SOS2), value and the range of the variable for which it applies. However, our example does not include any bounds defined for a variable, but the above screen is shown to illustrate how other variable types than the default (non-negative real) are specified . The next step in building a model, is to specify the constraint groups, this is achieved with the following screen.

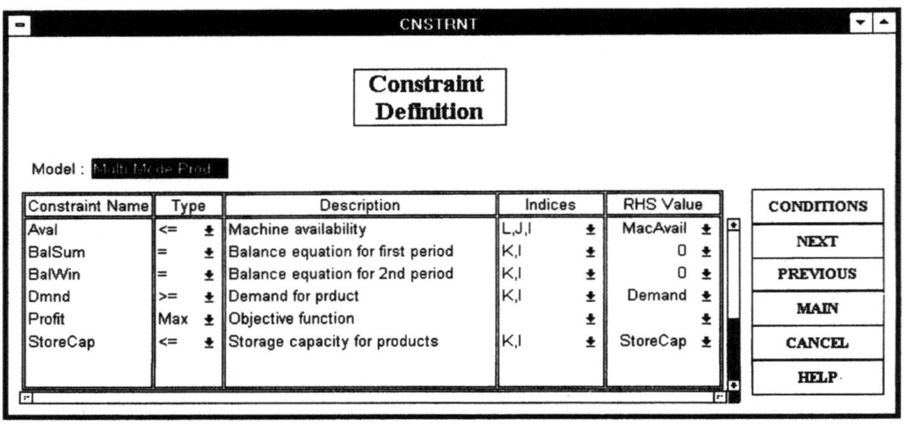

The constraint name, its type (max, min, >=, <=, =, !=, range), a brief description, indices over which the constraint is defined, and the RHS value are specified in the constraint definition. The indices and RHS are both chosen from pick-up lists where indices shows the available indices and RHS refers to data tables defined previously, or is a scalar value. Upper Limit and Lower Limit of a range constraint are specified on a pop-up screen. If the constraint group is not defined for all the range of a dimension, then the various conditions concerning its validity are specified in the following screen.

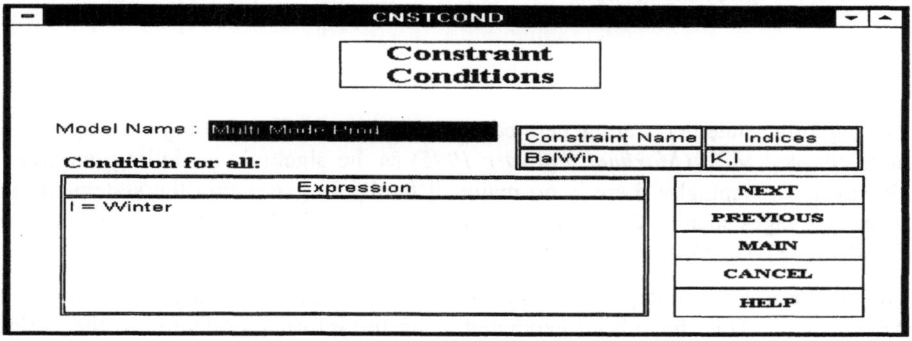

One or more conditions can apply to a constraint. The above figure specifies that the constraint group BalWin is defined only for i = Winter (winter period). The terms of this constraint are shown in the following screen.

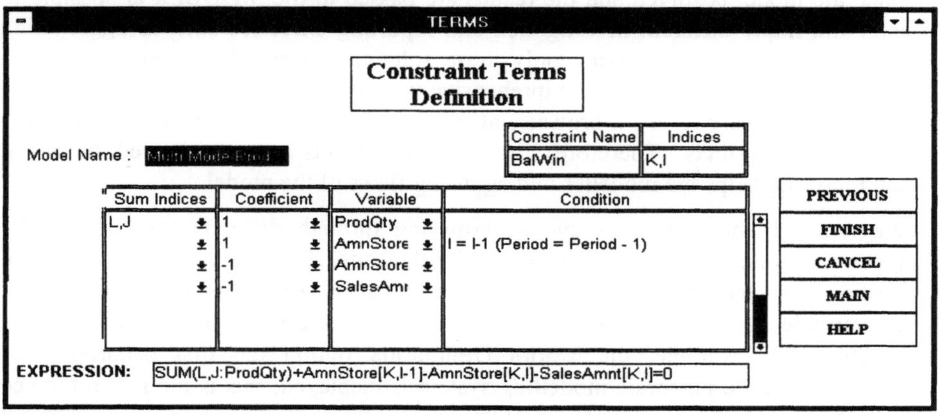

Individual constraint terms are defined as a product form $C*X$ where C refers to a coefficient (a data table already defined or a scalar value) and X refers to a variable already defined. As before, these two entries are chosen from pick-up lists. Indices that have been used to dimension a constraint group are not available as summation indices. As constraint terms are defined, the constructed algebraic expression is generated to make the constraint more readable and comprehensible; it also can be used by more experienced modeller to specify the constraint.

Once the model is constructed, the algebraic representation is used to generate the matrix and this is subsequently submitted to the solver for optimisation. This is shown in the following section.

4.3 Model Generation

The algebraic statements are generated for a text-based LP modelling language. We have adopted MPL (*Maximal Software 1995*) as the algebraic modelling language. However, conceptually there is no major difficulty in generating the statements in other languages such as GAMS (*Brooke et al, 1988*), AIMMS (*Bisschop & Entriken 1993*), AMPL (*R. Fourer et al 1993*).

The algebraic statements, in MPL syntax, generated for our example are shown in Appendix B. They have been generated using the information stored in the Script Driver and data instantiation is specified in the data section where tables are read from the MDDB. The model is now ready to be submitted to the solver.

4.4 Investigation, Reporting and Browsing

Once the model is solved and the results are passed to the MDDB, it is possible to carry out many different investigation and reporting exercises such as aggregating the solution values, displaying only those solution values which can be within a specified range. However, a comprehensive way to investigate a model is through a browser. In order to use a browser, all model components are connected together using hypertext links. Therefore, if the expected results are not achieved, browsing and reporting helps the modeller in diagnosing flaws of the model.

Diagram 4.1 shows an example of browsing through different modules of the hypertext system for the Multi-Mode-Prod (Nuts, Bolts and Washers) model. Starting from the Variable Definition Screen in the script driver and clicking on the variable AmnStore a menu is displayed listing the possible connections. By selecting, for example, the Model Definitions option the first occurrence of AmnStore in the algebraic modelling system is displayed; this is a direct reference link. Subsequently the Algebraic References within this module can be viewed showing all the constraints as well as variable type definitions (bounds) in which this variable appears.

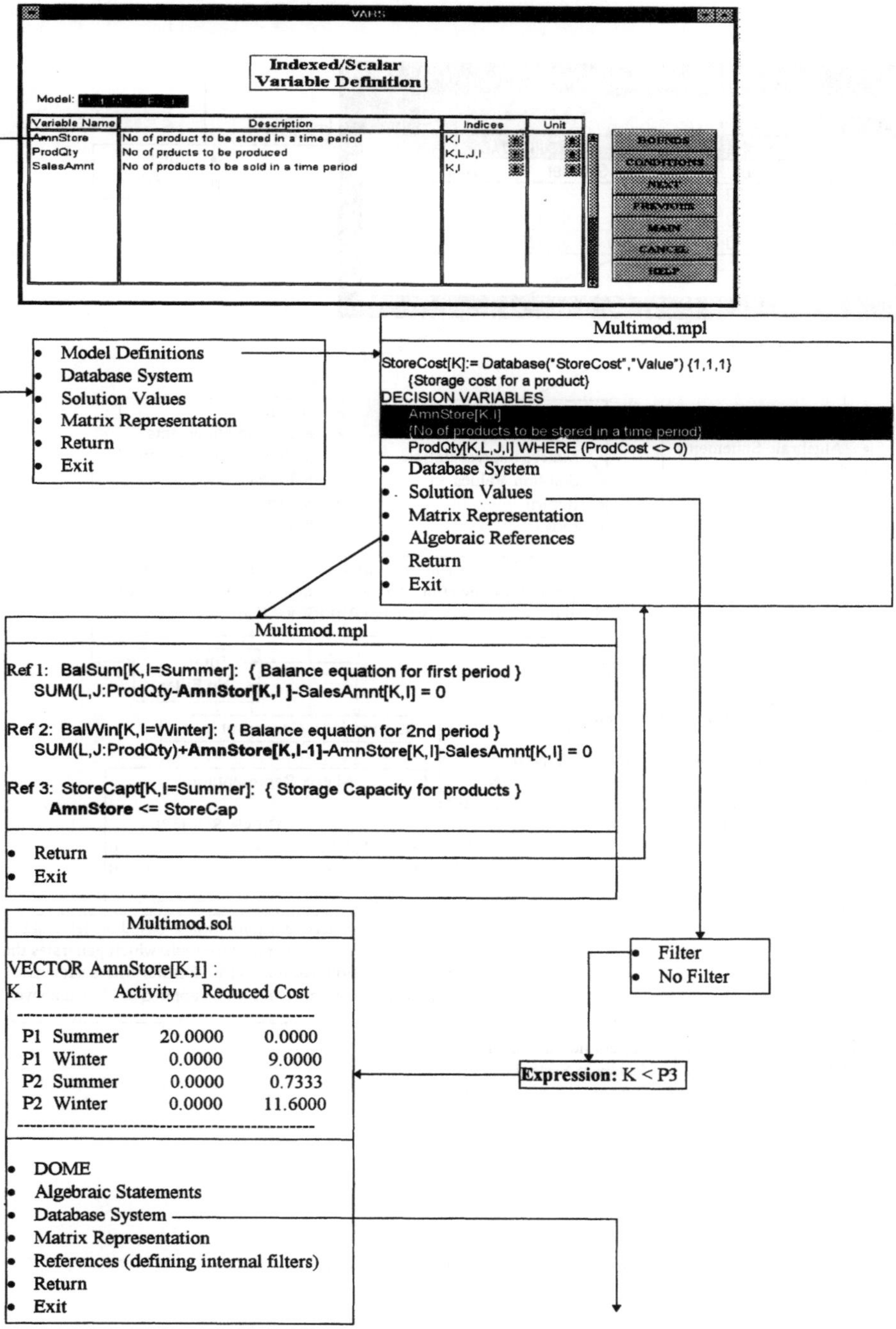
VARS
Indexed/Scalar Variable Definition
Model:
Variable Name | Description | Indices | Unit
AmnStore | No of product to be stored in a time period | K,I
ProdQty | No of prducts to be produced | K,L,J,I
SalesAmnt | No of products to be sold in a time period | K,I
BOUNDS
CONDITIONS
NEXT
PREVIOUS
MAIN
CANCEL
HELP
Model Definitions
Database System
Solution Values
Matrix Representation
Return
Exit
Multimod.mpl
StoreCost[K]:= Database("StoreCost","Value") {1,1,1}
{Storage cost for a product}
DECISION VARIABLES
AmnStore[K,I]
{No of products to be stored in a time period}
ProdQty[K,L,J,I] WHERE (ProdCost <> 0)
Database System
Solution Values
Matrix Representation
Algebraic References
Return
Exit
Multimod.mpl
Ref 1: BalSum[K,I=Summer]: { Balance equation for first period }
SUM(L,J:ProdQty-AmnStor[K,I]-SalesAmnt[K,I] = 0
Ref 2: BalWin[K,I=Winter]: { Balance equation for 2nd period }
SUM(L,J:ProdQty)+AmnStore[K,I-1]-AmnStore[K,I]-SalesAmnt[K,I] = 0
Ref 3: StoreCapt[K,I=Summer]: { Storage Capacity for products }
AmnStore <= StoreCap
Return
Exit
Multimod.sol
VECTOR AmnStore[K,I] :
K I Activity Reduced Cost
P1 Summer 20.0000 0.0000
P1 Winter 0.0000 9.0000
P2 Summer 0.0000 0.7333
P2 Winter 0.0000 11.6000
DOME
Algebraic Statements
Database System
Matrix Representation
References (defining internal filters)
Return
Exit
Filter
No Filter
Expression: K < P3

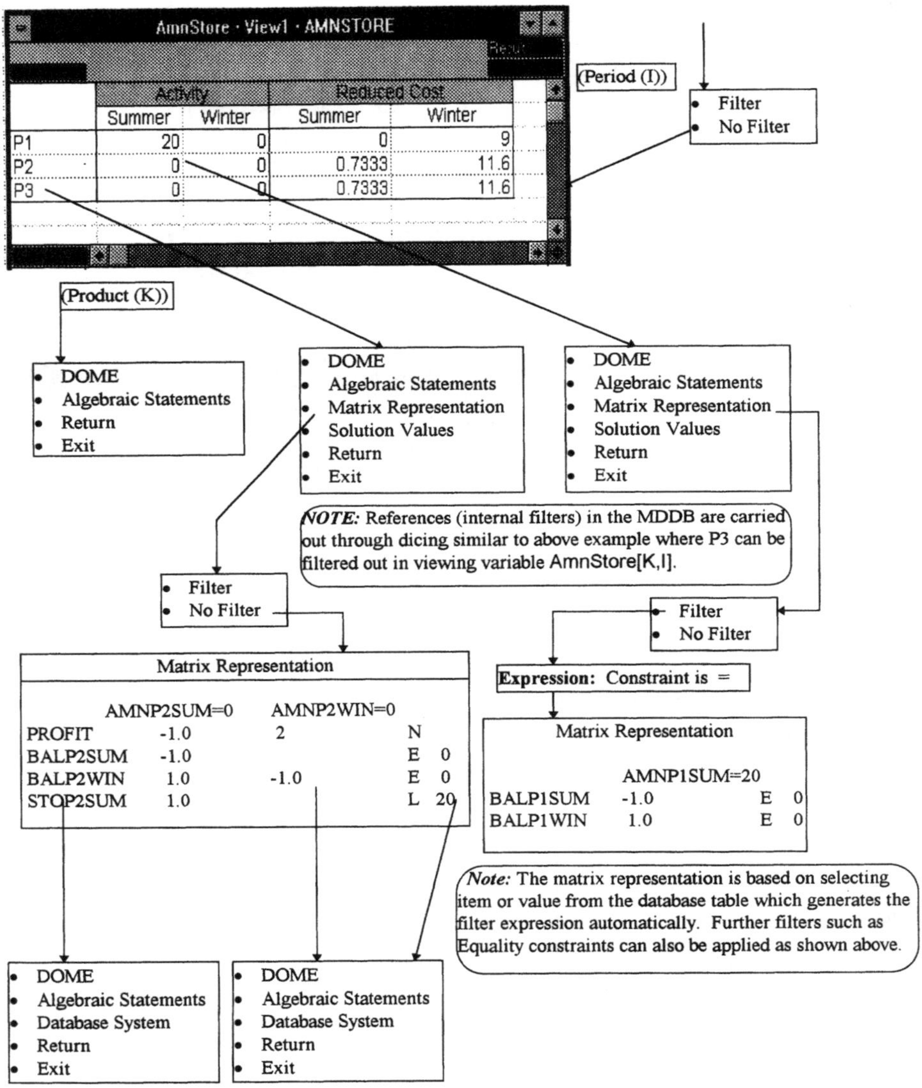

Diagram 4.1 Model and Solution Browsing

Choosing Solution Values prompts the user for specifying a filter, if any. The specified filter outputs solution values in which K < P3. This is shown in the expression box of Diagram 4.1. Then, the Database System option displays the corresponding values for the AmnStore variable within the MDDB shown in the screen labelled AmnStore.View1.AMNSTORE. It is worth noting that the filter K < P3 does not apply after leaving the solution module. However, the MDDB itself provides analytical display features such as dicing, roll-up and drill-down. Any of the categories (sets), items (set members) and solution values can be browsed and selecting any one of them displays its corresponding menu. The menus for items and values are similar, but when a value is selected the generated filter is more restrictive. In this example, clicking on the solution value 20 for k=P1 and i=Summer a filter (k=P1 AND i=Summer) is generated so that only the corresponding values from other modules can be viewed.

An additional filter, specifying only the equality constraints has resulted in displaying the corresponding Matrix Representation. However, clicking on item P2 generates the filter i=P2 and the corresponding Matrix Representation is displayed showing all the constraints and variables for AmnStore where i=P2. The user can carry on browsing to other modules or exit at any time.

5. ARCHITECTURE AND OVERVIEW OF THE SYSTEM

In this section the system architecture, its functional elements and their interaction as well as an overview of the system and its features are described.

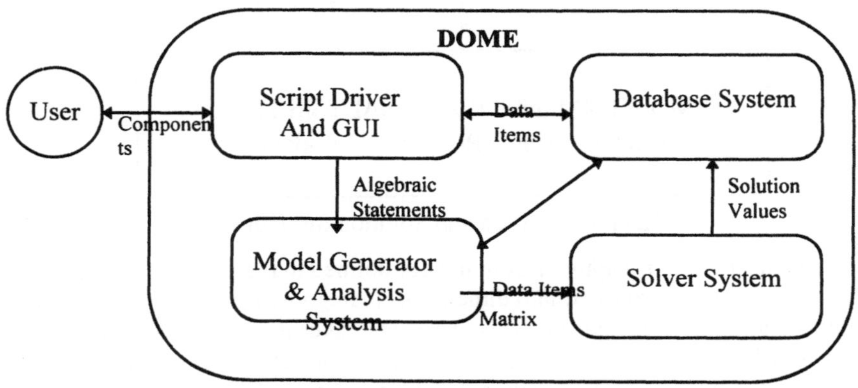

Diagram 5.1 DOME Architecture

5.1 Architecture

Diagram 5.1 illustrates the overall architecture of DOME and outlines the control and information flow between its components. DOME is divided into four major subsystems which interact with each other; the major functions cover construction through the script driver, generation, solving and analysis of LP models:

- **Script drive**: This is an enhanced version of CAMPS (*Lucas & Mitra 1988*) whereby the set of interactive menus and screens allow the modeller to build a model. It generates the algebraic statements which are submitted to the modelling system. Any model inconsistency is reported back and unit conversions are also carried out. When reporting solutions, it acts as a medium for presenting the solution to the user.
- **Database system**: This provides common structure for the detailed data tables and variables as shown in diagram 5.2:

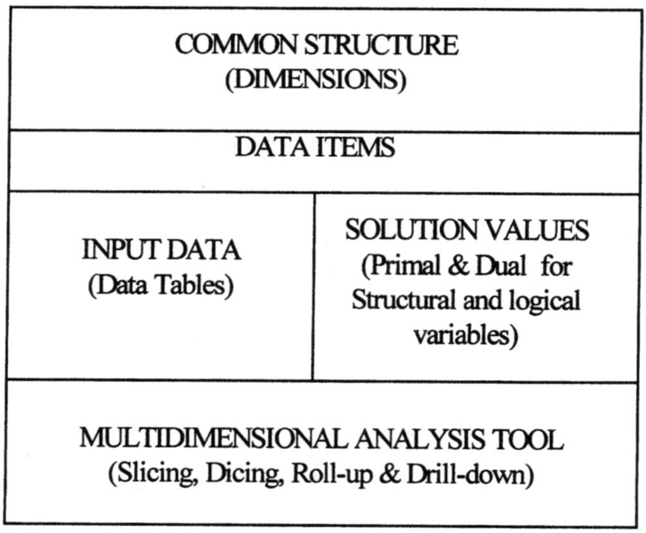

Diagram 5.2 Data and Solution Structures

The data structure is captured from indices defined in the model so that the indices (sets) are translated into dimensions (categories) and set members are items within these categories. This approach makes it possible for data to be separated from the model components. The MDDB table manipulation tool enables the analysis of both input data tables and solution data. The advantage of MDDB can be harnessed in this way so that by performing "on the fly" operations on data dimensions or values different views can be presented to the user.

- **Model generator and analysis system**: The generated algebraic statements from the script driver are transferred to a matrix representation through the model generator which is a text-based LP modelling language. The aim is to exploit facilities of the modelling system in constructing, investigating and analysing the LP models.
- **Solver system**: The solver carries out optimisation and computes the solution values for decision variables which are returned to the database system for presentation and analysis by the end user.

5.2 System Overview

The tasks carried out by the system are divided into three major areas: model construction, model hypertext system and model utilities.

- **Model construction**: The standard model construction tasks are performed by the system. These tasks are: model creation through interactive menus and screens; model revision or alterations; model generation of algebraic statements; model optimisation by submitting it to a solver. Model checking is also carried out interactively so that the model is always syntactically correct; for example if a variable is dimensioned by an undefined index such as *k*, the modeller is given the opportunity to either create a new set and define its members, or change this chosen index. Moreover, conversion knowledge for units in which the data is measured is stored in the system so that units of components are consistent with each other throughout the model. The measuring units are either basic or derived and are defined by the modeller.
- **Model hypertext system**: The hypertext system connects a number of data modules (Database System, MPS Data, Matrix Representation) and modules holding symbolic model definitions (DOME Screen, Algebraic Statements). These modules and their interconnection are shown in Diagram 5.3:

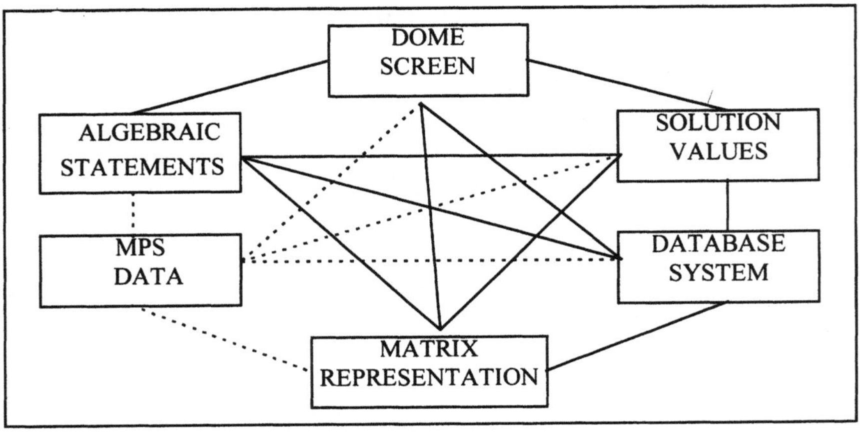

Diagram 5.3. Hypertext System
The dotted lines indicate that MPS external format model definition is a facility which may be dropped in the future. It is included in order to be compatible with legacy systems, but for implementation purposes it is of low priority.

- **General Browsing**: The hypertext system is used at a simple level for browsing. In order to browse model data, model statements, generated matrices and solution values external links are used to navigate across modules. Within each module further navigation is also possible. All browsing is controlled by either direct reference or by selection criteria formally defined as a filter. (See Appendix B).
- **Investigation and Analysis**: External and internal filters are used in order to view the required information. See appendix C for a complete list of possible links between different modules within the hypertext system. This approach is then extended to support more traditional investigations namely, model (data) analysis, solution report and solution analysis. Changes to data and their effect on the solution values can be observed as well by storing and navigating through multiple cases.
- **Reporting:** Data and solution of a model are analysed at two levels. The first level comprises plain reports, such as solution values, optimal ranges and totals. The second level utilises the analytical features of the MDDB. For example dicing is used as an internal filter to view only required information and slicing is used to give different views of the data. The MDDB also provides the

facility of graphically presenting solution and model data values through bar charts.

- **Model utilities**: Miscellaneous functions such as copying, deleting, renaming and documentation listing all major components of a model and their relationships, are also included in the system.

6. SUMMARY

DOME has been designed to provide an easy-to-use, integrated environment for model elicitation and investigation. This is achieved by providing different menus and screens through which a model can be constructed progressively. All relevant model, data and solution components are linked by a hypertext engine so that a range of useful analyses and investigation of the model and solution can be carried out. By incorporating MDDB and its features, we are able to support data modelling within the framework of structured algebraic LP models. As the new generation of decision support tools provide easy access to MDDB, our system is set out to make LP modelling more readily available to the DSS and EIS user community.

7. REFERENCES

Arbor Software Corporation, "Multidimensional Analysis: Converting Corporate Data into Strategic Information", UK, 1993.

Ashford, R.W. & R.C. Daniel, "LP-MODEL: XPRESS_LP's Model Builder", IMA Journal of Mathematics in Management, 1, pp 163-176, 1987.

Bisschop, J.J. & R. Entriken, AIMMS User Guide, Version 2.03, Paragon Decision Technology, The Netherlands, 1993.

Bisschop, J.J. & R. Stegeman, "Linking Graphical Objects to Index Sets in an Algebraic Modelling Language" presented at APMOD95, Brunel University, Uxbridge, Middx, UK, April 1995.

Bradley, G.H. & D.R. Clemence, "A Type Calculus for Executable Modelling Languages", IMA Journal of Math. in Management, 1, pp 227-291, 1987.

Brooke, A., D. Kendrick, A. Meeraus, GAMS: A User's Guide, The Scientific Press, 1988.

Business Intelligence Ltd, "External Information and EIS" , Business Intelligence Ltd., UK, 1992.

ButlerBloor, "EIS Software: An Evaluation and Comparison", ButlerBloor Ltd., UK, 1994.

Carnelley, P., A. Shamp & M. Sheina, "Ovum Evaluates: Second Generation EIS", Ovum Ltd, UK, 1994.

Choobineh, J., "SQLMP: A Data Sublanguage for Representation and Formulation of Linear Mathematical Models", ORSA Journal on Computing, Vol. 3, No 4, 1991.

Chung, Q.B. & R.M. O'Keefe, "Provision for Naive Users of Model Management Systems", European Journal of Information Systems, Volume 2, No. 3, pp 225-236, 1993.

Codd, E.F., S.B. Codd & C.T. Salley, Providing OLAP (On-Line Analytical Processing) to User-Analysts: An IT Mandate, E.F. Codd & Associates, 1993.

Collaud, G. & J. Pasquier-Boltuck, " gLPS: A Graphical Tool for the Definition and Manipulation of Linear Problems", European Journal of Operational Research (72), pp 277-286, 1994.

DBMS, "A Drill-down Analysis of Multidimensional Databases", pp 60-71, July 1994.

Ellison, E.F. & G. Mitra, "UIMP: User Interface for Mathematical Programming", ACM Transactions on Mathematical Software, Volume 8, No. 3, , pp 229-255, 1982.

Finklestein, R., Understanding the Need for On-Line Analytical Servers, Comshare, UK, 1994.

Fourer, R., "Modelling Languages versus Matrix Generators for Linear Programming", ACM Transactions on Mathematical Software, Vol. 9, No. 2, pp 143-183, 1983.

Fourer R., D.M. Gay, B.W. Kernighan, AMPL User Guide, The Scientific Press, 1993.

Geoffrion, A.M., "An Introduction to Structured Modelling", Management Science, Volume 33, No. 5, 1987.

Holtham, C., Executive Information System and Decision Support, Chapman & Hall and UNICOM, UK, 1992.

Jones, C.V., "An Introduction to Graph-based Modelling Systems, Part 1: Overview", ORSA Journal on Computing, 2, pp. 136-151, 1990.

Kall, P. & S.W. Wallace, Stochastic Programming, John Wiley and Sons, 1994.

Kenan Technologies, Multidimensional database technology, Cambridge, Massachusetts, USA, 1994.

Koutsoukis, N-S & G. Mitra, "Information Systems Support for Decision Making", Brunel University, UK, 1996 - Technical Report (TR/1/96).

Lotus Development Corporation, USA, Improv for Windows Handbook, Lotus Improv version 2.1, 1993.

Lucas, C. & G. Mitra, "Computer-Assisted Mathematical Programming (Modelling) System (CAMPS)", The Computer Journal, Vol. 31, No. 4, 1988.

Maturana, S.V., "Issues in the Design of Modelling Languages for Mathematical Programming", European Journal of Operational Research 72, pp 243-261, 1994.

Maximal Software Incorporation, USA, MPL Modelling System, Release 3.0, 1994.

Mitra, G.,ed., Mathematical Models for Decision Support, NATO ASI Series, Series F: Computer and System Sciences, Volume 48, 1988.

Mitra G., C. Lucas, S. Moody & B. Kristjansson, "Sets and Indices in Linear Programming Modelling and their Integration with Relational Data Models", Computational Optimisation and Applications, Vol. 4, No. 3, pp 263-283, 1995.

Mulvey, J.M., R.J. Vanderbei, S.A. Zenios, "Robust Optimisation of Large Scale Systems, Operations Research, Vol 43, No. 2, pp 264-281, 1995.

Murphy, F.H., E.A. Stohr, , "An Intelligent System for Formulating Linear Programs", Decision Support Systems, 3, 1986.

Savage, S.L., Fundamental Analytical Spreadsheet Tools for Quantitative Management (FASTQM), McGraw Hill, 1992.

Senne, E.L.F. & C. Lucas, Towards an Intelligent Graphical Interface for Linear programming Modelling, Brunel University, UK, 1994 - Technical Report (TR/21/94).

Sharda, R, Linear and Discrete Optimisation and Modelling Software, UNICOM, UK & Lionheart Publishing, USA., 1993.

Shneiderman, B. & G. Kearsley, Hypertext Hands-on: An Introduction to a New Way of Organising and Accessing Information, Addison-Wesley, 1989.

APPENDIX A: ROLE OF DIFFERENT CONSTITUENTS

The system considered in this paper needs positioning in respect of a number of constituents whom we have variously referred to as user, end user, modeller, MS/OR professional and problem owner. In this appendix we wish to formally define their roles.

- **Modeller**: The role of a modeller is to capture a real world decision problem as a mathematical model and translate it into a machine readable form in a particular modelling system language such as MPL, GAMS or AMPL. A modeller is also concerned with investigation of the model and its solution, in particular, how changes in different components of a model can affect the solution. Synonyms: *MS/OR Professional, Analyst*

- **End User**: The end user is only interested in the decision making and uses the results from the model. Hence, in an integrated modelling and solving environment, the solution values must be translated into description of decisions and expressed in units which make sense to the end user. Synonyms: *Problem Owner, Decision Maker*

APPENDIX B: ALGEBRAIC STATEMENTS

The algebraic statements are generated by the model generator using MPL syntax as shown below:

```
TITLE
        Multi Mode Prod
{ This is a model of producing NUTS, BOLTS and WASHERS in multi-period
production mode }
INDEX
        L = (M1,M2,M3)  {Machine}
        J = (Normal, Overtime)    {Mode}
        I = (Summer, Winter)      {Period}
        K = (P1,P2,P3)            {Product}
DATA
        Demand[K,I] := Improv("Demand", "Value") {25,30,30,30,25,25}
        { Demand for a prod in a time period}
        MacAvail[I,L,J] := Improv("MacAvail", "Value")
        {100,80,100,90,40,30,110,100,50,40}
```

```
        { No of hrs of machine availability}
        Price[I,K] := Improv("Price", "Value") {10,10,9,11,11,10}
        { Selling price of Product in time I}
        ProdCost[I,L,J,K] := Improv("ProdCost", "Value")
        {2,3,4,3,4,5,4,3,2,5,4,3,1,0,0,2,0,0,3,4,5,4,5,6,5,4,3,6,5,4,2,0,0,3,0,0}
        { Production Cost}
        ProdTime[I,L,J,K] := Improv("ProdTime","Value")
        {4,5,6,3,4,57,6,6,6,5,5,3,0,0,2,0,0,5,6,7,4,5,5,8,7,7,7,6,6,4,0,0,3,0,0}
   {No of hrs required to produce a prod in time I in mode J on machine L}
        ResalVal[K] := Improv("ResalVal", "Value") {2,2,1}
        { Resale value of a product}
        StoreCap[K]:= Improv("StoreCap","Value") {20,20,0}
        { Storage capacity for a product}
        StoreCost[K]:= Improv("StoreCost","Value") {1,1,1}
        {Storage cost for a product}
DECISION VARIABLES
        AmnStore[K,I]
        {No of products to be stored in a time period}
        ProdQty[K,L,J,I] WHERE (ProdCost <> 0)
        { No of products to be produced}
        SalesAmnt[K,I]
        { No of products to be sold in a time period}
MODEL
        Max Profit = SUM(I,K:Price*SalesAmnt)
                     - SUM(I,L,J,K:ProdCost*ProdQty)
                     - SUM(I=Summer,K:StoreCost*AmnStore)
                     +SUM(I=Winter,K,ResalVal*AmnStore)
        {Objective function}
SUBJECT TO
        Aval[L,J,I]:                {Machine Availability}
          SUM(K:ProdTime*ProdQty) <= MacAvail
        BalSum[K,I=Summer]:    { Balance equation for first period }
          SUM(L,J:ProdQty)-AmnStor[K,I]-SalesAmnt[K,I] = 0
        BalWin[K,I=Winter]:              { Balance equation for 2nd period }
         SUM(L,J:ProdQty)+AmnStore[K,I-1]-AmnStore[K,I]-SalesAmnt[K,I] = 0
        Dmnd[K,I]:                  { Demand for product}
          SalesAmnt >= Demand
        StoreCapt[K,I=Summer]:           { Storage Capacity for products }
          AmnStore <= StoreCap
END
```

APPENDIX C: FILTER SYNTAX

In order to follow a hypertext link a selection criterion may be invoked. This selection criterion is defined as a filter. A filter specifies a logical condition that has to be satisfied to follow a hypertext link. It can be a simple filter or a combination of other filters and is formally defined as follows:

<Filter> ::= <Term> | (<Term>) <LogOp> (<Term>) | NOT <Term> | {<Term>}
<LogOp> ::= AND | OR
<Term> ::= <RowRef> | <ColRef> | <ValRef>
<RowRef> ::= <RowSpec> | <RowIndex>
<RowSpec> ::= Constraint IS <Rtype>
<Rtype> ::= >= | <= | = | range
<RowIndex> ::= RIndex <IndOp> RIndexVal
RIndex ∈ Indices defined in the algebraic model and referenced (used) in the constraint definition.
RIndexVal ∈ RIndex
<IndOp> ::= > | < | = | != | >= | <=
<ColRef> ::= <ColSpec> | <ColIndex>
<ColSpec> ::= Variable IS <CType>
<CType> ::= Real | Integer | Binary | SOS1 | SOS2
<ColIndex> ::= CIndex <IndOp> CIndexVal
CInex ∈ Indices defined in the algebraic model and referenced (used) in the varaiable definition
CIndexVal ∈ CIndex
<ValRef> ::= <Element> <ValOp> Value
<ValOp> ::= >= | <= | = | !=
<Element> ::= RHS | Cost | AMat | Solution Value | Primal | Dual | Up Bnd | Lo Bnd | Fx Bnd
Value ∈ R

Interpretation of the Signs:
( ) Precedence
{ } Repetition
| Selection
< > Non terminal terms
::= Is Syntactically defined as

APPENDIX D: HYPERLINKS IN DOME

Diagram 4.3 provided a broad view of hypertext links amongst the hypertext module in DOME. In this appendix we establish the possible connections from each module to others. Type of connections and any exceptions are also specified.

TO FROM	*DOME Screen*	*Algebraic Statements*	*Database System*	*Solution Values*	*Matrix Rep. (MPS)*
DOME Screen	N/A	N	F	F	F
Algebraic Statements	N/A	N	F	F	F**
Database System	N	N	F	F*	F*
Solution Values	N	N	F*	F	F
Matrix Rep. (MPS)	N	N	F	F	F

Key:
F: Filter Reference
N: Name Reference
N/A: Not Applicable
*: Only applied to decision variables
**: Only applied to decision variables and constraint names

GLOSSARY of Information Systems and Mathematical Programming Acronyms

CATEGORY in MDDB	Sets in LP
DSS	Decision Support System
EIS	Executive Information System
IP	Integer Programming
IT	Information Technology
ITEM in MDDB	Set members in LP
LP	Linear Programming
MDDB	Multidimensional Database System
MIS	Management Information System
MP	Mathematical Programming
OLAP	On-Line Analytical Processing
RDBS	Relational Database System

14

ENHANCING USER UNDERSTANDING VIA MODEL ANALYSIS IN A DECISION SUPPORT SYSTEM

David M. Steiger

Bryan School of Business and Economics
University of North Carolina at Greensboro
Greensboro, NC 27215
steigerd@iris.uncg.edu

ABSTRACT

This paper focuses on the analysis of solved, model instances in a decision support system (DSS). Specifically, we provide a unique taxonomy for the existing model analysis systems and approaches based on the type of logic employed, deductive or inductive. Then, we expand the taxonomy based on the premise that the primary purpose of a DSS is to enhance the decision maker's understanding of the modeled business environment. We then propose a framework for the next evolution of analysis systems and discuss the potential application of knowledge bases and artificial intelligence to enhance a user's understanding of mathematical models in a DSS environment.

1 INTRODUCTION

A Decision Support System (DSS) is a computer-based system whose objective is to help the decision maker understand semi-structured and ill-structured problems by providing access to, and interaction with, data and models. That is, a DSS provides the decision maker with a means of perceiving the intricacies, tradeoffs and key relationships of the modeled system; e.g., the controllable parameters that are most important, how these parameters interact with each other to impact the model results, or what accounts for some unexpected difference in what-if cases. Historically, most of the research and development efforts in DSS, especially in the management science literature, have focused on the solution techniques of modeling. Only recently have we seen research efforts that are devoted to model formulation aids such as modeling languages, graphical user interfaces and knowledge-based tools intended to simplify the formulation process. Relatively less effort has been devoted to model analysis (see Section 2 for notable exceptions to this general statement).

The focus of this paper is on model analysis in a DSS; specifically, the computer-aided analysis of one or more model instances to enhance the understanding by the decision maker of the business environment represented in the model. Within the model analysis process, there are several potential sources of enhanced understanding. One potential source is the deductive analysis of a single model instance within the framework of the knowledge inherent in the associated mathematical model. For example, if the model is formulated and solved with linear programming (LP), the LP sensitivity analysis may help the decision maker understand the relative importance of key controllable parameters; e.g., how an (incremental) change in a single parameter impacts the solution [16]. Another potential source of understanding is the inductive analysis of multiple, related solved model instances. For example, suppose several model instances that have different parameter values are solved. Then, an analysis of the resulting multiple, solved model instances may provide new knowledge, and enhanced understanding, of the relative importance of the individual parameters and how the key parameters interact to affect the model solution [40, 45].

The purpose of this paper is twofold: 1) to propose a taxonomy of the current model analysis systems and approaches, and 2) to expand this taxonomy to include a new evolution of model analysis systems. This evolution calls for the integrated use of 1) deductive and inductive model analysis, 2) the implicit or explicit knowledge-base of the model, and 3) the arguments that enhance understanding of the modeled environment. For example, a simplified auxiliary model (or metamodel) generated from the inductive analysis of multiple model instances [41] might be enhanced and interpreted in view of the paradigm-specific knowledge-base of the model and the associated deductive sensitivity analysis [16].

This evolution in model analysis will be encouraged by: 1) the needs of the decision maker in his/her search for understanding of the complex problem represented by the model, 2) the innovative advances in computer science (machine learning, database mining [21], and other data analysis technologies) and their application to insightful model analysis, and 3) the synergistic combination of cognitive science (theory of understanding), artificial intelligence (knowledge extraction, machine learning, expert systems) and DSS (model management systems).

In Section 2, we examine model analysis in light of a taxonomy that not only classifies the current analysis systems, but leads naturally to the next evolution. In Section 3, we provide a framework for the development of the next evolution of model analysis. And in Section 4, we discuss several conclusions and research directions.

2 TAXONOMY AND REVIEW OF THE CURRENT MODEL ANALYSIS SYSTEMS

Model analysis systems, an integral component of model management systems [9, 15], are systems that 1) interpret and manipulate the output of model solvers, 2) analyze existing knowledge and/or extract new knowledge concerning the complex business environment represented by the model, and 3) enhance the decision maker's insight into the business environment represented by the model [6, 23]. The two existing classes of model analysis are based on the predominant type of logic employed; i.e., deductive and inductive model analysis. Both of these classes are defined and discussed below, complete with descriptions of existing and prototype analysis systems.

2.1 Deductive Model Analysis Systems

Deductive model analysis systems (DMAS) typically apply paradigm- or model-specific knowledge to a single instance of the model, addressing such questions as "Why is this the solution?", "Why do the solutions of two model instances differ so much?" or, in the case of linear programming models, "Why is this instance infeasible?" There are DMAS for each of the three major modeling paradigms: linear programming (LP), simulation and spreadsheet models.

One example of a linear programming-based DMAS is ANALYZE [16, 17, 18], a successor of PERUSE [31]. ANALYZE is a computer-assisted analysis system that provides interactive query of the LP matrix and the solution values to: 1) document and verify the model, 2) trace causation via network path analysis, 3) analyze causes of infeasibility, and 4) simplify the model via substructure elimination or imbedded network identification. ANALYZE provides "an artificially intelligent environment, with English translation of results automatically obtained" that is suitable for both expert and non-expert users of linear programming [18].

In addition to LP-based DMAS, there is a prototype DMAS which helps analyze simulation models named I-KBS [37]. I-KBS is an intelligent system which helps managers conceive and develop simulation models as well as learn and verify interactions between model entities. I-KBS provides run-time graphics to aid decision makers in their understanding of causal relations (i.e., events that cause other events) and detect errors in logical relationships.

Finally, there is a DMAS for spreadsheet-based models named IFPS/Plus [11]. IFPS/Plus is a commercially-available system based on the ROME/ERGO research system developed at Carnegie-Mellon University [29, 30, 46]. This DMAS features a set of explanation commands for interpreting and explaining the difference between a specified variable in two model instances (spreadsheets) or in two different periods of the same spreadsheet. These commands allow the user to "trace the path of influences in a model or consolidated structure so you can see not only which variables are most important from a definition viewpoint, but also which ones had the most influence on the change in values" [11]. For example, the WHY command lets the user ask questions about why a value changes, providing the answer in terms of the smallest subset of variables that accounts for most, say 80%, of the change. Variables in this subset are further categorized based on whether they have a positive or negative (counteractive) effect on the variable in question. Unfortunately, the WHY command is limited to those variables or parameters that are computed by the same spreadsheet formulas.

In addition, the submodel/surrogate model analysis capabilities of Candle-lighting [22, 23, 24] fall into the DMAS category. In this approach, a submodel consists of an equation (i.e., model) that, when executed, determines the value for the parameter of the main model; e.g., an objective function cost coefficient may really be a function of labor hours, cost of labor and an inflation factor. A surrogate model represents a rule of thumb about some underlying relationship between the main model parameter and other, more elementary, components. Candle-lighting uses these user-specified submodels and surrogate models to answer such questions as: Do any submodels have common variables, and if so, how much can these common variables change before the basis changes? and What is the cost of changing the results of the model given that we can alter some assumptions in the model?

2.2 Inductive Model Analysis Systems

Inductive model analysis systems (IMAS) operate on a set of related model instances that represent historical situations familiar to the decision maker and/or what-if cases specified by him/her. The primary goal of IMAS is to help the decision maker develop insight(s) into the business environment represented by the model. Geoffrion [13] states that such insight is seldom evident from the output of a single model run, but is instead generated from the study of many 'what-if' cases; i.e., the decision maker must go beyond knowing "what" the solution is for a given set of input data, but must discern "why" the solution is what it is. Developing insight(s) is ultimately a process of finding trends and surprising model behavior based on comparing multiple model instances and/or comparing the behavior of the model to that which is expected or observed in the real world [20]

Inductive model analysis systems (IMAS) are distinguished from deductive analysis systems (DMAS) by both the required input and the type of processing logic employed. That is, IMAS operate on many model instances and apply inductive analysis technologies to extract new knowledge, whereas DMAS operate on one or two model instances and apply deductive analysis techniques based on known, paradigm- or model-specific knowledge. Also, IMAS are independent of the modeling paradigm, matching the generality and analysis requirements of state-of-the-art modeling languages such as Structured Modeling [14].

There are several model analysis systems known to the authors that qualify as IMAS. The first of these [26] applies regression analysis to multiple runs of a (simulation) model to determine the impact of changes in the model's input parameters on its output; i.e., it uses regression to build a metamodel of the simulation model. For example, in an M/M/s queuing model we might want to determine the impact of changes in three input parameters (the arrival rate λ , the service rate μ, and the number of servers, s) on the output variable average queue length, v. To do so, we might select 10 unique combinations of values for λ, μ and s, with each combination replicated 20 times generating 20 v values for different pseudorandom number seeds. The resulting 200 combinations of λ, μ, s, and v would then be used as input for the regression analysis, with the independent variables being λ, μ, and s, and the dependent variable being v.

The authors of this analysis technique rely heavily on statistical experimental design to determine the optimal number of simulation runs that provide statistically significant results and yet keep simulation runs to a minimum. For example, they recommend an orthogonal experimental design, specifically a fractional factorial design, when considering main effects only; i.e., assuming a first order polynomial regression model with no interaction between independent variables. They also suggest that it is better to assume significant interactions between pairs of factors, as well as quadratic effects; i.e., a second order polynomial regression model. Such designs require many more simulation runs. To reduce the number of possible dependent variables in such cases, the authors suggest the sequential bifurcation screening method [3, 4].

Such regression analysis applied to multiple simulation runs makes several assumptions: 1) that the analyst knows, a priori, the structure of the resulting metamodel (i.e., the order of the approximating polynomial and the statistically significant interaction terms), 2) that the error terms are well-behaved (i.e., normally independently distributed), 3) and that prior knowledge is available concerning the direction of change in the dependent variable caused by changes in each independent variable and/or the interaction term.

Another IMAS is LISA [38, 39]. LISA accepts, as input, a set of model instances generated via Monte Carlo simulation and uses standardized rank regression coefficients and partial rank correlation coefficients to determine the key parameters in a probabilistic model; i.e., those parameters that are most important in determining variations in the output variable.

A third IMAS is Global Sensitivity Analysis [45], the goal of which is to provide extended sensitivity analysis. The input for Global Sensitivity Analysis consists of: 1) the marginal distributions of up to ten stochastic model parameters or groups of parameters, where a parameter might be an objective function coefficient, a right-hand side value, or a matrix coefficient, if the model is a linear programming model, 2) a pre-specified set of nonlinear, multinomial terms involving the stochastic parameters, with the set of terms based on prior human knowledge or model-specific expertise, and 3) a large set of perhaps 2000 solved model instances obtained by repeatedly varying (via Monte Carlo simulation) one or more of the stochastic parameters and resolving the model. Processing of the solved model instances consists of using backward, stepwise, least-squares regression to find the "best fit" between some subset of multinomial terms and the set of instance solutions. The output is the best fit polynomial (consisting of an additive model of 20 - 40 or more multinomial terms) and the associated regression coefficients, along with the coefficient of determination, R2. The output polynomial is then used to determine the sensitivity of the original model to changes in a given parameter and to determine which stochastic parameter(s) are most influential in the model solution. The primary advantages of Global Sensitivity Analysis include the following: 1) least-squares fits of multinomial terms can be used to provide good approximations for "explaining" the variations in solutions, and 2) all software used in this system is commercially available as add-ins for standard spreadsheet packages.

The fourth IMAS is called INSIGHT [40, 41]. INSIGHT is an prototype analysis environment whose goal is to generate, using artificial intelligence, a simplified auxiliary model that helps the decision maker develop insight(s) into the business environment being modeled. The input for INSIGHT consists of a small set of perhaps 10-50 solved model instances generated either by the decision maker through a series of what-if instances or by Monte Carlo simulation if the marginal distributions of the stochastic parameters are known or estimated. Processing consists of two sequential steps: first, the Group Method of Data Handling (GMDH) algorithm [12, 36] is utilized to generate an additive model of linear and/or nonlinear multinomial terms that explains variations in the instance solutions based on variations in the input parameters, and second, stepwise regression is utilized to form a "best fit" simplified auxiliary model from a subset of the multinomial terms produced by GMDH. This simplified auxiliary model helps the decision maker develop insight(s) into the interactions and tradeoffs between key model parameters, as well as the sensitivity of model solutions to changes in parameter values.

GMDH, a cross between neural networks and nonparametric statistical regression, provides the following advantages in this application: 1) no prior knowledge of the model's structure, or pre-established set of multinomials, is required, 2) a smaller set of samples is required as compared to statistical regression (e.g., GMDH requires as few as 100 sample points or model instances to construct a multinomial of degree 8 in 20 variables), 3) no assumptions with respect to linearity or continuity of the solutions values (e.g., objective function values in the case of LP models) or normality of residuals are required, and 4) chances of overfitting are reduced [2, 36]. INSIGHT has been used to duplicate the insight-generating simplified auxiliary model for a classical facility location model without requiring the human expertise or mathematical manipulation used in Geoffrion's [13] analysis.

A final IMAS approach is the inductive analysis approach of Candle-lighting [24, 25]. Candle-lighting uses a genetic algorithm to generate a set of model instances which promises to closely satisfy or optimize the goals represented as part of the model's construction.

2.3 Limitations of Deductive/Inductive Model Analysis Systems

There are two primary shortcomings of current deductive and inductive model analysis systems. First, no mechanism exists to promote, or even allow, the integration of existing knowledge inherent in a DMAS (or the model itself) with new knowledge extracted via an IMAS. Such integration of knowledge could provide a significant enhancement to the decision maker's understanding of the business environment.

Second, there is no methodology or basis for evaluating multiple metamodels, either with respect to each other or with respect to validating the sufficiency and necessity of individual models and their components. Such metamodels can be generated in several ways: 1) by applying different machine learning technologies, such as neural networks, GMDH, etc., 2) by changing the set of instances that are input to inductive technologies, 3) by changing the input order of those instances (which may lead to different models in neural network-based technologies), and/or 4) by changing the appropriate technology parameters (e.g., the step size in neural network techniques, complexity parameter in GMDH-based techniques, or the threshold criteria for the removal/addition of terms in stepwise regression).

3 A FRAMEWORK FOR ENHANCING UNDERSTANDING VIA MODEL ANALYSIS

The primary purpose of the analysis systems is to enhance the decision maker's understanding of the complex business environment represented in the DSS. We assume in the following discussion that the decision maker uses the analysis of multiple what-if cases to explore the thing s/he is trying to understand; e.g., the key factors and relations underlying the business environment.

The two primary components of this framework include metamodel generation and evaluative argument generation. Each framework component is discussed individually below.

3.1 Metamodel Generation

One method of explaining a model, and especially the effects of changes in input parameters on the model, is by deriving and presenting one or more metamodels [1, 26, 41]. Such metamodels can be generated as outputs of the inductive model analysis systems discussed in Section 2 above, as relations specified by the decision maker based on human expertise, and/or as new models generated from a combination of deductive and inductive technologies. Multiple metamodels can also be generated by applying different AI-based technologies to the generation and analysis of multiple, solved model instances (e.g., neural networks, GMDH, genetic algorithms) and/or by varying the sequence or contents of the set of model instances input to these inductive technologies. Note that evaluative arguments (see below) can be used to eliminate inapplicable and relatively inferior metamodels resulting from less appropriate technologies, insufficient data, too few model instances, and/or designs that are not generalizable (due perhaps to over learning). Further, since such elimination occurs at an early stage in the analysis there is little penalty for having too many metamodels or applying different technologies.

3.2 Evaluative Argument Generation

Evaluative arguments compare different models and metamodels to determine the strengths and weaknesses of each, evaluate the necessity and sufficiency requirements of an individual metamodel and its components, and provide verification for a given model. In comparing different models, the evaluative arguments can rely on internal evaluation measures such as the accuracy of a model over some set of instances, or model complexity with respect to the number of terms, degrees of polynomial terms, etc. The system can also query the decision maker for other measures of comparison along with their definitions and relative evaluation weights. In addition, the various technologies can be utilized to help the decision maker rank order the models with respect to one or more criteria; e.g., the analytical hierarchy process.

Evaluative arguments also include a hypothesis testing capability for all potential models. This includes the ability to evaluate potential models based on a) rules presented to, and validated by, the decision maker, b) internally generated validity checks based on the substitution of extreme parameter values into the model to see if reasonable/feasible results occur, and c) substitution of multiple values of non-key parameters into the original model to verify that the solution does not change significantly [34].

In evaluating a model's sufficiency and necessity, one must use specific measures that capture the explanatory power of the model and its components. For example, the coefficient of determination, R^2, provides a measure of sufficiency for an inductive model that supposedly explains the variations in output due to changes in input parameters. Likewise, the necessity requirement of model components can be measured by Wagner's [45] all-save-one R^2-based measure.

In evaluating metamodels and their component terms, dimensional analysis should be included for all mathematical expressions in the model to insure dimensional consistency, as suggested by several researchers [35, 43, 44]. Bhargava reduces the dimensional analysis problem to numeric arithmetic by a "suitable encoding of fundamental units as primes," deriving units as "products of primes" [5].

As an example of the application of evaluative arguments, consider the following proposed model whose purpose is to calculate the area of a parallelogram:

$$A = (bh) + (m * c^2) + [b + l_1 - (t + l_2)]$$

where b = length of the parallelogram's base, h = height of the parallelogram, l_1 = length of one of the sides connected to the base, l_2 = length of the side opposite to l_1, t = length of the parallelogram's top, m = mass of the parallelogram, c = speed of light (in m/sec2), A = area of the parallelogram (in m^2), and all lengths are measured in meters.

In this example, the second term is eliminated due to dimensional inconsistency with the area, A. In addition, the third term is eliminated due to the necessity condition; i.e., it can be omitted without affecting the accuracy of the model. Further, both the base and the height in the first term are necessary, since omitting one or the other reduces the accuracy of the calculated area when compared to given instances. Finally, the first term provides an accurate solution when compared to several given instances; i.e., it is a valid model.

4 CONCLUSIONS AND RESEARCH DIRECTIONS

It has been stated that the purpose of modeling is insight, not numbers [13]. Given that modeling is the central component of every DSS [7, 42] and that insight is a special type of (holistic) understanding [19, 27, 28, 32, 33], then it follows that the purpose of DSS is understanding; specifically, understanding by the decision maker of the business environment depicted in the DSS. We have suggested that one source of such understanding is the analysis of one or more model instances using deductive or inductive logic, or both. Also required are various types of arguments that show how the model and model instances enhance the decisionmaker's understanding.

In the DSS literature, especially the management science aspects of it, the focus of research has historically been on modeling techniques and solution algorithms. It is time now to give the analysis of solutions the importance it deserves. An early effort in this regard is to articulate the tasks of model analysis. Several DSS researchers [8, 10] have developed some theory in this vein, but it is still an area needing further refinement. Specifically, we need to identify a "minimal spanning set" of tasks that leads to successful model analysis, and validate this set of tasks through experimentation.

Another research area could explore and evaluate technologies that are potentially applicable to analysis and understanding. Initial evaluation could match the input, processing, output and feedback characteristics of these technologies against the corresponding requirements of the prime analysis tasks mentioned above. The results could provide a research agenda for the application of the technologies to the analysis tasks, along with empirical testing of their effectiveness.

REFERENCES

[1] Agrawal, S. C. (1985). Metamodeling: A Study of Approximations in Queueing Models. The MIT Press, Cambridge, MA.

[2] Barron, R. L., A. N. Mucciardi, F. J. Cook, A. R. Barron and J. N. Craig (1984). Adaptive Learning in Networks: Development and Applications in the U.S. of Algorithms related to GMDH. In S. J. Farlow (Ed.) Self-Organizing Methods in Modeling: GMDH Type Algorithms. Marcel Dekker, New York, 25-66.

[3] Bettonvil, B. (1990). Detection of Important Factors by Sequential Bifurcation. Tilburg University Press, Tilburg, Netherlands.

[4] Bettonvil, B. and Kleijnen, J.P.C. (1994). Identifying the Important Factors in Simulation Models with Many Factors. Working Paper, Department of IS and Center, Tilburg, University, Tilburg, Netherlands.

[5] Bhargava, H. K. (1993). Dimensional Analysis in Mathematical Modeling Systems: A Simple Numerical Method. ORSA Journal on Computing, 5:1, 33-39.

[6] Blanning, R. W., A. B. Whinston, A.M. Chang, V. Dhar, C. W. Holsapple, M. Jarke, S. O. Kimbrough, J. Leren, and M. J. Prietula (1992). Model Management Systems. In (E. A. Stohr and B. R. Konsynski, Editors) Information Systems and Decision Processes. IEEE Computer Society Press, Los Alamitos, CA, 205-206.

[7] Bonczek, R. H., C. W. Holsapple and A. B. Whinston (1980). Future Directions for Developing Decision Support Systems. Decision Sciences, October, 1980.

[8] Brennan, J. J. and J. J. Elam (1986). Understanding and Validating Results in Model-based Decision Support Systems. Decision Support Systems. 2, 49-54.

[9] Dolk, D. R. (1986). A Generalized Model Management System for Mathematical Programming. ACM Transactions on Mathematical Software. 12:2, 92-126.

[10] Elam, J. J. and B. Konsynski (1987). Using Artificial Intelligence Techniques to Enhance the Capabilities of Model Management Systems. Decision Sciences, 18:3. 487-501.

[11] EXECUCOM (1992). Interactive Financial Planning System: User's Manual. EXECUCOM, Austin, TX.

[12] Farlow, S. J. (Ed.) (1984). Self-Organizing Methods in Modeling: GMDH type Algorithms. Marcel Dekker, New York.

[13] Geoffrion, A. M. (1976). The Purpose of Mathematical Programming Is Insight, Not Numbers. Interfaces. 7:1, 81-92.

[14] Geoffrion, A. M. (1987). An Introduction to Structured Modeling. Management Science. 33:5, 547-588.

[15] Greenberg, H. J. (1982). A Tutorial on Computer-Assisted Analysis. In H. J. Greenberg, F. H. Murphy and S. H. Shaw (Eds.) Advanced Techniques in the Practice of Operations Research. Elsevier Science Publishing Company, New York, 212-249.

[16] Greenberg, H. J. (1993a). A Computer-Assisted Analysis System for Mathematical Programming Models and Solutions: A User's Guide for ANALYZE. Kluwer, Boston, MA.

[17] Greenberg, H. J. (1993b). Enhancements of ANALYZE: A Computer-Assisted Analysis System for Linear Programming. ACM Transactions on Mathematical Software. 19:2 223-256.

[18] Greenberg, H. J. (1994). Syntax-directed Report Writing in Linear Programming Using ANALYZE. European Journal of Operational Research. 72, 300-311.

[19] Hilgard, E. R. (1956). Theories of Learning,(2nd Ed). Appleton-Century-Crofts, Inc., New York.

[20] Jones, C. V. (1992). "User Interfaces and Operations Research," in E. G. Coffman, J. K. Lenstra and A. Y. Kan (Eds.) Handbook of Operations Research, Vol. 13, North-Holland/Elsevier, Amsterdam, 603-668.

[21] Kaufman, K.A., R. S. Michalski, and L. Kerschberg (1991). Mining for Knowledge in Databases: Goals and General Description of the INLEN System. In (G. Piatetsky-Shapiro and W. J. Frawley, Editors) Knowledge Discovery in Databases, AAAI Press, Menlo Park, CA, 449-462.

[22] Kimbrough, S. O., C. W. Pritchett, M. P. Bieber and H. K. Bhargava (1990). The Coast Guard's KSS Project, Interfaces 20:6, 5-16.

[23] Kimbrough, S. O., S. A. Moore, C. W. Pritchett and C. A. Sherman (1992). On DSS Support for Candle-Lighting Analysis, Transactions of DSS-92. 118-135.

[24] Kimbrough, S. O., J. R. Oliver and C. W. Prichett (1993). On Post-Evaluation Analysis: Candle-Lighting and Surrogate Models. Interfaces, 23:3, 17-28.

[25] Kimbrough, S. O. and J. R. Oliver (1994). On Automating Candle-Lighting Analysis: Insight from Search with Genetic Algorithms and Approximate Models. Proceedings of the Twenty-seventh Annual Hawaii International Conference on System Sciences, 1994. 536-544

[26] Kleijnen, J. P. C. (1994). Sensitivity Analysis and Optimization of Simulation Models. Working Paper, Center and Department of I.S., Tilburg University, Tilburg, Netherlands.

[27] Kohler, W. (1925). The Mentality of Apes. (English translation by E. Winter). Harcourt, Brace and Co., New York.

[28] Kohler, W. (1969). The Task of Gestalt Psychology. Princeton University Press, Princeton, NJ.

[29] Kosy, D. W. and B. P. Wise (1984). Self-explanatory Financial Planning Models. Proceedings of the National Conference of Artificial Intelligence, August 1984, 176-181.

[30] Kosy, D. W. and B. P. Wise (1986). Overview of Rome: A Reason-Oriented Modeling Environment in L.F. Psu, editor). Artificial Intelligence in Economics and Management. Elsevier Science Publishers, North-Holland, 21-30.

[31] Kurator, W. G. and R. P. O'Neill (1980). PERUSE: An Interactive System for Mathematical Programs. ACM Transactions on Mathematical Software. 6:4 489-509.

[32] Lee, D. L. (1965). "Perception, Intuition and Insight" in (W. R. Niblett, Editor) How and Why Do We Learn, Faber and Gaber, London.

[33] Logan, F. A. and D. P. Ferraro (1978). Systematic Analyses of Learning and Motivation. John Wiley & Sons, New York.

[34] Nickerson, R. S., D. N. Perkins and E. E. Smith (1985). The Teaching of Thinking, Lawrence Erlbaum Associates, Hillsdale, N.J.

[35] Piela, P.C., T. Epperly, K. Westerberg and A. Westerberg (1991). An Object-Oriented Computer Environment for Modeling and Analysis, Part 1 - The Modeling Language. Computers and Chemical Engineering, 15:1, 53-72.

[36] Prager, M. H. (1988). Group Method of Data Handling: A New Method for Stock Identification. Transactions of the American Fishery Society. 117, 290-296.

[37] Reddy, Y. V. (1985). The Role of Introspective Simulation in Managerial Decision Making. DSS-85 Transactions, IADSS, University of Texas at Austin, 18-32.

[38] Saltelli, A. and T. Homma (1992). Sensitivity Analysis for Model Output. Computational Statistics and Data Analysis. 13 (1992) 73-94.

[39] Saltelli, A. and J. Marivoet (1990). Non-parametric Statistics in Sensitivity Analysis for Model Output: A Comparison of Selected Techniques. Reliability Engineering and Systems Safety. 28 (1990) 229-253.

[40] Sharda, R. and D. M. Steiger (1995). Using Artificial Intelligence to Enhance Model Analysis. In (Nash and Sofer, editors) The Impact of Emerging Technologies on Computer Science and Operations Research, Kluwer Publishing 263-280.

[41] Sharda, R. and D. M. Steiger (1996). Inductive Model Analysis Systems: Enhancing Model Analysis in Decision Support Systems. Information Systems Research (to appear).

[42] Sprague, R. H., Jr. and E. D. Carlson (1982). Building Effective Decision Support Systems, Prentice-Hall, Englewood Cliffs, NJ.

[43] Stephanopoulos, G., G. Henning and H. Leone (1990). MODEL.LA. A Modeling Language for Process Engineering - I. The Formal Framework. Computers and Chemical Engineering, 14:8, 813-846.

[44] Stephanopoulos, G., G. Henning and H. Leone (1990). MODEL.LA. A Modeling Language for Process Engineering - II. Multifaceted Modeling of Processing Systems. Computers and Chemical Engineering, 14:8, 847-869.

[45] Wagner, H. M. (1995). Global Sensitivity Analysis. Operations Research, (November/December, 1995) 948 - 969.

[46] Wise, B. P. and D. W. Kosy (1986). Model-Based Evaluation of Long-Range Resource Allocation Plans in. (L. F. Pau, editor) Artificial Intelligence in Economics and Management. Elsevier Science Publishers, North-Holland, 93-102

15

BANK FAILURE PREDICTION USING DEA TO MEASURE MANAGEMENT QUALITY

Richard S. Barr and Thomas F. Siems*

Department of Computer Science and Engineering
Southern Methodist University
Dallas, Texas 75275
email: barr@seas.smu.edu.

** Financial Industry Studies*
Federal Reserve Bank of Dallas
Dallas, Texas 75201
email: siems@seas.smu.edu.

ABSTRACT

Presented are new failure-prediction models for detecting a bank's troubled status up to two years prior to insolvency using publicly available data and a new category of explanatory variable to capture the elusive, yet crucial, element of institutional success: management quality. Quality is assessed using data envelopment analysis (DEA), which views a bank as transforming multiple inputs into multiple outputs, focusing on the primary functions of attracting deposits and making loans. The new models are robust and accurate, as verified by in-depth empirical evaluations, cost sensitivity analyses, and comparisons with other published approaches.

Bank failures have increased at an alarming rate in the past decade. The number of commercial bank failures averaged less than seven per year from 1950 to 1980, but has since escalated to an annual average of 175 from 1986 through 1991. This rapid increase raises many questions regarding the safety and soundness of the banking industry: Why has there been a sudden increase in the number of bank failures? What can be done to slow the bank failure rate? How can the likely collapse of an institution be anticipated and prevented?

Often the failure of a bank can be avoided, or the bailout costs minimized, by early detection of a bank's troubled status and subsequent intervention by regulatory authorities. Key to this effort is the identification of the bank's potential for failure. To this end, researchers have sought mathematical models that predict institutional failure in an accurate and timely manner.

Interest in predicting bank failures dates from Secrist's classic study [21]. It examined 741 national banks that failed in the late 1920s and early 1930s and 111 banks that did not fail prior to 1933 to "secure indications of likely survival or of death." This comparative analysis was the first of its kind and sought to discover the symptoms of failure and non-failure.

Not until the early 1970s did interest in predicting bank failures return. This renewed interest stemmed from the development of "newer" classification techniques, such as discriminant analysis and nonlinear regression, for separating "potential failures" from "probable survivors." Now, with the number of failures surging, the need for more effective prediction models has become evident. While an early-warning system cannot replace the on-site examination, which allows for personal interaction with the bank's management and employees and permits first-hand evaluation of operating procedures, levels of risk-taking, and long-range strategic planning. However, an effective warning system can complement the on-site examination process by identifying troubled institutions that need early examination or possible intervention.

As outlined in [25], for an early-warning (failure-prediction) model to be useful to regulators it must have the following ingredients:

- *Understandability.* For the model to be accepted and used by the examination staff, it must be understandable and fit their intuitive perceptions of how banks fail; the model will not be trusted if it is considered a "black box" where banks are rated according to some highly complex process.
- *Ease of use/quickness.* To yield the greatest benefits, an early-warning system's data must be timely and accurate . Fortunately, banks are already required to submit quarterly balance sheet and income statement information to banking regulators, and models built around this database can rely on up-to-date, reliable inputs.

One means of enhancing the regulators' understanding of a model is to build it around familiar factors. When examiners evaluate a bank's health, they develop an overall rating based on *C*apital adequacy, *A*sset quality, *M*anagement quality, *E*arnings ability, and *L*iquidity position—hence the term *CAMEL rating.* Federal regulators developed the numerical CAMEL rating system in the early 1970s to help structure their examination process. Since then, the use of CAMEL factors has become widespread, due to its simplicity and use by regulators. Financial data and relationships are the principal ingredients for scoring capital, asset quality, earnings, and liquidity. Assessing management quality, however, requires professional judgment of a bank's compliance to policies and procedures, aptitude for risk-taking, development of strategic plans, and the degree of involvement by the bank's officers and directors in making decisions.

Most failure-prediction models include variables that can be categorized under four of the five CAMEL factors; the variable that is typically missing is the one which assesses management quality. This is a rather curious paradox since the quality of management is often cited as the *leading* reason for failure.

In this paper, we present a bank failure prediction model that is developed around a new paradigm for assessing a bank's management quality. This paradigm views a bank as processing multiple inputs to produce multiple outputs, and focuses on its key financial intermediation functions of acquiring deposits and making loans and investments. Using *data envelopment analysis* (DEA), a management quality metric is established that is designed as a proxy for the 'M' in the CAMEL rating.

For failure prediction, the management quality metric is combined with variables representing the other four factors in the CAMEL rating, as well as a proxy for local economic conditions. The resultant probit-regression model is compared with the leading alternative approaches and evaluated for sensitivity to prediction error. The empirical testing not only shows the new model to be dramatically superior to all other approaches, but underscores the validity and importance of the management quality metric in forecasting. This study not only advances the state-of-the-art in institutional failure prediction, but pioneers the use of DEA as a predictive factor that has far-reaching applications.

1 THE ROLE OF MANAGEMENT IN BANKING

Clearly, the quality of a bank's management is key to its long-run survival, and one of management's greatest challenges is coping with the industry's increasing uncertainties and accompanying risks. Internally, this means that bankers must effectively allocate scarce resources, implement controls and procedures to minimize risks and control costs, and be open to the use of new technologies to increase operating efficiencies. Externally, they must keep pace with new regulatory actions, economic fluctuations, societal trends, technological advances, and changes taking place in the global economy.

Kaufman [13] postulated that "to survive in a risky world, banking firms must cope with risk and manage it." This management of risks seems to be the key factor in bank failure. Consider the following statements from recent research studies:

> "...apart from fraud, a single event rarely cripples a bank fatally. Instead, the culprit is consistent mismanagement or risky strategies." [6]
>
> "...bank failure, which affects only a handful of banks, is caused by mismanagement; mismanagement, furthermore, of the basic, old-fashioned risks of banking like lending, liquidity, and controls." [9]
>
> "Management factors generally explain why one bank survives while another fails when facing almost identical circumstances." [7]
>
> "It is the management of the bank that determines success or failure." [18]
>
> "...management incompetence in its broad sense is a major cause of bank failure." [15]
>
> "...the ultimate determinant of whether or not a bank fails is the ability of its management to operate the institution efficiently and to evaluate and manage risk." [20]

Many studies of bank performance and bank failure cite management quality as the most important factor to long-run survival. For a bank to continue to survive, its management must understand, manage, and control the increased risks inherent in today's financial environment. This means that bankers must effectively allocate resources and efficiently control the bank's operations.

2 HOW CAN MANAGEMENT QUALITY BE MEASURED?

In 1993, Barr, Seiford, and Siems [2] (BSS) presented an approach to measuring a bank's managerial quality using only its financial statements. Their model considers the essential financial intermediation functions of a bank and computes a scalar measure of efficiency. This measure of transformational efficiency is considered a proxy for management quality and is derived using a multiple-input, multiple-output model as follows.

2.1 Data Envelopment Analysis: an Overview

In standard microeconomic theory, an enterprise can be viewed as a *decision-making unit* (DMU) that is concerned with transforming a set of m different inputs into s different outputs. For example, BSS models a bank as a DMU with six inputs (full-time-equivalent employees, salary expense, premises and fixed assets, other noninterest expense, total interest expense, and purchased funds) and three outputs (core deposits, earning assets, and total interest income). Decision units of the same kind can be compared for their relative transformational efficiency—this is often performed in traditional analyses using a series of simple ratios of individual outputs to individual inputs.

DEA is a non-parametric method initiated by Charnes, Cooper and Rhodes[10], which uses a generalized ratio to consider multiple inputs and outputs simultaneously. Specifically, if DMU k consumes amounts X_{ik} of inputs, $i = 1, \ldots, m$, and produces amounts Y_{jk} of outputs, $j = 1, \ldots, s$, then its overall efficiency can be given as: $E_k = \sum_{j=1}^{s} Y_{rk} u_{rk} / \sum_{i=1}^{m} X_{ik} v_{ik}$ where u_{jk} and v_{ik} are positive scaling factors, or *weights*, that are to be determined by the DEA process. Rather than using preset weights to evaluate all banks, an optimal set of weights is computed for each bank that makes its ratio as large as possible. These weights are found for an individual bank k by solving the following fractional linear programming problem:

$$\begin{aligned} &\text{Maximize } E_k \\ \text{subject to: } E_d &\leq 1, d = 1, \ldots, n \\ u_j, v_i &\geq 0, i = 1, \ldots, m; j = 1, \ldots, s \end{aligned}$$

By solving the above problem for each decision unit, an empirical production surface (or efficient frontier) is constructed that shows the maximum amount of outputs that can be achieved by combining various quantities of the inputs, based on the observed behavior of a population. DEA uses linear programming with observed data to locate these frontiers and determine the efficiency of each organizational unit. It determines which of the n DMUs define the surface and, hence are most efficient. Efficient DMUs lie on the surface and have a relative efficiency measure of 1; inefficient DMUs are not on the surface, have an $E_k < 1$.

Note that, unlike regression analysis, DEA is not limited to a single output, and is directed to identify the "best practice" units, not central location of the data. See [23] for a survey of DEA methods, and [22] for a 500-entry bibliography on DEA. Standard computer programs are available to solve such models (see [11]).

2.2 Using DEA as a Management Quality Metric

As described above, BSS model a bank as a transformer of six inputs into three outputs. Since such a process reflects the key activities of management, the DEA efficiency metric was expected to be a good proxy for managerial quality and empirical evidence in the BSS study supported this view. Specifically, the research revealed that significant statistical differences in average management quality scores existed up to three years prior to failure. Figure 4 illustrates the study's finding that from December 1984 to June 1987 the average DEA scores for a sample of surviving banks ranged between 0.8142 and 0.8315; whereas, for banks which failed the first half of 1988, the average scores dropped from 0.7986 in December 1984 to 0.6694 in June 1987.

Furthermore, when pooling the data to compute average scores for one year prior to failure, 94.6% of the survivors had scores above 0.7, but only 50% of the failed banks. Additionally, over 50 percent of the survivors scored above 0.8, while only 13.9 percent of the failures did. The average DEA score for failures one year prior to failure was 0.7032 and for survivors it was 0.8203.

Similarly, for banks that are two years prior to failure, there is again a demarcation between surviving banks and failed institutions. However, the differentiation in DEA scores is not as pronounced. For failures, the average DEA score two years prior to failure was 0.7637 and for survivors it was 0.8394.

The empirical results of the BSS study "confirm that the quality of management is crucial to a bank's survival. Scores for surviving institutions are statistically higher than the scores for failed banks. Furthermore, banks that are nearer failure are found to have lower efficiency scores." These results are significant in that banks that survive can be statistically differentiated from banks that fail based on the management quality scores generated by the DEA model. Hence, the use of these scores as a variable in a bank-failure prediction model seems promising.

3 EARLY-WARNING MODELS

An early-warning model is simply an established procedure (usually statistical) for classifying banks into groups (usually *failure* and *non-failure*). This is typically done using only financial characteristics of candidate institutions. The goal of an early-warning model is to identify an institution's financial weakness at the initial stage in its process of deterioration so as to warn interested parties of its potential failure. Banking regulators have recognized the usefulness of early-warning models and have conducted intensive research efforts in this area for nearly two decades so that they can classify problem institutions before they have a major financial impact on the federal deposit insurance fund.

Modern studies of institutional failure began with Beaver [4] who used financial ratios to predict bankruptcy of non-financial firms; this was also the first model to forecast solvency from accounting data . Altman [1] and Sinkey [25] subsequently developed discriminant models for predicting corporate bankruptcy.

Early-warning models for banks began with Meyer and Pifer [17]; since then, various researchers have used multivariate techniques to explain past closures and predict future failures. While discriminant models have evolved over the years, the variables used to predict failure have largely remained constant. For financial institutions, they generally fall into the CA_EL categories and, more recently, included variables to capture operating efficiencies and local economic conditions. (For other readings on early-warning models, see [8, 12, 17, 21, 26].)

Generally, the measures used to evaluate operating efficiencies are regarded as proxy variables for the quality of management. The major complication, however, is that it takes many ratios to encompass the operating characteristics of an organization. Hence, to measure management quality, most recent studies have focused on the ratio of total operating income to total operating expense.

Three prominent models are compared with the new models developed in this study: those appearing in [12, 16, 18]. These were selected because they use economic and publicly-available financial data, none of which require on-site examination. While all three studies did not define "failure" the same way, the

Martin [16], 1970-76 Data: 5642 Survivors, 58 Failures		
Logit Model Variables	*Coeff*	*Sig*[a]
Constant	-5.33	**
Net income / total assets	-120.86	**
Gross charge-offs / net operating income	2.20	**
Commercial & industrial loans / total loans	7.89	**
Gross capital / risk assets	-35.63	**

Hanweck [12], 1973-75 Data: 177 Survivors, 32 Failures		
Probit Model Variables	*Coeff*	
Constant	-4.14	*
Net operating income / total assets	-69.49	**
Equity capital / total assets	14.86	
% change in net operating income / total assets	-0.01	
% change in total assets	-1.18	
Loans / capital	0.26	**
Size: log(assets)	0.02	

Pantalone & Platt [18], 1983-84 Data: 226 Survivors, 113 Failures		
Logit Model Variables	*Coeff*	
Constant	-0.01	
Net income / total assets	-71.39	**
Equity capital / total assets	-11.79	**
Total loans / total assets	7.71	**
Commercial & industrial loans / total loans	3.72	**
|%| change in residential construction	0.10	

[a]Coefficient is significant: * = at the 0.05 level; ** = at the 0.01 level

Table 1 Summary of the Test Set of Bank-Failure-Prediction Models

traditional view that failure occurs through a declaration of insolvency by one of the regulatory agencies is adhered to for this test. All models utilize probit or logit regression as the statistical classification technique and the data used for the survivors in each analysis were randomly selected. Table 1 provides a comparison of the three models using information presented in their respective articles.

These three models have many other similarities: (1) all reflect capital adequacy with a single ratio of capital to assets; (2) each contains two asset quality variables; and (3) each represents earnings with a net-income-to-total-assets ratio. (Hanweck includes an additional earnings variable to capture the percentage change in net operating income.) The models are differentiated by the fact that Hanweck included an institutional scale variable and Pantalone and Platt included a local economic conditions term.

4 DEVELOPING NEW EARLY-WARNING MODELS TO PREDICT BANK FAILURE

For an early-warning model to be reliable and useful, other facets of a bank should be analyzed in harmony with the quality of management. To date, the most effective early-warning models attempt to replicate the regulators' CAMEL rating scheme. Thus, a model that identifies variables to represent the factors in the CAMEL rating should not only prove to be among the most accurate, but have a greater chance of acceptance by regulatory bodies. Additionally, since it appears that local economic conditions play an important role in the success or failure of financial institutions, a variable to proxy a bank's local economic conditions is also considered.

The models in this paper are developed from the regulator's viewpoint and designed to operate on current data. In other words, for a model devised to predict failure t periods ahead, the variables considered in the model must be available at least t periods prior to the time of failure.

The new models embody two forecast lead-times: prediction 12 to 18 months prior to failure (referred to as a "one-year-ahead" model) and 24 to 30 months prior (referred to as a "two-year-ahead" model). For both lead-times, models based on the probit regression classification methodology are implemented.

We hypothesize that the quality of management is important to a bank's survival. Hence, the DEA efficiency measure should be a highly significant variable in the failure-prediction equation. We expect that our new model will classify survivors and failures more accurately when the management quality metric is included, and will out-perform previously published classification schemes.

4.1 Data Sets Used for Model Construction and Validation

For both the one-year-ahead (1YA) and two-year-ahead (2YA) models, commercial banks which failed from 1986 through 1988 and a random sample of banks which survived through 1989 composed the population. The following restrictions were in place: (1) banks had to have been in operation for at least three years, and (2) only banks with total asset size of between $20 and $300 million were included.

The age limitation was intended to keep *de novo* institutions from entering into the failure equations. New institutions often have anomalous balance sheets and income statements that could negatively affect the reliability and usefulness of a failure-prediction model. Also, since the regulatory agencies charter new financial institutions, they already have controls in place to meet their goal of fostering a safe and sound banking system by restricting entry to the industry to only those with strong prospects for success.

The size limitations were invoked because of the tremendous differences in the managerial operations between small, medium-sized, and large banks. The largest banking institutions (i.e., those with over $300 million in total assets) were eliminated from the sample population because they are generally watched more closely and examined more frequently by banking regulators. In general, the larger the bank, the more frequent and thorough the examination. In fact, it is common for bank examiners to be occupied in an on-site examination of a large bank for months, sometimes years. With near-continuous monitoring, there is no real need for an early-warning model.

The smallest banks (those under $20 million in assets) were dropped from the population for the following reasons.

- Managing a small bank is much different than managing a larger bank. Research has identified major differences in performance, markets, and operating costs (see [5, 14, 19]). Smaller banks have internal operating disadvantages, primarily related to the absence of economies of scale.
- These smaller banks often behave different from "traditional" banks. By accepting deposits and buying securities, some act as investment houses or mutual funds. Others use borrowed money predominantly to make loans, relying little on core deposits.

Time Period	Statement Date	Failure Date	No. of Survivors	No. of Failures	Total Banks
1	Dec. 31, 1984	1/86–6/86	71	40	111
2	June 30, 1985	7/86–12/86	74	46	120
3	Dec. 31, 1985	1/87–6/87	75	40	115
4	June 30, 1986	7/87–12/87	74	48	122
5	Dec. 31, 1986	1/88–6/88	75	51	126
6	June 30, 1987	7/88–12/88	76	69	145
	Model Construction Totals:		**445**	**294**	**739**
7	Dec. 31, 1987	1/89-6/89	76	54	130
8	June 30, 1988	7/89–12/89	76	65	141
	Holdout Sample Totals:		**152**	**119**	**271**

Table 2 Summary of Test Data for One-Year-Ahead Models

- The deposits for the smallest banks are small relative to larger institutions: only 1.3% of total banking deposits. One could argue that on-site examinations may not be worth their cost and effort for the smallest institutions.
- As banks grow through increased mergers and acquisitions, the smallest banks are gradually disappearing, while the percentage of medium-sized and large banks has increased.

The population used in this study includes over 70% of the financial institutions in operation in the United States, and most of the "at risk" banks. Economic and financial statement data were employed for each six-month period beginning with December 31, 1984 and ending with June 30, 1988.

As shown in Table 2, the data for failing institutions were divided into eight time periods, based on their failure date. Thus, if failure occurred during the last six months of 1988, economic and financial figures for June 30, 1987 were used as the one-year-ahead data. Data for time periods 1 through 6 were then pooled and, together with data from the survivors, used to build each model.

Data for the survivors were divided into the same eight time periods and grouped so that no bank would be represented in two different time periods. In other words, of the 611 survivors in the population, banks were sampled at random without replacement and placed into one of the eight time periods.

This grouping was done to ensure that the data for the surviving banks would be independent and identically distributed. This random selection scheme reduced the total number of survivors represented in the model to 597 because of missing observations for some of the banks. (Due to the size boundaries, some banks move into and out of the population at various dates in their time series. This could result in a bank not being represented in the data used to build and test the model. Because this predicament exists when using the model in practice, the slight reduction in the number of survivors represented in the population was not seen as a problem.)

To develop the 1YA failure-prediction model, the data were aggregated for time periods 1 through 6. Hence, there were 445 survivors and 294 failures in the population with all observations independent and identically distributed random variables. The holdout sample consisted of time periods 7 and 8, with a total of 152 survivors and 119 failures.For the 2YA model, a similar selection process was used, but it resulted in a slightly different number of banks in the population. To construct the 2YA model, there were 391 survivors and 211 failures. The holdout sample consisted of 202 survivors and 126 failures.

4.2 Variable Selection

Our new models include only six variables: one for each factor of the CAMEL rating and another to proxy local economic conditions. This restriction was designed to keep them parsimonious and understandable to regulators and bankers. A large number of variables were analyzed statistically (see Barr and Siems [3] for details] before selecting those described in Table 3.

While some of the variables have appeared in earlier models, this is the first study in which the DEA efficiency score has been used in a bank forecasting model, or any regression model in the literature. All values—including DEA—are based on publicly available data, and many are from the Federal Reserve System's Call Report Database. Table 3 shows, for each variable included in the 1YA and 2YA models, the mean and standard deviation for both the survivors and failures. Also given are the results of a t-test on each variable, testing the hypotheses that the true survivor and failure means are equal: the t-statistic and the significance level, or probability of identical means.

All of the selected variables show significant differences in means between the survivor and failure groups, in both 1YA and 2YA cases. All of the selected variables have significantly different means for survivors versus failures and the

Variable	One Year Ahead		Two Years Ahead	
	Survivors	Failures	Survivors	Failures
Equity Capital/Total Loans	18.85%	8.56%	18.65%	10.83%
Nonperforming Loans/Total Assets	1.42%	5.63%	1.28%	3.27%
DEA Efficiency Score	0.8203	0.7033	0.8394	0.7637
Net Income/Total Assets	0.89%	-2.18%	0.93%	-0.62%
Large Deposits/Total Assets	10.03%	24.97%	9.98%	26.12%
% Change In Residential Construction	-1.68%	-15.55%	0.33%	-12.55%

Table 3 Comparison of Variable Averages for One-Year and Two-Years Prior To Failure

differences in mean values follow natural expectations: survivors had higher capital, better quality assets, more efficient management, higher earnings, and more liquidity.

In summary, the six variables selected for our failure-prediction models were:

Variable	*Proxy for*
Equity Capital/Total Loans	Capital Adequacy
Non-performing Loans/Total Assets	Asset Quality
DEA Efficiency Score	Management Quality
Net Income/Total Assets	Earnings Ability
Large Dollar Deposits/Total Assets	Liquidity
Percentage Change in Residential Construction	Local Economic Conditions

A comparison of the 1YA and 2YA mean values gives insight into differences between the surviving and failing populations. Except for the liquidity measure, the gap between the mean values for survivors and failures widens as failure approaches. Also, there is little difference in mean values for the survivors, but a noticeable deterioration for failures (except for the liquidity variable). Thus, one might expect the 1YA model to predict failure with greater accuracy than the 2YA model.

4.3 Empirical Testing: The One-Year-Ahead Model

The standard probit methodology was used to develop models that would classify banks as either survivors or failures. The models were configured so that the dependent variable took on the value of zero (0) for survivors and one (1) for failures. Hence, a negative (positive) coefficient means that the variable is inversely (directly) related to failure and directly (inversely) related to survival. When used for classification, a fractional *cut-off value* for the computed dependent variable—usually 0.5—assigns an observation to either the survivor or failure group.

The 1YA model, with the coefficients and *t*-ratios, is presented in Table 4. To develop this model (and others in the study), a pooled time-series cross-sectional dataset was constructed. Tests for aggregation bias concluded that a structural shift in the data was detected in time period 6. However, because the shift did not classify banks better than the holdout sample, the structural shift model was discarded.

For this model, all variables were significant at the 0.01 level and the signs on the variables' coefficients are appropriate. Specifically, for the variables with positive coefficients—ratios of non-performing loans to total assets, and large dollar deposits to total assets—one would expect that the higher the variable level, the higher the probability of failure (because the expected score becomes larger). Similarly, for the variables with negative coefficients—equity-capital-to-total-loans ratio, the DEA efficiency score,ratio of net income to total assets, and the percentage change in residential construction—one would expect that the higher the variable level, the lower the probability of failure.

The variable with the highest *t*-ratio, in absolute value terms, was the DEA efficiency score with a *t*-ratio of -8.457. Thus, it appears that the quality of management added the most information to the new 1YA bank-failure prediction model.

The *likelihood ratio index* (identified as the LR index in the table) measures the goodness of fit as a pseudo-r^2. For this model the likelihood ratio index was 0.72—a much better fit than the Martin [16], Hanweck [12], and Pantalone and Platt [18] models with LR indices of 0.42, 0.40, and 0.51, respectively, on the same dataset.

	One Year Ahead (LR Index: .72)		Two Years Ahead (LR Index: .58)	
Variable	Coeff.	t-ratio	Coeff.	t-ratio
Constant	5.1395	6.747†	2.9100	3.901†
Equity Capital/Total Loans	-9.6993	-4.982†	-9.4078	-4.854†
DEA Efficiency Score	-7.7682	-8.457†	-4.6943	-5.500†
Nonperforming Loans/ Total Assets	17.8065	5.442†	24.5388	4.990†
Net Income/Total Assets	-22.0646	-4.080†	-3.9871	-.635†
Large Deposits/Total Assets	5.8907	7.060†	5.9031	8.209†
% Change in Residential Construction	-2.7024	-4.567†	-3.0982	-4.850†
Percent Correctly Classified	In-Sample	Out-Sample	In-Sample	Out-Sample
Survivors	94.4%	96.1%	92.1%	92.1%
Failure	89.5%	96.6%	82.9%	94.4%
Total	92.4%	96.3%	88.9%	93.0%

†Significant at the 0.01 level.

Table 4 Full Models and Classification Results

The classification results for this new 1YA bank failure-prediction model were exceptionally accurate. As shown in the lower portion of Table 4, using a classification cut-off of 0.5, the model correctly identified 94.4% of the survivors and 89.5% of the failures for a total classification accuracy of 92.4% (the total number of correct classifications divided by the total number of banks in the sample).

The results for the holdout sample were even stronger. For the two holdout time periods combined, the new model correctly classified 96.1% of the survivors and 96.6% of the failures for an overall classification accuracy of 96.3%. The classification results for all of the models discussed in this paper were statistically significant at the 0.01 level as compared to the results of a naive forecast based on sample proportions.

4.4 Empirical Testing: The Two-Year-Ahead Model

For the models which predict bank failure 24 to 30 months ahead, the proxy variable for earnings was no longer significant, as shown in Table 4. Additionally, the most significant variable (in terms of greatest absolute t-ratio) is no longer the management quality term but, rather, liquidity: large-dollar deposits over total assets.

The 2YA model had an LR index of roughly 0.58 compared to 0.72 for the 1YA model. This is lower because by using data that are further from the actual failure date, the two groups (survivors and failures) tend to look more similar.

Also, as expected, the classification results of the 2YA model were not as accurate as those for the 1YA model. Using a classification cut-off of 0.5, the percentage of correctly identified banks for the in-sample group was 88.9%, with 92.1% of the survivors and 82.9% of the failures correctly classified. However for the holdout sample, the 2YA model correctly identified 92.1% of the survivors, and 94.4% of the failures, for a strong 93.0% overall accuracy.

4.5 Effect of Removing the Management Quality Variable

Of concern in this study is the impact of the DEA efficiency score on the model. Since it is the most significant variable in the 1YA instance, one might expect that the fit and classification results would deteriorate if this variable were eliminated.

To test the impact of the management quality variable, it was removed from the models and a second analysis performed. The coefficients and classification result are given in Table 5 for the "reduced" 1YA and 2YA models. For the 1YA case, the original variables all remain significant with the expected signs. The LR index of fit drops from 0.72 for the full model to 0.62 for the reduced model, and the classification accuracy is lower. With a 0.5 classification cut-off, 89.0% of all in-sample banks were correctly classified and the out-of-sample accuracy dropped from 96.3% to 94.8%. Removing the DEA efficiency variable from the 2YA model has a similar effect: the significance of each variable remains roughly the same as before, the fit metric drops, and the classification results are lower.

These results demonstrate the importance of bank management in the success or failure of a bank. The models which included the management quality variable fit the original data much better and classified failures and survivors more accurately for both in-sample and holdout-sample tests.

4.6 Classification Accuracy Analysis

Table 6 summarizes the classification results on our data sets for the Martin, Hanweck , Pantalone and Platt, and all of our new models (with and without the DEA variable). For each model, the table shows the LR index of fit, in-sample classification accuracy rates, and holdout sample classification accuracy rates for both survivors and failures. All classifications are based on a 0.5 logit/probit cut-off point (discussed further in the next section).

For the 1YA data, the two new models classified both failures and survivors more accurately than the previously developed models from the literature. In addition, the model which includes the management quality variable was the most accurate of all.

	One Year Ahead (LR Index: .62)		Two Years Ahead (LR Index: .54)	
Variable	Coeff.	t-ratio	Coeff.	t-ratio
Constant	-.8165	-3.037†	-.7913	-2.515†
Equity Capital/Total Loans	-9.4796	-5.575†	-9.8316	-5.369†
Nonperforming Loans/ Total Assets	19.9399	6.747†	23.2073	5.114†
Net Income/Total Assets	-18.1565	-3.931†	-2.8936	-.490†
Large Deposits/Total Assets	5.4826	7.752†	6.0231	8.785†
% Change in Residential Construction	-2.3222	-4.520†	-3.1322	-5.247†
Percent Correctly Classified	In-Sample	Out-Sample	In-Sample	Out-Sample
Survivors	92.4%	96.1%	91.3%	90.6%
Failure	84.0%	93.3%	80.6%	95.2%
Total	89.0%	94.8%	87.5%	92.4%

†Significant at the 0.01 level.

Table 5 Model Results without the Management Quality Metric

Model	LR index	In-Sample Correct Classification			Out-of-Sample Correct Classification		
		Survivors %	Failures %	Total %	Survivors %	Failures %	Total %
One Year Ahead:							
Martin	.42	93.0	71.1	84.3	89.5	94.1	91.5
Hanweck	.40	92.8	70.1	83.8	94.7	86.6	91.1
Panalone & Platt	.51	89.7	77.9	85.0	90.1	90.8	90.4
New model w/o DEA	.62	92.4	84.0	89.0	**96.1**	93.3	94.8
New model with DEA	**.72**	**94.4**	**89.5**	**92.4**	**96.1**	**96.6**	**96.3**
Two Years Ahead:							
Martin	.22	**94.1**	46.0	77.2	96.5	61.9	83.2
Hanweck	.18	93.9	41.7	75.6	**97.0**	54.8	80.8
Panalne & Platt	.38	89.8	71.1	83.2	92.6	84.1	89.3
New model w/o DEA	.54	91.3	80.6	87.5	90.6	**95.2**	92.4
New model with DEA	**.58**	92.1	**82.9**	**88.9**	92.1	94.4	**93.0**

Table 6 Comparative Classification Accuracy Results for All Models

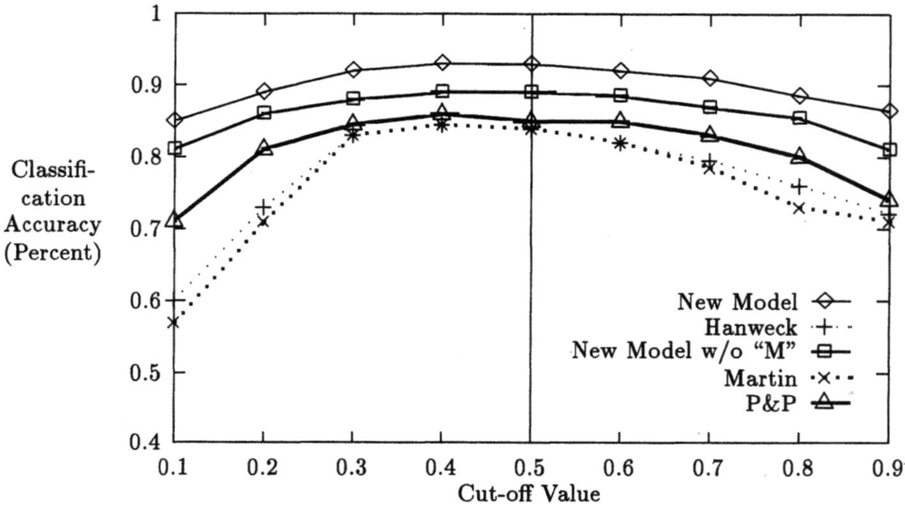

Figure 1 Effect of the cutoff value on the 1YA models' classification accuracy.

Similarly, classification accuracy comparisons are shown for the 2YA models. Again, the new bank-failure prediction model which incorporates the management quality variable has the best fit and classified banks more accurately both in-sample and out-of-sample.

To demonstrate the robustness of the new models, Figure 1 compare the overall in-sample classification accuracy rates at cut-off points ranging from 0.1 to 0.9 for all of the models for the 1YA and 2YA models, respectively. In both cases, our models that include the managerial quality variable out-perform all of the other models analyzed for overall accuracy. In fact, the worst case for our 1YA model is more accurate than the best cases of the Hanweck, Martin, and Pantalone and Platt models.

Cut-off Point	Number of Misclassifications		Relative Cost Magnitude Type I Error:Type II Error			
	Type I	Type II	1:1	2:1	5:1	10:1
0.1	7	104	111	118	139	**174**
0.2	12	66	78	90	126	186
0.3	15	45	60	**75**	**120**	195
0.4	23	31	**54**	77	146	261
0.5	31	25	56	87	180	335

Table 7 Relative Costs Of Misclassifications for the One-Year-Ahead Model

5 COSTS OF MISCLASSIFICATIONS

Most published studies assert that the cost of misclassifying a bank that fails (a Type I error) is greater than the cost of misclassifying a bank that continues to survive (a Type II error). The reasoning is that the cost to perform an on-site examination that results in significant operating improvements is less than the cost to bail out the same bank if it was not examined and had failed. Although this is the conventional wisdom, researchers have not identified the actual costs of misclassifications to permit minimization of the total expected costs.

The likelihood of both types of errors is affected by the selection of the probit cut-off point. A lower cut-off will decrease the probability of a Type I error, while increasing the likelihood of a Type II. Table 7 shows the number and type of in-sample misclassifications for different cut-off points applied to the 1YA model. Hence most researchers have taken the stance, advocated by Pantalone and Platt, that a cut-off less than 0.5 should be used but have not justified their chosen value.

The "best" probit classification cut-off point minimizes the total cost of misclassifying banks. We analyzed four different scenarios, where the ratios of Type I error costs to Type II error costs are: 1:1, 2:1, 5:1, and 10:1. For example, the 10:1 ratio assumes the cost of misclassifying a bank that in fact fails is ten times the cost of misclassifying a bank that survives; 1:1 assumes that the two costs are equal.

Shown in Table 7 is the total misclassification cost, defined as the relative cost of a Type I error multiplied by the number of failures misclassified, plus the cost of a Type II error multiplied by the number of survivors misclassified. For the 1:1 scenario, the total cost is the sum of the number of banks misclassified. When misclassifying failures is twice as costly as examining survivors (column 2:1), the total cost is equal to twice the number of failures misclassified (Type I error) plus the number of survivors misclassified (Type II error).

Under the four cost magnitudes, a minimum value can be found that corresponds to the cut-off value that should be chosen. For the 1:1 case, a cut-off of 0.4 will yield the minimum total cost of 54. For the 2:1 and 5:1 scenarios, a cut-off point of 0.3 should be used, and a value of 0.1 should be used for the 10:1 scenario. Search procedures could be applied to find the optimal cut-off point.

6 POLICY IMPLICATIONS

More accurate bank failure-prediction models, such as those developed in this study, have four key implications for policy makers. First, more accurate models allow banking regulators to deploy their examination resources more efficiently. An effective early-warning model will detect and classify the weakest financial institutions. With banks rated on a survival/failure likelihood scale, regulators can focus on those that are the greatest threat to the deposit insurance fund. The impacts not only include a reduction in expenses for the regulatory agencies (through more efficient resource allocation) but a potential savings to the taxpayers through early identification of weak banks that can be "turned around" before threatening the deposit insurance fund.

Second, the models developed in this study can be used by regulators and bankers to gain a better understanding of the reasons for bank failure. The 1YA model shows that management quality has the greatest impact on bank failure; banks receiving the highest DEA efficiency scores are much more likely to survive than banks which have relatively low scores. The 1YA model identifies five additional variables that are significant in differentiating between failures and survivors: low capital (as a percentage of loans), a high level of non-performing loans, low net income, a high level of large dollar deposits, and slow growth (or rapid decline) in the annual rate of residential construction in the bank's home state.

If regulators have a better understanding of why banks fail, they will be better able to offer bankers suggestions on how to become more stable. In addition, managers will be alerted to threatening conditions so that action can be taken before more serious problems arise.

A better bank failure-prediction model strengthens the entire examination process by identifying banks more objectively. Bankers can have a better understanding when examiners suggest changes in their operations. Additionally, the value of the quarterly call report data would be greatly enhanced.

Finally, the new early-warning models developed in this study could be used to develop a variable-rate deposit-insurance-premium structure. The banks at the greatest risk of failure would be required to pay the highest premiums for deposit insurance, much like smokers pay higher life insurance premiums than nonsmokers.

7 CONCLUSIONS

In this paper, two new bank-failure prediction models were developed. Both the one-year-ahead and two-year-ahead models use proxy variables for each factor in the CAMEL rating plus a variable to capture local economic conditions. For the first time, bank failure-prediction models were developed using a DEA efficiency variable to proxy management quality.

The results of these models indicate that management is, indeed, important to the successful operation of a bank. When the management variable was removed from the full model, the results were worse in terms of the model's fit to the data and its classification accuracy. The newly-developed models also show superior results to leading published approaches, as analyzed using a standardized dataset.

From a policy standpoint, these early-warning models can be used to detect and classify the weakest financial institutions, while providing regulators and bankers insight into the reasons for bank failure. Their high accuracy should allow banking regulators to allocate more efficiently their resources to perform on-site examinations, and should strengthen the entire examination process by providing a simple and objective detection mechanism for failure-prone banks. Such new tools will ably assist those striving to ensure a safe and secure future for the nation's banking system.

REFERENCES

[1] Altman, E. I. Financial Ratios, Discriminant Analysis and the Prediction of Corporate Bankruptcy, *The Journal of Finance,* 1968, *23,* 589-609.

[2] Barr, R. S., Seiford, L. M., and Siems, T. F. An Envelopment-Analysis Approach to Measuring the Management Quality of Banks, *Annals of Operations Research*, 1993, *38*, 1-13.

[3] Barr, R.S. and Siems, T.F. Predicting Bank Failure Using DEA to Quantify Management Quality, working paper, Federal Reserve Bank of Dallas (Dallas, TX), 1993.

[4] Beaver, W. H. Financial Ratios as Predictors of Failure, Empirical Research in *Accounting: Selected Studies* (The Institute of Professional Accounting, University of Chicago), 1966, 71-127.

[5] Bentson, G. J., Hanweck, G. A., and Humphrey, D. B. Operating Costs in Commercial Banking, *Economic Review*, 1982, November, 6-21.

[6] Booker, I. O.Tracking Banks from Afar: A Risk Monitoring System, *Economic Review*, 1983, November, 36-41.

[7] Bovenzi, J. F., and Nejezchleb, L. Bank Failures: Why Are There So Many?, *Issues in Bank Regulation*, 1985, *8*, 54-68.

[8] Bovenzi, J. F., Marino, J. A., and McFadden, F. E. Commercial Bank Failure Prediction Models, *Economic Review*, 1983, *68*, November, 14-26.

[9] Cates, D. C. Bank Risk and Predicting Bank Failure,*Issues in Bank Regulation*, 1985, *9*, 16-20.

[10] Charnes, A., Cooper, W.W., and Rhodes, E. Measuring the Efficiency of Decision Making Units,*European Journal of Operational Research*, 1978, *2*, 429-444.

[11] Charnes, A., Cooper, W.W., Lewin, A.Y., and Seiford, L. *Data Envelopment Analysis: Theory, Methodology and Applications*, 1994, Kluwer Academic Publishers, Boston.

[12] Hanweck, G. A. Predicting Bank Failure, *Research Papers in Banking and Financial Economics*, 1977, (Financial Studies Section, Division of Research and Statistics, Board of Governors of the Federal Reserve System), November.

[13] Kaufman, G. G. Banking Risk in Historical Perspective, *Bank Structure and Competition*, 1986, May, 231-249.

[14] King, B. F. Changes in Large Banks' Market Shares, *Economic Review*, 1982, November, 35-40.

[15] Looney, S. W.,Wansley, J. W., and Lane, W. R. An Examination of Misclassifications with Bank Failure Prediction Models, *Journal of Economics and Business*, 1989, *41*, 327-336.

[16] Martin, D. Early-warning of Bank Failure: A Logit Regression Approach, *Journal of Banking and Finance*, 1977, *1*, 249-276.

[17] Meyer, P. A., and Pifer, H. W. Prediction of Bank Failures, *The Journal of Finance*, 1970, *25*, 853-868.

[18] Pantalone, C. C., and Platt, M. B. Predicting Commercial Bank Failure Since Deregulation, *New England Economic Review*, 1987, Jul/Aug, 37-47.

[19] Rhoades, S. A., and D. T. Savage Post-Deregulation Performance of Large and Small Banks,*Issues in Bank Regulation,* 1990, Winter, 20-31.

[20] Seballos, L. D. and Thomson, J. B. Underlying Causes of Commercial Bank Failures in the 1980s, *Economic Commentary*, 1990, Federal Reserve Bank of Cleveland, September.

[21] Secrist, H.*National Bank Failures and Non-Failures: An Autopsy and Diagnosis*, 1938, Principia Press, Bloomington, IN.

[22] Seiford, L.M. A Data Envelopment Analysis Bibliography (1978-1992), in nes, A., Cooper, W.W., Lewin, A.Y., and Seiford, L. *Data Envelopment Analysis: Theory, Methodology and Applications*, 1994, Kluwer Academic Publishers, Boston.

[23] Seiford, L.M. and Thrall, R.M. Recent Developments in DEA: The Mathematical Approach to Frontier Analysis,*Journal of Econometrics*, 1990,*46*, 7-38.

[24] Siems, T. F. Quantifying Management's Role in Bank Survival, *Economic Review*, 1992, First Quarter, 29-41.

[25] Sinkey, J. F., Jr. *Problem and Failed Institutions in the Commercial Banking Industry*, 1979, JAI Press, Greenwich, CT.

[26] West, R. C. A Factor-Analytic Approach to Bank Condition, *Journal of Banking and Finance*, 1985,*9*, 253-266.

16

A COOPERATIVE MULTI-AGENT APPROACH TO CONSTRAINED PROJECT SCHEDULING

Dan Zhu and Rema Padman*

College of Business Administration
University of Iowa
Iowa City, IA 52242
**The Heinz School of Public Policy and Management*
Carnegie Mellon University
Pittsburgh, PA 15213

ABSTRACT

Different approaches to a problem generally produce distinct algorithms that have significantly different performance characteristics in varying problem environments. In most instances, a single algorithm is chosen to be applied throughout the solution process of a given problem, thus losing the opportunity to exploit the desirable features of other methods. An Asynchronous Team (A-Team) is a software organization that facilitates cooperation amongst multiple algorithms so that together they produce better solutions than if they were acting alone. This paper explores the feasibility of applying the A-Team approach to the practical problem of resource-constrained project scheduling with cash flows (RCPSPCF). This problem investigates the scheduling of precedence-related and resource-constrained activities to maximize the Net Present Value of the expenses incurred and payments made over the duration of a project. This is a complex combinatorial optimization problem which precludes the development of optimal schedules for projects of practical size. We embed several simple heuristics for solving the RCPSPCF within the iterative, parallel structure of A-Team which provides a natural framework for distributed problem solving. Preliminary results on several randomly generated project networks show that the combination of multiple, simple heuristics perform very well when compared to many single pass, complex optimization-based heuristics proposed in the literature.

1 INTRODUCTION

Resource-constrained project scheduling problems (RCPSP) are widely encountered in practice, in situations ranging from construction to software project management, and research and development. This problem is concerned with the scheduling of a project consisting of a number of activities which are linked together by precedence relations and multiple resource restrictions. When cash flows exist in the form of expenses for initiating activities and payments for completed work, the development of project schedules which maximize the net present value (NPV) of the project is of considerable practical importance [1].

This is a difficult combinatorial optimization problem which precludes the development of optimal schedules. In particular, as the size of project increases, or as parameters such as duration and resource requirements of the project change, it is harder to generate good schedules in reasonable time. Many heuristics have been proposed in the literature to solve the resource-constrained project scheduling problem with cash flows [5, 11, 12, 13]. They have been extensively tested on projects varying in size, complexity, level of resource-constrainedness, pattern of cash flows, and so on. These studies have shown that no single heuristic dominates in all the project environments [5, 10, 17]. This raises an interesting question as to whether multiple heuristics acting in parallel can generate better solutions than each one acting alone.

Some early proposals for applying multiple heuristic processes (agents) to populations of solutions occur in [4, 6, 7]. In [7], for example, auxiliary strategies are introduced to modify the choice rules of the agents based on history. These strategies include reinforcing consistent features of good solutions and alternately driving the solutions across or away from regions where they cluster. More recently, these proposals have become embodied in tabu search strategies that use varied forms of memory to assist the underlying processes, as in [8], where the learning procedure called target analysis is suggested as a supplement to the strategy of ranging through different types of move structures and evaluation alternatives. Instances of this multiple agent strategy for solving combinatorial problems have been applied in parallel processing studies of tabu search [2, 3]. These studies report that asynchronous uses of multiple move types and decision rules are superior to synchronous uses. As we will show, our findings reinforce these conclusions and further demonstrate that appropriate coordination of simple agents can prove superior to the application of more complex agents. Moreover, this outcome applies even without using memory to guide the operation of simple agent s beyond conditions of local optimality, though memory may provide further advantages.

In this paper, we thus investigate the collaborative efforts of multiple heuristic procedures acting as agents that guide the search for a better schedule. In previous studies on the resource-constrained project schedulng problem, heuristics have generally been applied in a one-at-a-time, single pass, forward sequencing mode that exploits the capabilities of the specific heuristic strategy. However, this approach also precludes the ability to improve solutions by taking advantage of the good features of other strategies in conjunction with the current one. Talukdar et al. [15] discuss the construction and dynamics of asynchronous software organizations, called Asynchronous Team (A-Team), that facilitate the cooperation of multiple algorithms that can be applied in parallel to solve a problem.

Given the number of heuristics that have been developed for the RCPSPCF that use several different approaches, this paper investigates the potential of the A-Team approach to generate better solutions by using some of the simplest heuristic procedure s. We conduct some experiments on randomly generated project networks to evaluate the effectiveness of our approach and report our preliminary results. These results are compared with the performance of three complex, optimization-based heuristics from the literature. The experiments indicate that the cooperative approach has considerable promise, performing significantly better than the three optimization-based heuristics and has highest average NPV.

The next section briefly describes the RCPSPCF with an illustrative example and we also present some details of the A-Team aproach. In section 3, we discuss the design of the system, followed by experimental results in section 4 and conclusions in section 5.

2 BACKGROUND

2.1 Resource-Constrained Project Scheduling Problem With Cash Flows

The resource constrained project scheduling problem with cash flows addresses the scheduling of a number of activities subject to constraints on precedence requirements and resource limitations. A series of cash flows occur over the duration of the project in the form of cash outflows for project expenditures and cash inflows as payments for completed activities.

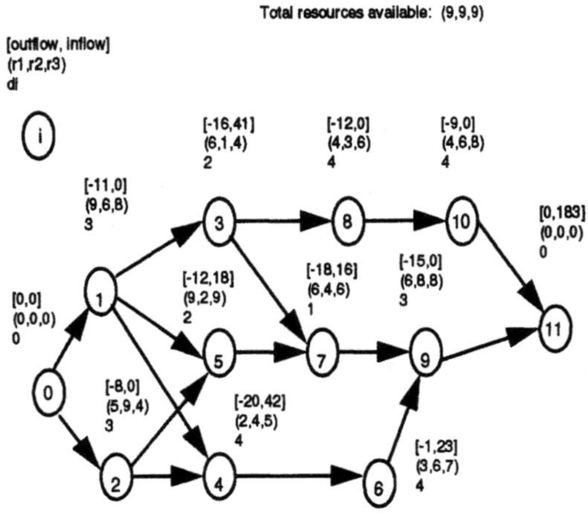

Figure 1 Example Project Network

Figure 1 depicts an example project network. Each node represents an activity and the directed arcs represent the precedence relationships among activities. Each activity i ($i = 1, \ldots, m$) has a fixed duration d_i, resource requirements $(r_{i1}, \ldots, r_{iK})$ of K types (in this example K=3), and cash flows F_i^{out} and F_i^{in} representing expenses and payments.

Given a project network with m activities, and a discount rate of α for the cash flows, the RCPSP with the NPV objective can be formulated, as shown below, as a mixed integer nonlinear program where x_{it} ($i = 1, \ldots, m$) are binary variables that determine the start time of activity i, i.e., $x_{it} = 1$ if activity i starts at time t, and 0 otherwise.

Maximize

$$\sum_{i=1}^{m} \sum_{t=E_i}^{L_i} F_i^{out} e^{-\alpha(t)} x_{it} + \sum_{i=1}^{m} \sum_{t=E_i}^{L_i} F_i^{in} e^{-\alpha(t+d_i)} x_{it}$$

Subject to

$$\sum_{t=E_i}^{L_i} x_{it} = 1 \quad i = 1, \ldots, m \tag{16.1}$$

$$\sum_{t=E_j}^{L_j} t x_{jt} - \sum_{t=E_i}^{L_i} t x_{it} \geq d_i \quad \forall\ (i,j) \in A \tag{16.2}$$

$$\sum_{i=1}^{m} \sum_{q=t}^{t+d_i-1} r_{ik} x_{iq} \leq R_{kt} \quad k = 1, \ldots, K; t = 1, \ldots, D \tag{16.3}$$

Where:

m	:	Number of activities
F_i^{out}	:	Cash outflow for activity i
F_i^{in}	:	Cash inflow for activity i
α	:	Discount rate
d_i	:	Duration of activity i
E_i	:	Earliest start time for activity i
L_i	:	Latest start time for activity i
K	:	Number of resource types in the project
r_{ik}	:	Requirement of resource type k by activity i
R_{kt}	:	Amount of resource type k available in period t
A	:	Set of precedence relationships
D	:	Deadline for the project

Constraint (1) ensures a unique start time for each activity. Constraint (2) describes the precedence relationships. Constraint (3) represents the resource utilization requirements. The objective is to maximize the NPV of the cash flows.

Obtaining optimal schedules for projects of reasonable size is impractical since the problem is NP-complete [9]. Hence heuristic procedures have found extensive application. Many of these heuristics use information generated by the Critical Path Method such as slack, early and late start times, and so on [5]. Russell [13] and Padman et al. [11, 12] developed and tested several optimization-based heuristics that use information from a transshipment network relaxation of the RCPSPCF such as dual prices and target scheduling dates. The heuristics are embedded in a single-pass, forward sequencing procedure. Once a heuristic is selected, it is applied throughout the scheduling process. Previous studies show that no single heuristic dominates in all the project environments [5, 10, 11, 12, 13, 17]. However, there has been no investigation of the benefits of combining multiple strategies on an instance of the RCPSPCF. As the characteristics of the project change over the duration of the project, different heuristics may adapt to the new conditions to generate better solutions. In this research, we explore the collaborative efforts of a number of heuristics that cooperate as a team of agents that synergistically help each other to solve this complex problem. This raises many issues about the selection of appropriate heuristics and the order in which they should be applied. The Asynchronous Team is an organization that provides the mechanisms to explore many combinations through cooperation of the heuristics within an iterative framework.

2.2 Asynchronous Team

An Asynchronous Team (A-team) is a type of software organization in which a collection of software objects, each acting as an autonomous agent, cooperate with each other to effectively tackle a problem [14, 15, 16]. A class of agents form the team. Inside the team, each agent works asynchronously on a population of solutions. The premise of this method is that the combined efforts of multiple agents that perform construction and iterative modification generally lead to a solution better than those generated by any individual agent acting alone. Throughout the execution of multiple autonomous agents, there is no central supervisory control. New agents can be added or old ones deleted without needing any managerial structure. The only communication channel for these agents to interact is through the shared memory.

In this organization, there are construction agents which are heuristics that generate initial feasible schedules, and modification agents which are heuristics that improve these schedules. Each heuristic generates its best solution that can then be applied as the input to another modification heuristic procedure. All solutions are stored in the working memory. The organization also allows the elimination of bad solutions by appropriate application of destruction agents. The destruction agents ensure that, over a number of iterations, the evolved solutions are better than the original set. The destruction mechanism eliminates a new solution if its performance on a given criterion is the worst among those in its set. Otherwise, it picks a solution from the working memory using a linear probability distribution whereby the best solutions have the least likelihood of being selected for destruction. Thus newer and better solutions populate the working memory. In order to ensure good solutions in reasonable time, it is desirable to include simple heuristics in this organization.

In the most general case, all three types of agents - construction, modification, and destruction - are continually active; new schedules are under construction, existing schedules are being modified, and inferior schedules are being removed from the system. Thus the working memory may contain both partial and completed schedules, being processed by construction and modification agents, respectively, at any given time. Schedules can be selected randomly or using problem specific information for these operations.

A-team has been successfully applied to the traveling salesman problem (TSP), reconfigurable modular manipulator systems (RMMS) and power system fault diagnosis [14, 15, 16]. The complex nature of the RCPSPCF, the availability of a number of algorithms and heuristics in the literature, and the potential advantages of multiple heuristics working in parallel to generate better solutions provide the motivation for exploring the applicability of the A-team approach for this problem. We have adapted some simple heuristics for use within the A-team organization and measured the system's performance against that of three complex, optimization based heuristics from the literature that have been shown to work well.

3 DESIGN OF THE SYSTEM

In order to assemble the A-team to solve the resource constrained project scheduling problem with cash flows, a number of heuristics need to be identified. These heuristics include both heuristics for generating schedules (construction agents) and heuristics for repairing schedules (modification agents).

3.1 Construction Agents

At any given time during the scheduling process, a number of activities in the project may become eligible for scheduling when their predecessor activities are completed. Given the constraints on the availability of resources, some of the activities in this queue can start immediately while others have to be delayed. Heuristics are used to prioritize among the activities from the eligible activity queue. In our initial study, we consider the following heuristics as construction agents:

- Random Selection (Random): Choose activities randomly from the eligible activity queue and schedule them till available resources are exhausted.
- Shortest Processing Time (SPT): Select activities from the eligible activity queue in ascending order of their durations.
- Longest Processing Time (LPT): Select activities from the eligible activity queue in descending order of their durations.
- Minimum Slack (MINSLK): Select the activity with the minimum slack. The slack is the difference between the late finish time and early finish time computed using the critical path method. Slack is continuously updated.
- Cash Flow Weight (CFW): Select activity with the highest cash flow weight. The cash flow weight of each activity is computed as the sum of the cash flows of all its successors.

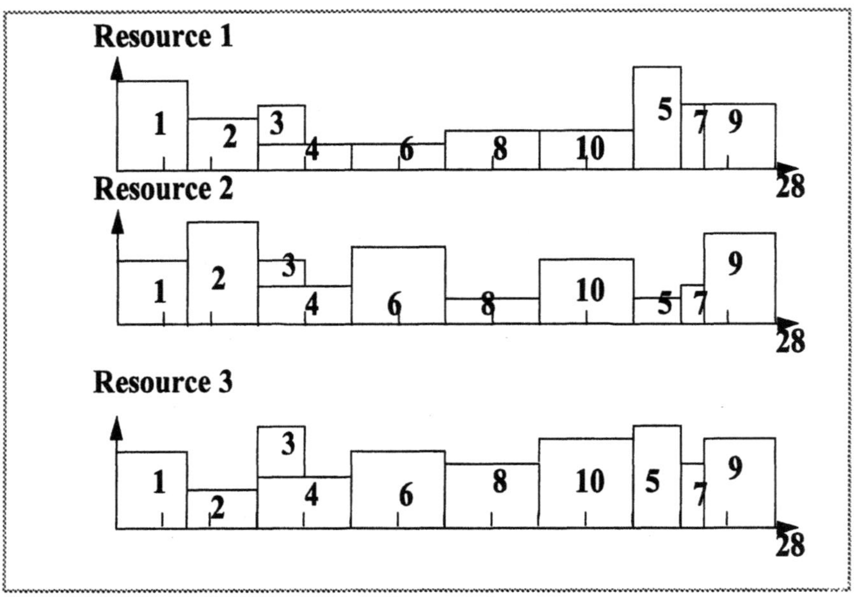

Figure 2 An Example Schedule

To illustrate, given a resource limit vector of [9,9,9] in our example problem, a gantt chart representation of the schedule produced by a Longest Processing Time (LPT) heuristic rule is shown in Figure 2. With a discount rate of $\alpha = 0.01$, the NPV of this schedule is \$15378.

3.2 Modification Agents

Once some initial solutions are obtained, a number of general and domain specific tactics can be employed to improve the schedule. In the RCPSPCF, the NPV can be improved if activities with higher payments can be scheduled as early as possible and activities with higher expenses can be scheduled as late as possible. Using this motivation to identify appropriate activities, we apply the method of iterative repair that includes the following modification agents:

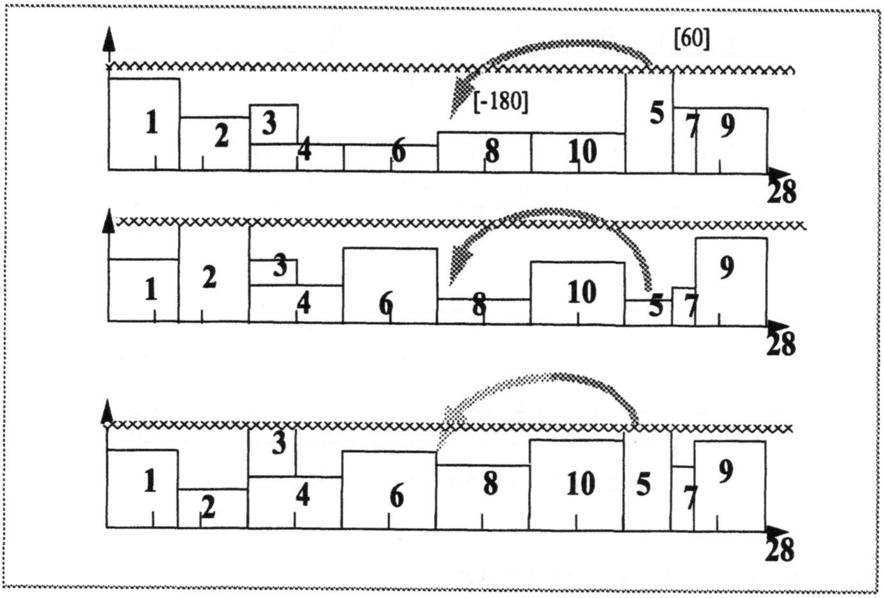

Figure 3 An Illustration of Repair Heuristic

- Adjacent Pairwise Interchange (API): Swap two adjacent activities.
- Left and Right Shift (LRS): Shift an activity one or more time units to the left/right.

- Positive Left Insertion (PLI): Insert the activity with net positive cash flows earlier in the schedule and delay the activity with net negative cash flows.

To illustrate, in Figure 3, we apply the positive left insertion(PLI) heuristic to work on the example schedule shown in Figure 2. It can be observed that activity 5 can bring in a positive net cash flow of \$60 while activity 8 leads to a negative net cash flow of \$180. The PLI heuristic thus inserts activity 5 ahead of activity 8. The repaired schedule after this modification is shown in Figure 4. The NPV of this repaired schedule is \$15455. One insertion thus results in a net gain of \$77.

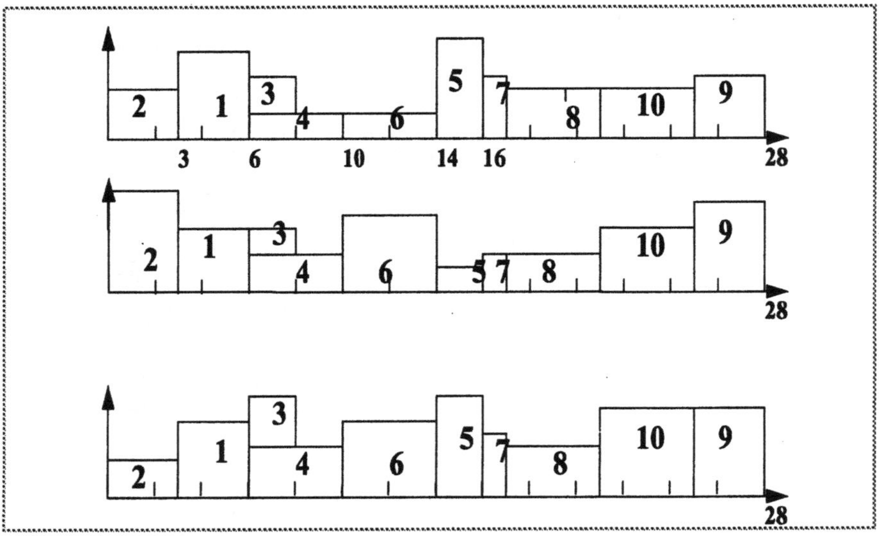

Figure 4 The Schedule After Repair

3.3 Solution Procedure

Figure 5 presents a system design for the A-team approach for RCPSPCF. The problem specification is stored in the long term memory. A number of construction heuristics independently generate some initial feasible schedules. Thus the initial population for the example problem consists of five feasible schedules. These initial solutions are then stored in the working memory for selection by the modification heuristics. At any iteration, a selection procedure

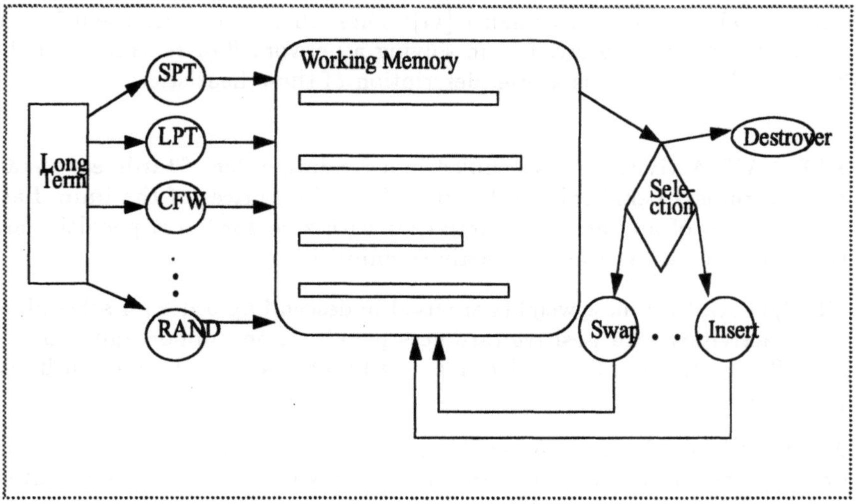

Figure 5 System Design

is initiated to choose a schedule for repair; in this study, we use random selection from the population of solutions. The modification agent to be applied to this schedule is also chosen randomly. During repair, schedules maintain precedence feasibility, but may violate resource constraints. Constraint propagation is used to restore feasibility of the new schedule. When the population of solutions overwhelm the memory capacity, a destroyer automatically eliminates some bad solutions; these are identified based on the NPV of the schedule. The procedure halts after a specified number of iterations.

4 EXPERIMENTS AND PRELIMINARY RESULTS

The above section provides the framework of a cooperative, multi-agent system for solving the RCPSPCF. We conducted some initial experiments to test this approach on a set of nine projects. Each project consists of 12 activities, with each activity requiring three types of resources. Table 1 presents the Net Present Value (NPV) for each problem. The first three heuristics in the table are optimization-based heuristics identified from the literature that have been shown to have good performance [11]. These three heuristics use information such as dual prices derived from solving a network flow-relaxation of the RCPSPCF. The following is a brief description of these heuristics:

- LTP/LAN: Activities are scheduled in ascending order of tardiness penalties. Tardiness penalties are the sum of the dual prices of the immediate predecessors of any activity. Activities with zero tardiness penalties are scheduled according to lowest activity number.
- CFW/OCC: Cash flow weights are used in descending order for scheduling those activities with positive tardiness penalties, and opportunity cost of cash flows, in ascending order, for scheduling those with zero tardiness penalties.
- MTP/ET: Activities are selected from the schedule queue in descending order of their tardiness penalties. Activities with zero tardiness penalties are scheduled according to an early time rule.

As can be observed from Table 1, the simple repair heuristic has a higher mean NPV when compared to the three complex optimization-based heuristics. In addition, in seven out of the nine projects examined, the repair heuristic performs better than all the optimization-based heuristics. CFW/OCC, which performed the best among sixteen heuristics in an extensive study reported in [11], is also a reasonable heuristic as indicated by its superior performance in about half the projects.

Table 1 Comparison of Heuristics on Small Projects

Problem	LTP/LAN	CFW/OCC	MTP/ET	REPAIR
1	19062	19006	19066	19073
2	20579	20599	20585	20610
3	14946	15015	14990	15017
4	15729	15872	15763	15917
5	15952	16020	15977	16080
6	19668	19783	19635	19786
7	7919	7905	7874	7905
8	10951	11015	10944	10986
9	8961	9017	8971	9132
Mean NPV	14863.0	14914.7	14867.2	14945.1

An expanded experiment using 30 randomly generated project networks from [11] (called the PSD problems) compared the performance of the repair heuristic against CFW/OCC, LTP/LAN, as well as the best heuristic out of the set of sixteen reported for each problem in [11]. Each problem in this set consists of 46 activities, with low levels of three project characteristics such as profit margin (less than 30%), frequency of progress payments (on average, a payment at the seventh activity), and level of resource-constrainedness (average utilization factor of 1.0).

Table 2 summarizes the NPVs of the four approaches for the 30 problems. The third column of NPVs, as mentioned earlier, reports the best of sixteen heuristics for the given problem. It can be observed that the repair heuristic dominates in performance over all problems except one (problem 377). In addition, the average percentage improvement is also a significant 8.8% over the best heuristic. These results indicate that iterative repair strategies using the A-team framework have considerable potential for generating good solutions for the RCPSPCF.

5 CONCLUSION

The preliminary results show that a combination of multiple simple heuristics perform very well when compared to many of the complex, single pass heuristics considered in this study. Iterative repair heuristics not only improve scheduling performance, but are also able to deal with changes in problem parameters. Extensive testing needs to be conducted to evaluate the performance of the approach over all significant project environments in terms of both solution quality and time. Other important aspects for further investigation include the identificaton of rules for switching from one agent to another during schedule construction and modification, and the impact of learning on the selection of agents for the different operations of construction, modification, and destruction. The parallel structure of A-team provides a natural scheme suitable for exploring these dynamics in a distributed problem solving environment such that good solutions can be generated for this difficult scheduling problem.

Acknowledgements

This research is partially supported by a grant from the Central Investment Fund for Research Enhancement at the University of Iowa, a grant from the Faculty Development Fund at Carnegie Mellon University and a grant from the US Army (DAAH04-94-G-0239). We would like to thank Professor Fred Glover of the University of Colorado at Boulder and two anonymous referees for many helpful comments on an earlier draft of the paper.

REFERENCES

[1] R. Bey, R.H. Doersch and J.H. Patterson, 1981. The Net Present Value Criterion: Its Impact on Project Scheduling, *Project Management Quarterly*, Vol.12, No.2, 35-45.

[2] T.G. Crainic, M. Toulouse and M. Gendreau, 1993a. A Study of Synchronous Parallelization Strategies for Tabu Search, Publication 934, Centre de Recherche sur les Transports, Universite de Montreal.

[3] T.G. Crainic, M. Toulouse and M. Gendreau, 1993b. Appraisal of Asynchronous Parallelization Approaches for Tabu Search, Publication 935, Centre de Recherche sur les Transports, Universite de Montreal.

[4] W. Crowston, F. Glover, G. Thompson and J. Trawick, 1964. Probabilistic and Parametric Learning Methods for the Job Shop Scheduling Problem, GSIA Working Paper, Carnegie Mellon University.

[5] E.W. Davis and J.H. Patterson, 1975. A Comparison of Heuristic and Optimal Solutions in Resource-Constrained Project Scheduling, *Management Science*, Vol.21, No.8, 944-955.

[6] H. Fisher and G.L. Thompson, 1963. Probabilistic Learning Combinations of Local Job-shop Scheduling Rules, In *Industrial Scheduling*, Prentice-Hall, Inc., 225-251.

[7] F. Glover, 1977. Heuristics for Integer Programming Using Surrogate Constraints, *Decision Sciences*, Vol.8, No.1, 156-166.

[8] F. Glover, 1990. Tabu Search: A Tutorial, *Interfaces*, Vol.20, No.1, 74-94.

[9] M.R. Garey and D.S. Johnson, 1979. Computers and Intractability: A Guide to the Theory of NP-Completeness, W.H. Freeman and Co., New York.

[10] I.S. Kurtulus and E.W. Davis, 1982. Multi-Project Scheduling: Categorization of Heuristic Rules Performance, *Management Science*, Vol. 28, No. 2, 161-172.

[11] R. Padman, D.E. Smith-Daniels and V.L. Smith-Daniels. Heuristic Scheduling of Resource-Constrained Projects with Cash Flows, forthcoming in *Naval Research Logistics.*

[12] R. Padman, D.E. Smith-Daniels, 1993. Early-Tardy Cost Trade-Offs in Resource Constrained Projects with Cash Flows: An Optimization-based Approach, *European Journal of Operational Research*, Vol. 65, 295-311.

[13] R.A. Russell, 1986. A Comparison of Heuristics for Scheduling Projects with Cash Flows and Resource Restrictions, *Management Science*, Vol. 32, No. 10, 1291-1300.

[14] S.N. Talukdar, 1993. Asynchronous Teams, *Fourth Symposium on Expert Systems Applications to Power Systems.*

[15] S.N. Talukdar and P.S.de Souza, 1992. Scale Efficient Organizations, *Proceedings of the 1992 IEEE International Conference on Systems, Man and Cybernetics*, Vol.2, 1458-1463.

[16] S.N. Talukdar, P.S. de Souza, and S. Murthy, 1993. Designing Organizations for Computer-based Agents, *International Journal of Engineering Intelligent Systems for Electrical Engineering and Communications.*

[17] D. Zhu and R. Padman, 1995. Neural Networks for Heuristic Selection: An Application in Resource-Constrained Project Scheduling, *The Impact of Emerging Technologies on Computer Science and Operations Research*, eds. S.G. Nash and A . Sofer, Kluwer Academic Publishers, 297-312.

Table 2 Comparison of Heuristics on PSD Problems

Problem	LTP/LAN	CFW/OCC	Best NPV (Hno)	REPAIR
361	747.81	741.09	758.87 (11)	792.11
362	743.19	763.61	787.41 (6)	827.71
363	810.60	815.38	821.99 (13)	833.00
364	812.06	793.00	812.06 (12)	860.14
365	968.54	979.15	984.66 (2)	1011.38
366	808.64	805.11	814.50 (8)	856.45
367	916.58	916.95	935.50 (6)	974.66
368	952.24	978.53	1008.59 (1)	1021.00
369	849.85	846.99	849.85 (12)	881.00
370	903.64	917.71	926.30 (1)	951.00
371	705.77	724.76	729.92 (6)	761.68
372	892.24	910.57	910.57 (13)	910.00
373	784.10	796.34	796.34 (14)	834.52
374	911.17	950.44	961.35 (1)	980.00
375	717.69	757.30	769.53 (1)	777.00
376	889.68	938.25	938.25 (13)	983.18
377	721.95	747.35	751.32 (15)	742.27
378	964.82	959.06	968.11 (15)	995.75
379	903.94	903.45	903.94 (12)	919.51
380	906.89	916.11	916.11 (14)	1004.72
381	874.86	899.83	906.16 (4)	957.00
382	843.64	843.64	874.23 (15)	914.7
383	668.13	668.50	676.00 (15)	743.90
384	758.13	761.49	802.70 (9)	834.33
385	791.29	798.98	799.51 (1)	820.54
386	688.97	688.66	695.05 (15)	744.33
387	829.12	810.93	829.12 (12)	875.00
388	843.40	837.74	848.90 (6)	866.46
389	994.35	1030.62	1030.67 (13)	1054.56
390	883.79	895.01	900.48 (8)	948.44
Mean NPV	836.24	846.55	856.93	932.83

17

SCHEDULING A FLOW SHOP TO MINIMIZE THE MAXIMAL LATENESS UNDER ARBITRARY PRECEDENCE CONSTRAINTS

Joanna Józefowska and Arkadiusz Zimniak

Institute of Computing Science
Poznań University of Technology

ABSTRACT

A flow shop consisting of m stages with arbitrary precedence constraints between tasks is considered. The optimization criterion is maximum lateness. Heuristic algorithms have been developed for this NP-hard problem and solutions generated by them have been compared with the optimal ones for reasonably sized problems. A polynomially solvable special case has been presented. The conclusions concerning further use and development of heuristics are also presented.

1 INTRODUCTION

Let us consider a printed circuit board (PCB) assembly system consisting of three stages, where each task (mounting a component on a board) consists of the following three operations: selection of a component, fixing the component on a board and testing the board. Some components may be mounted in an arbitrary order but for some of them the order is fixed. Thus arbitrary constraints between tasks may occur. In that case the successive task can start only after all three operations of the preceeding taks are completed. A board may return after testing to the first stage for mounting the next component or it may leave the system being sent to the next production stage. Different boards leave the system after completion of different tasks. The boards have to be delivered to the next production stage before given deadlines. Any time a deadline is exceeded, the production at the next stage is being delayed causing a given penalty cost. The goal of scheduling the assembly system is to avoid this penalty cost, i.e. to minimize the maximal lateness of all tasks (or ensure a nonpositive value of the maximum lateness). Similar assembly systems are often met in practice also for other technologies. This problem is modelled as a problem of scheduling an m-machine flow shop with arbitrary precedence constraints between tasks with the objective to minimize maximum lateness. It is proved in Section 2 that for two machines and unit processing times the problem can be solved in time $O(n^2)$, where n is the number of tasks to be scheduled. However, in general, the problem is NP-hard and thus it can be effectively solved only with the use of heuristic algorithms. These algorithms are usually specialized for particular types of problems. Several such proposals are presented in this paper. On the other hand some general heuristic approaches are developed with the aim to support a wide range of combinatorial problems, regardless of their specifity. One of such approaches is simulated annealing which has been investigated by other authors (e.g., in the context of scheduling a permutation flow shop to minimize the schedule length [6, 7, 8, 10]). A simulated annealing algorithm designed for the considered problem has been implemented and tested as well.

2 PROBLEM DESCRIPTION AND FORMULATION

In this paper an m-machine flow shop is considered. It consists of m stages, one machine at each stage. Every task consists of a sequence of m operations. Operation j $(j = 1, \ldots, m)$ has to be completed on machine j. Moreover, precedence constraints between tasks are allowed. Task T_i precedes task T_k $(T_i \prec T_k)$ means that the last operation on task T_i has to be completed before the first operation on task T_k can start. Precedence constraints are represented by an arbitrary acyclic directed graph G with nodes representing tasks. Every task T_i has to be completed by a given deadline d_i. The problem is to find a schedule minimizing the maximum lateness (i.e., $\max_i (C_i - d_i)$, where C_i denotes the completion time of task T_i, $i = 1, \ldots, n$).

The computational complexity of scheduling a flow shop under arbitrary precedence constraints has been examined earlier for the minimization of makespan C_{max} with the following results. For two machines, unit processing times and arbitrary precedence constraints the problem can be solved in time polynomial in n [4, 9]. Assuming arbitrary processing times, the problem becomes NP-hard even for two machines. Knowing that transformation from C_{max} to L_{max} leads to a problem of at least the same computational complexity [5], we can easily see that our problem $(Fm \mid prec \mid L_{max})$ is also NP-hard. Thus it is justified

Table 1 Computational complexity of scheduling problems in a flow shop under arbitrary precedence constraints.

No.	Number of machines	Processing times	Optimization criterion	Computational complexity	References
1	2	$p_{ij} = 1$	C_{max}	$O(n)$	[9]
2	2	arbitrary	C_{max}	str. NP-hard	[3]
3	m	$p_{ij} = 1$	C_{max}	open	
4	2	$p_{ij} = 1$	L_{max}	$O(n^2)$	Theorem 1
5	m	arbitrary	L_{max}	str. NP-hard	[5]

to solve it using heuristic algorithms. The complexity results are summarized in Table 1.

3 AN OPTIMAL ALGORITHM FOR A SPECIAL CASE

First the two machine flow shop is considered. Many manufacturing systems consist of two machines only and for this configuration and unit processing times an optimal solution can be found in $O(n^2)$ time.

The problem under consideration is the following: $F2 \mid prec, p_{ij} = 1 \mid L_{max}$

We will use the following notation:

$M1, M2$ - machines constituting the flow shop

$J = \{T_1, T_2, ..., T_n\}$ - set of tasks to be executed; there are two unit-time operations to be executed on each task: the first one on machine $M1$ and the second one on machine $M2$

$d_i > 0$ - deadline of task T_i

$\prec$ - partial order on J; whenever $T_i \prec T_k$, the execution of the last operation on T_i must be completed before the execution of the first operation on T_k is begun.

Definition 1

A task T_i is called a predecessor of task T_k (and T_k is a successor of T_i) whenever there exists a sequence of tasks $T'_1, T'_2, ..., T'_l$ such that $T_i \prec T'_1 \prec T'_2 \prec \ldots \prec T'_l = T_k$.

Definition 2

Given an n-element set J, $d_i, i = 1 \ldots, n$ and the partial order $\prec$ in J, a valid schedule is a total function $f \ : \ J \longrightarrow [0, \infty)$ that satisfies the following two conditions:

(1) for all $t \in [0, \infty), || \{T_i \in J : f(T_i) \leq t < f(T_i) + 1\} || \leq 1$

(2) whenever $T_i \prec T_k$, $f(T_i) + 2 \leq f(T_k)$.

Function f assigns a starting time to each task in J, property (1) states that no more than one task can start at any time and property (2) ensures that the partial order constraints are met.

Let us call a *modified deadline* of task T_i a minimum of its original deadline and the latest completion time of task T_i allowing all its successors to be completed before their deadlines. The following algorihtm assigns modified deadlines to all tasks in J.

ALGORITHM 1

1. Select a task T_i that has not been assigned a modified deadline d_i' and all of whose successors have been assigned modified deadlines.
2. For each modified deadline d' assigned to a successor of T_i determine the number $g(i, d')$ of successors of T_i having modified deadline less than or equal to d' and set

$d_i' = min\,\{d' : min(d_i, d' - (g(i,d') + 1))\}$

d_i' is the modified deadline of task T_i.

3. If not all tasks have been assigned modified deadlines, return to step 1. Otherwise stop.

It is easy to see that the computational complexity of the Algorithm 1 is $O(n^2)$. Let us now prove some properties of the modified deadlines.

Lemma 1

A valid schedule meets all modified deadlines if and only if it meets all the original deadlines.

Proof: Since $d_i' \leq d_i$ for each $1 \leq i \leq n$, any valid schedule meeting the modified deadlines obviously meets the original deadlines. Suppose a valid schedule f meets all the original deadlines, but fails to meet some modified deadline. Choose a task T_i, with f(T_i) as large as possible, such that $d_i' < f(T_i) + 2 \leq d_i$. By the computation of the modified deadlines there must exist d' such that the number $g(i,d')$ of successors of T_i with modified deadlines not exceeding d_i satisfies $d_i' = d' - [g(i,d') + 1]$. Since each of those $g(i,d')$ successors of T_i is completed by time d' and cannot begin execution until $f(T_i) + 2 > d_i'$, there must be $g(i,d')$ tasks executed in the time interval of

$$[d' - (f(T_i) + 2)] < d' - d_i' = g(i,d') + 1$$

which is impossible. Thus f must meet all the modified deadlines.

Theorem 1

Let $L = T_1, T_2, ..., T_n$ be any priority list such that the modified deadlines satisfy $d_i' \leq d_{i+1}'$, for $1 \leq i \leq n-1$. If a valid schedule obtained using L fails to meet the original deadlines, then there exists no valid schedule which meets all the deadlines.

Proof: Suppose a valid schedule f obtained using L fails to meet the original deadlines. Then by Lemma 1, f must also fail to meet the modified deadlines. Let T_i be the task with minimum $f(T_i)$, which fails to meet its modified deadline. Since $T_l \prec T_k$ implies $d_l' < d_k'$, the task with the smallest modified deadline must be ready for execution not later than the other tasks. Thus the choice of L ensures that at any time $t < f(T_i)$ at least one task with a modified deadline less than or equal to d'_i is being executed (on either machine). We now consider two cases.

Case 1. Suppose there exists an integer time $t < f(T_i)$ such that no one task started at time $t+1$. We may assume that t is the largest such integer time and let T_k be the task with $f(T_k) = t$ and $d'_k \leq d'_i$. By this choice of t, one task with deadline less than or equal to d'_i begins at each time $t', t+2 \leq t' \leq f(T_i)$. Together with T_i these tasks create a set of $[f(T_i) - (t+2)] - 1 = f(T_i) - t - 1$ tasks having deadlines less than or equal to d'_i and all having started execution after the completion of T_k. Since T_k was the task with deadline less than or equal to d'_i starting execution at time t and no task started execution at time $t+1$, the choice of L and the priority list scheduling method imply that T_k must be a predecessor of all these $f(T_i) - t - 1$ tasks. Therefore T_k has at least $f(T_i) - t - 1$ successors with modified deadline less than or equal to d'_i. Thus by computation of the modified deadline we have:

$$d'_k \leq d'_i - [g(i,d') + 1] = d'_i - [(f(T_i) - t - 1) + 1] = d'_i - f(T_i) + t$$

and since $f(T_i) > d'_i - 2, d'_k < t + 2$.

But this says that T_k failed to meet its modified deadline, contradicting the choice of T_i as the first such task.

Case 2. Suppose that exactly one task with modified deadline less than or equal to d'_i begins at each integer time $t < f(T_i)$. Then there are at least $f(T_i)$ tasks with deadline less than or equal to d'_i. No schedule can ever complete more than $(t-1)$ tasks by any time t. Thus since $f(T_i) \geq d'_i - 1$ we have $f(T_i) \geq (d'_i - 1) > d'_i - 2$ and there cannot exist any valid schedule meeting all the modified deadlines. This proves the theorem.

Conclusion 1

The problem of scheduling n tasks with arbitrary precedence constraints and unit processing times of all operations in a two-machine flow shop with the objective to minimize the maximal lateness can be solved in time $O(n^2)$.

Proof: It follows from Theorem 1 that a list schedule obtained for a list on which tasks are ordered according to the nondecreasing order of their modified deadlines is optimal with respect to the maximal lateness. Calculation of modified deadlines takes $O(n^2)$ time.

4 HEURISTIC ALGORITHMS

In the general case, we are not able to find the latest completion time of a task allowing all its successors to be completed before their deadlines. Thus we calculate an upper bound of this value using various formulas. We will still call the minimum of the original deadline and this upper bound - a modified deadline. Five list scheduling heuristics are proposed for the general m-machine case. They combine in different ways the idea of modified deadlines and the Coffman-Graham algorithm [1] to create a list of tasks. In Algorithm 2 the modified deadlines are calculated and tasks are labeled according to the Coffman-Graham algorithm, solving ties by applying the priorities following from the modified deadlines. To apply the Coffman-Graham idea of labeling tasks let us recall the definition of the lexicographical order.

Definition 3

Let $\mathbf{l} = (n_1, \ldots, n_k)$ and $\mathbf{l}' = (n'_1, \ldots, n'_k)$ be sequences of integers, and $k, k' \geq 0$. If $k = 0$ then $\mathbf{l}$ is an empty sequence. We say that $\mathbf{l}$ is *lexicographically smaller* than $\mathbf{l}'$, written $\mathbf{l} <^{\cdot} \mathbf{l}'$, if
(i) the two sequences agree up to some index j, but $n_{j+1} < n'_{j+1}$ (i.e. there exists j, $1 \leq j \leq k$, such that for all i, $1 \leq i \leq j$, $n_i = n'_i$ and $n_{j+1} < n'_{j+1}$), or if
(ii) sequence $\mathbf{l}$ is shorter, and the two sequences agree up to the length of $\mathbf{l}$ (i.e. $k \leq k'$ and $n_i = n'_i$ for all i, $1 \leq i \leq k$).

ALGORITHM 2

1. Calculate the modified deadlines of tasks in the following way

   $$d'_i := min\{d_i, T_k \in suc(T_i) : d'_k - \sum_{j=1}^{m} p_{ij}\},\ i = 1, \ldots, n$$

   where $suc(T_i)$ denotes the set of all successors of task T_i in graph G (i.e. edge (T, T_i) belongs to graph G). Assign label 1 to an arbitrary task without successors in G; $l := 2$;

2. Assume that labels $1, \ldots, l-1$ have been already assigned. Let U denote the set of tasks without labels, successors of which are already labeled. For every task T from U a sequence $l(T)$ is built in the following way. Let $T_1, T_2, \ldots, T_k$ denote all successors of T (i.e. the edge (T, T_m) belongs to G for $m = 1, 2, \ldots, k$). Labels $e(T_1), \ldots, e(T_k)$ in a decreasing order form $l(T)$. Choose all tasks T such that for each task $T' \in U$ inequality $l(T) <^{\cdot} l(T')$ holds and denote the set of these tasks as V. Order tasks in V according to the non-decreasing order of their modified deadlines. Assign the next label to the last task in V. Repeat Step 2 until all tasks are labeled.

3. Schedule tasks according to the list, starting with the task with the greatest label. If there is no task ready, schedule an empty task on $M1$ until the next operation on Mm is finished.

The computational complexity of Algorithm 2 is clearly $O(n^2)$.
In Algorithm 3 tasks are ordered according to non-decreasing modified deadlines and then scheduled on the basis of this order.

ALGORITHM 3

1. Calculate the modified deadlines of tasks in the following way

   $$d'_i := min\{d_i, min\{T_k \in suc(T_i) : d'_k - \sum_{j=1}^{m} p_{ij}\}\},\ i = 1, \ldots, n$$

   where $suc(T_i)$ denotes the set of all successors of task T_i in graph G (i.e. edge (T, T_i) belongs to graph G). Assign label 1 to an arbitrary task without successors in G; $l := 2$;

2. Assume that labels $1, \ldots, l-1$ have been already assigned. Let U denote the set of tasks without labels, successors of which are already labeled. Choose a task with the largest modified deadline from set U. Assign the next label to the chosen task. Repeat step 2 until all tasks are labeled.

3. Schedule tasks according to the list, starting with the task with the largest label. If no task is ready, schedule an empty task on $M1$ until the next operation on Mm is finished.

The computational complexity of Algorithm 3 is $O(n^2)$. It follows from the complexity of calculating the modified deadlines.

Algorithm 4 is similar to Algorithm 3, but calculates the modified deadlines in a more sophisticated way.

ALGORITHM 4

1. Calculate the modified deadlines of tasks in the following way

   $$d_i' := min\{d_i, min\{T_k \in suc(T_i) : d_k' - \sum_{\substack{T_l \in suc(T_i): \\ d_l' < d_k'}} \sum_{j=1}^{m} p_{lj}\}\},$$

   $i = 1, \ldots, n$

   where $suc(T_i)$ denotes the set of all successors of task T_i in graph G (i.e. edge (T, T_i) belongs to graph G). Assign label 1 to an arbitrary task without successors in G; $l := 2$;

2. Assume that labels $1, \ldots, l-1$ have been already assigned. Let U denote the set of tasks without labels, successors of which are already labeled. Choose a task with the largest modified deadline from set U. Assign the next label to the chosen task. Repeat step 2 until all tasks are labeled.

3. Schedule tasks according to the list, starting with the task with the greatest label. If there is no task ready, schedule an empty task on $M1$ until the next operation on Mm is finished.

Complexity of Algorithm 4 is $O(n^2)$.
Algorithms 5 and 6 calculate a modified deadline of a task on the basis of a partial schedule of the set of successors of this task. They differ in the formula used for calculating the modified deadlines. Their complexity is $O(n^2)$.

ALGORITHM 5

1. Select a task T_i that has not been assigned a modified deadline d_i' and all of whose succesors have been assigned modified deadlines.

2. Let $suc(T_i)$ be a set of all succesors of T_i in graph G. Find a schedule $S1$ of tasks from $suc(T_i)$ according to the list of their modified deadlines starting at time 0. Calculate the modified deadline of task T_i in the following way

   $d_i' := min\{d_i, min\{T_k \in suc(T_i) : d_k' - c_k\}\},\ i = 1, \ldots, n$

   where c_k denotes the completion time of task T_k in schedule $S1$.

3. Assume that labels $1, \ldots, l-1$ have been already assigned. Let U denote the set of tasks without labels, whose successors are already labeled. Choose a task with the largest modified deadline from set U. Assign the next label to the chosen task. If there are still tasks without labels go to step 2.

4. Schedule tasks according to the list, starting with the task with the greatest label. If no task is ready, schedule an empty task on $M1$ until the next operation on Mm is finished.

ALGORITHM 6

1. Select a task T_i that has not been assigned a modified deadline d_i' and all of whose succesors have been assigned modified deadlines.

2. Let $suc(T_i)$ be a set of all succesors of T_i in graph G. Find a schedule $S1$ of tasks from $suc(T_i)$ according to the list of their modified deadlines starting at time 0. Calculate the modified deadline of task T_i in the following way

   $$d_i' := min\{d_i, min\{T_k \in suc(T_i) : d_k' - s_k\}\} - \sum_{j=1}^{m} p_{ij}, \ i = 1, \ldots, n$$

   where s_k denotes the starting time of task T_k in schedule $S1$.

3. Assume that labels $1, \ldots, l-1$ have been already assigned. Let U denote the set of tasks without labels, whose successors are already labeled. Choose a task with the largest modified deadline from set U. Assign the next label to the chosen task. If there are still tasks without labels go to step 2.

4. As in Algorithm 5.

5 SIMULATED ANNEALING APPROACH

Simulated Annealling is one of the widespread randomized iterative improvement methods, that accepts with some probability, decreasing in time, solutions having the value of cost function worse than the current one. The probability is controlled by a parameter called temperature, which is gradually reduced from a high value, at which most moves are accepted to a low one at which only moves improving cost function are enabled. Simulated Annealing was introduced as an efficient algorithm to solve combinatorial problems in the early 1980's. A standard simulated annealing algorithm can be written as follows (we assume that the objective is to minimize the goal function).

SIMULATED ANNEALING ALGORITHM

1. Choose an initial solution x.
2. Get an initial temperature T, such that virtually all transitions are accepted for almost all cost-increasing transitions.
3. Perform the following loop until N transitions are accepted or the number of transitions reaches L: 1) generate a solution y from the neighborhood $S(x)$ of the current solution x 2) let $df = f(y) - f(x)$, where $f()$ is goal function 3) if $df \leq 0$ (improving move) replace x by y 4) if $df > 0$ replace x with y, with the probability $p = exp(-df/T)$
4. Set T=k*T (reduce temperature)
5. If stopping conditions are satisfied then stop, otherwise go to Step 3.
6. Return the best solution found.

In order to apply the standard simulated annealing algorithm to the problem under consideration we have estabilished the following parameters:

- **initial solution** is a random list schedule;

- **initial temperature** $T = 30/log(1/0.9)$, which means that move increasing goal function by about 30 is accepted with probability 0.9; from experiments we know that this temperature lets almost all transitions to be accepted;

- **generating a neighbour** (i.e. a solution y from the neighborhood $S(x)$ of the current solution x) is performed in the following way:
1) choose randomly a task from all the tasks which precede the task with the maximal lateness (that's because only moving those tasks may improve the goal function) and denote it as *zad*,
2) denote position of last predecessor of task *zad* in the schedule as *nrpred*,
3) denote the position of the first successor of task *zad* in the schedule as *nrsucc*,

4) choose randomly a position from the range $(nrpred, nrsucc)$ and move task zad to this position;

- **the maximal number of accepted transitions** (for each value of T): $N = n$, where n is the number of tasks to be scheduled;

-**the maximal number of transitions** (both accepted and rejected): $L = (n/2) * (n/2)$. Notice, that N and l were chosen experimentally, respecting the fact that the average size of the neighbourhood is $n/2$;

- **temperature** is decreasing by factor k=0.9;

- **stopping condition:** Simulated Annealing stops when the current temperature is smaller than 0.5 which means that the probability of accepting an increase of the goal function by 1 (the smallest possible value) is less than 0.1.

6 COMPUTATIONAL RESULTS

Several groups of problems have been distinguished according to the number of machines in a flow shop $m \in \{2, 3, 5, 10\}$, number of tasks $n \in \{20, 50, 100\}$ and the density of the precedence graphs (that is the proportion of edges representing precedence relations between nodes to all such possible edges) $Q \in \{0.1, 0.2, 0.5, 0.7\}$. For each combination of (n, m, Q) one hundred instances of the problem have been generated as follows. Processing times were obtained as integer values from the range [1,100] by means of a standard randomizing procedure. The deadline for task T_i is a random integer from the range $\left[\sum_{T_k \in pred(T_i)} \sum_{j=1}^{m} \frac{p_{kj}}{m}, c\right]$, where $pred(T_i)$ denotes the set of all predecessors of task T_i, and c denotes the length of a random list schedule of the generated instance. The algorithms have been implemented in C and experiments have been performed on a Silicon Graphics Power Challenge computer in the Supercomputing and Networking Center in Poznań. The numerical results of the experiments are presented in Tables 2-7. Each table contains results for a given density (Q) of the precedence graph. For all combinations of (n, m, Q) four values obtained by all tested algorithms are reported:

max - maximum relative difference between the best solution and the solution generated by the considered algorithm:

$$max \left\{ i \in I, \frac{L_{max}(i) - L^*_{max}(i)}{L^*_{max}(i)} \right\}$$

where I is the set of all generated instances, $L_{max}(i)$ is the value of the cost function for instance i obtained by the considered algorithm and $L^*_{max}(i)$ is the best value of the cost function obtained for instance i by all applied algorithms.

mean - is the mean value of relative differences between the best solution and the solution generated by the considered algorithm over all instances:

$$\frac{1}{100} \sum_{i=1}^{100} \frac{L_{max}(i) - L^*_{max}(i)}{L^*_{max}(i)}.$$

best - is the number of instances for which the considered algorithm found the best solution;

and cpu - is the mean computational time of the algorithm over all instances.

The experiments show that for small density of the precedence graph (0.1 and 0.2) and small number of machines H5 proved to be the best heuristic. But with the increase of graph density and the number of machines its advantage over heuristics H3 and H6 falls down, so that for 10 machines the results produced by H3 and H6 are better than those obtained by H5. For small density of the precedence graph heuristic H2 is the worst one in every case. Heuristic H6 is always slightly better than H3 but its cpu is 50% longer. Similarly H5 is better than H4 and has longer computational times.

For medium and large graph density (0.5 and 0.7) the best heuristics are H3 and H6. They produce similar results while H6 is more time consuming. Again H5 is always slightly better than H4 and needs more cpu time. In this case, H2 is slightly worse than H5 for a small number of tasks and better for a large number of tasks.

7 CONCLUSIONS

The problem of scheduling unit time tasks in a two machine flow shop to minimize the maximal lateness has been shown to be polynomially solvable. Some

heuristics have been developed for the general, NP-hard, case and they have been computationally compared with a simulated annealing algorithm. The results of the experiment have been presented.
The simulated annealing algorithm gives results close to the optimal solutions. Unfortunately it is very time-consuming in comparison with other heuristics. So simulated annealing is preferable when we have enough time to look for a good solution. When this is not the case we should use one of the other heuristics.
Algorithm H5 is recommended for small density of the precedence graph and small number of tasks, while for large problems Algorithms H3 and H6 perform better than the other.

8 ACKNOWLEDGEMENTS

This research has been partially supported by the KBN grant No. 8T11F 010 08p02. The computational experiment has been carried out at the Poznań Supercomputing and Networking Center.

REFERENCES

[1] Coffman E.G.Jr., Graham L.R., "Optimal scheduling for two processor systems", Acta Informatica 1, 1972, pp. 200-213.

[2] Garey M.R., Johnson D.S., "Scheduling tasks with nonuniform deadlines on two processors", J.ACM, vol.23, No. 4, 1976, pp. 461-467.

[3] Johnson S.M.,. "Optimal two- and three-stage production schedules", Naval Research Logistics Quarterly 1, No.1, 1954, pp. 61-68.

[4] Józefowska J., "Scheduling unit-length two-machine flow shop under arbitrary precedence constraints", JORBEL, vol. 30, No. 2, 1990, 35-48.

[5] Lenstra J.K., Rinooy Kan A.H.G., Brucker P., "Complexity of machine scheduling problems", Ann. Discr. Math., 1, 1977, pp. 343-362.

[6] Osman I.H., Potts C.N., "Simulated annealing for permutation flow shop scheduling", Omega, No. 17, 1989, pp. 551-557.

[7] Ogbu F.A., Smith D.K., "The application of the simulated annealing algorithm to the solution of the $n/m/C_{max}$ flow shop problem", Computers and Opera. Research, No. 17, 1990, pp. 243-253.

[8] Ogbu F.A., Smith D.K., "Simulated annealing for the permutation flow shop problem", Omega, No. 19, 1991, pp.64-67.

[9] Steiner G., "The connection between the bump number problem and flow shop scheduling with precedence constraints", Computers Math. Applic., Vol. 20, No. 11, 1990, pp. 13-18.

[10] Zegordi S.H., Itoh K., Enkawa T., "Minimizing makespan for flow shop scheduling by combining simulated annealing with sequencing knowledge", Europ. J. Oper. Res., Vol. 85, No. 3, 1995, pp. 515-531.

Table 2 The results of the computational experiment for Q=0.1

H	m	n=20			n=50			n=100		
		max	mean	L	max	mean	L	max	mean	L
H1	2									
	3									
	5									
	10									
SA	2	0.80	0.030	79	0.16	0.01	81	0.47	0.03	79
	3	0.36	0.001	92	0.5	0.01	92	0.37	0.01	93
	5	0.12	0	95	0.1	0	94	0	0	100
	10	0.19	0	98	0	0	100	0	0	100
H2	2	6.64	1.27	1	8.96	1.45	0	4.37	1.43	0
	3	45.56	1.97	0	4.1	1.28	0	3.29	1.24	0
	5	614	11.97	1	2.31	0.98	0	2.55	1.43	0
	10	3.60	0.57	3	1.70	0.73	0	1.74	0.72	0
H3	2	2.71	0.47	0	1.34	0.42	1	1.28	0.59	0
	3	56.22	1.25	1	2.47	0.6	1	1.67	0.74	0
	5	340	6.5	5	1.18	0.43	1	1.61	0.93	0
	10	1.77	0.24	4	0.97	0.37	0	1.21	0.49	0
H4	2	1.83	0.35	11	2.08	0.35	5	1.17	0.44	2
	3	8.78	0.64	4	2.27	0.57	1	1.35	0.65	1
	5	466	7.36	2	1.71	0.61	0	1.95	1.22	0
	10	1.44	0.27	7	1.19	0.63	0	1.62	0.94	0
H5	2	1.86	0.32	11	1.4	0.26	12	1.34	0.25	19
	3	12.67	0.58	7	1.55	0.4	6	1.5	0.4	6
	5	354	5.4	5	1.21	0.41	5	1.47	0.84	0
	10	1.23	0.20	12	0.93	0.38	0	0.94	0.53	0
H6	2	2.71	0.47	7	1.44	0.41	1	1.3	0.58	0
	3	56.22	1.25	1	23	0.59	0	1.6	0.72	0
	5	340	6.5	5	1.16	0.43	0	1.57	0.92	0
	10	1.77	0.24	4	0.97	0.37	0	1.07	0.48	0

Table 3 The results of the computational experiment for Q=0.2

H	m	n=20			n=50			n=100		
		max	mean	L	max	mean	L	max	mean	L
H1	2									
	3									
	5									
	10									
SA	2	0.34	0.01	96	0.05	0	99	0	0	100
	3	0.55	0.01	92	0	0	100	0	0	100
	5	0.20	0	95	0	0	100	0	0	100
	10	0.08	0	96	0	0	100	0	0	100
H2	2	79.22	2.33	0	3.56	1.39	0	2.74	1.43	0
	3	5.94	1.12	0	2.84	1.1	0	2.95	0.9	0
	5	2.82	0.77	1	2.11	0.67	0	1.18	0.64	0
	10	1.97	0.34	7	1.13	0.38	0	0.84	0.27	0
H3	2	35.22	0.88	2	2.3	0.89	0	2.95	1.35	0
	3	3.46	0.52	5	1.44	0.72	0	1.65	0.94	0
	5	1.41	0.33	1	1.34	0.46	0	1.12	0.61	0
	10	0.85	0.16	5	0.48	0.23	0	0.44	0.24	0
H4	2	31.33	0.8	4	2.3	0.79	0	2.75	1.23	0
	3	3.58	0.54	3	2.01	0.87	0	2.23	1.12	0
	5	1.64	0.43	0	1.7	0.72	0	1.78	0.96	0
	10	1.08	0.26	2	1.14	0.48	0	0.93	0.52	0
H5	2	21.89	0.66	5	2.38	0.83	1	2.54	1.22	0
	3	2.88	0.45	3	1.74	0.73	0	2.16	0.83	0
	5	1.41	0.3	4	1.04	0.48	0	1.38	0.67	0
	10	0.72	0.16	5	0.67	0.29	0	0.53	0.30	0
H6	2	35.22	0.88	1	2.15	0.85	0	2.87	1.3	0
	3	2.66	0.51	5	1.32	0.69	0	1.6	0.92	0
	5	1.41	0.32	1	1.04	0.44	0	1.06	0.6	0
	10	0.85	0.15	6	0.45	0.23	0	0.40	0.23	0

Table 4 The results of the computational experiment for Q=0.5

H	m	n=20			n=50			n=100		
		max	mean	L	max	mean	L	max	mean	L
H1	2	0	0	100						
	3	0	0	100						
	5	0	0	100						
	10	0	0	100						
SA	2	0.23	0.02	72	0	0	100	0	0	100
	3	0.17	0.01	84	0	0	100	0	0	100
	5	0.06	0	89	0	0	100	0	0	100
	10	0.04	0	93	0	0	100	0	0	100
H2	2	2.36	0.55	0	1.24	0.39	0	0.79	0.35	0
	3	1.98	0.28	1	0.72	0.24	0	0.47	0.23	0
	5	0.77	0.17	1	0.53	0.16	0	0.3	0.15	0
	10	0.39	0.09	7	0.24	0.09	0	0.15	0.07	0
H3	2	1.48	0.39	1	1.32	0.39	0	0.85	0.38	0
	3	1.07	0.23	0	0.72	0.24	0	0.43	0.23	0
	5	0.64	0.13	2	0.41	0.13	0	0.28	0.12	0
	10	0.20	0.05	4	0.17	0.06	0	0.11	0.06	0
H4	2	2.11	0.42	0	1.13	0.45	0	1.17	0.46	0
	3	2.38	0.31	3	0.77	0.31	0	0.49,	0.29	0
	5	0.61	0.2	0	0.77	0.2	0	0.42	0.19	0
	10	0.78	0.11	2	0.26	0.12	0	0.25	0.12	0
H5	2	1.7	0.45	1	1.17	0.41	0	0.78	0.38	0
	3	3.02	0.27	1	0.92	0.27	0	0.5	0.24	0
	5	0.68	0.16	3	0.47	0.15	0	0.32	0.15	0
	10	0.51	0.08	5	0.18	0.09	0	0.13	0.08	0
H6	2	1.48	0.38	0	0.86	0.38	0	1	0.38	0
	3	0.65	0.24	0	0.73	0.24	0	0.39	0.22	0
	5	0.44	0.12	2	0.31	0.12	0	0.22	0.12	0
	10	0.35	0.06	4	0.14	0.06	0	0.11	0.06	0

Table 5 The results of the computational experiment for Q=0.7

H	m	n=20			n=50			n=100		
		max	mean	L	max	mean	L	max	mean	L
H1	2	0	0	100						
	3	0	0	100						
	5	0	0	100						
	10	0	0	100						
SA	2	0.13	0	89	0	0	100	0	0	100
	3	0.03	0	93	0	0	99	0	0	100
	5	0.02	0	98	0	0	100	0	0	100
	10	0.02	0	98	0	0	97	0	0	100
H2	2	0.64	0.18	7	0.55	0.15	0	0.29	0.15	0
	3	0.92	0.1	8	0.32	0.1	0	0.22	0.09	0
	5	0.18	0.06	8	0.2	0.06	0	0.12	0.05	0
	10	0.31	0.05	6	0.11	0.03	2	0.13	0.04	0
H3	2	1.36	0.15	7	0.49	0.15	0	0.25	0.14	0
	3	0.54	0.08	12	0.22	0.09	0	0.15	0.08	0
	5	0.23	0.05	13	0.12	0.05	0	0.1	0.06	0
	10	0.11	0.03	22	0.08	0.03	0	0.05	0.03	0
H4	2	0.58	0.18	9	0.62	0.16	0	0.34	0.16	0
	3	0.55	0.12	7	0.35	0.11	0	0.2	0.11	0
	5	0.21	0.07	5	0.16	0.07	0	0.16	0.08	0
	10	0.19	0.05	5	0.11	0.04	0	0.08	0.04	0
H5	2	0.61	0.16	8	0.49	0.16	0	0.33	0.14	0
	3	0.92	0.11	9	0.33	0.11	1	0.18	0.1	0
	5	0.19	0.05	15	0.18	0.07	1	0.13	0.06	0
	10	0.25	0.04	14	0.07	0.03	2	0.08	0.04	0
H6	2	1.29	0.15	5	0.4	0.14	0	0.3	0.14	0
	3	0.52	0.08	12	0.23	0.09	0	0.18	0.08	0
	5	0.19	0.05	13	0.14	0.05	0	0.1	0.05	0
	10	0.13	0.02	23	0.08	0.03	0	0.06	0.03	0

Table 6 The comparison of computational times for Q=0.1 and Q=0.2

H	m	Q=0.1			Q=0.2		
		n=20	n=50	n=100	n=20	n=50	n=100
SA	2	0.139	1.667	8.657	0.129	1.264	7.241
	3	0.201	2.487	13.202	0.163	2.004	11.721
	5	0.3	4.009	161.871	0.251	3.108	111.155
	10	2.898	23.959	277.826	2.786	23.079	157.752
H2	2	0.001	0.006	0.023	0.001	0.007	0.028
	3	0.002	0.006	0.026	0.002	0.008	0.031
	5	0.001	0.008	0.031	0.001	0.009	0.036
	10	0.002	0.011	0.043	0.003	0.012	0.049
H3	2	0.001	0.005	0.02	0.001	0.006	0.023
	3	0.001	0.006	0.023	0.002	0.007	0.026
	5	0.001	0.007	0.027	0.001	0.008	0.031
	10	0.002	0.010	0.040	0.002	0.011	0.044
H4	2	0.001	0.006	0.024	0.001	0.008	0.029
	3	0.001	0.007	0.026	0.002	0.008	0.032
	5	0.001	0.008	0.032	0.002	0.009	0.038
	10	0.002	0.011	0.045	0.002	0.013	0.052
H5	2	0.001	0.006	0.026	0.002	0.009	0.037
	3	0.001	0.007	0.029	0.004	0.01	0.043
	5	0.002	0.008	0.037	0.002	0.012	0.057
	10	0.002	0.013	0.055	0.003	0.017	0.087
H6	2	0.002	0.007	0.027	0.002	0.009	0.04
	3	0.002	0.008	0.03	0.002	0.011	0.045
	5	0.001	0.009	0.038	0.002	0.013	0.061
	10	0.002	0.014	0.058	0.003	0.019	0.092

Table 7 The comparison of computational times for Q=0.5 and Q=0.7

H	m	Q=0.5			Q=0.7		
		n=20	n=50	n=100	n=20	n=50	n=100
H1	2	1.287			0.041		
	3	1.725			0.056		
	5	3.663			0.058		
	10	7.589			0.107		
SA	2	0.078	0.831	5.169	0.062	3.465	3.942
	3	0.111	1.204	7.193	0.086	0.896	5.236
	5	0.155	1.868	48.617	0.131	1.404	41.002
	10	2.583	17.484	93.323	2.551	16.492	80.982
H2	2	0.001	0.011	0.046	0.002	0.016	0.058
	3	0.002	0.012	0.049	0.003	0.015	0.060
	5	0.002	0.010	0.054	0.003	0.016	0.066
	10	0.003	0.016	0.067	0.003	0.019	0.078
H3	2	0.002	0.008	0.036	0.020	0.011	0.043
	3	0.002	0.009	0.037	0.002	0.011	0.045
	5	0.002	0.011	0.042	0.002	0.013	0.050
	10	0.002	0.014	0.056	0.003	0.015	0.064
H4	2	0.002	0.012	0.045	0.003	0.013	0.056
	3	0.002	0.012	0.049	0.003	0.015	0.060
	5	0.002	0.015	0.054	0.003	0.017	0.064
	10	0.002	0.017	0.071	0.004	0.020	0.081
H5	2	0.003	0.017	0.088	0.003	0.023	0.138
	3	0.002	0.020	0.113	0.004	0.031	0.199
	5	0.003	0.027	0.160	0.005	0.040	0.269
	10	0.004	0.043	0.276	0.005	0.069	0.482
H6	2	0.003	0.019	0.099	0.003	0.027	0.156
	3	0.003	0.021	0.121	0.004	0.033	0.199
	5	0.003	0.028	0.174	0.005	0.044	0.291
	10	0.005	0.046	0.289	0.006	0.072	0.505

18

A GENETIC PROGRAMMING APPROACH FOR HEURISTIC SELECTION IN CONSTRAINED PROJECT SCHEDULING

Rema Padman and Stephen F. Roehrig

The Heinz School of Public Policy and Management,
Carnegie Mellon University,
Pittsburgh, PA 15213

ABSTRACT

The resource-constrained project scheduling problem (RCPSP) with cash flows investigates the scheduling of activities that are linked by precedence constraints and multiple resource restrictions. Given the presence of cash flows which represent expenses for initiating activities and payments for completed work, maximizing the Net Present Value of the project is a practical problem. This is a complex combinatorial optimization problem which precludes the development of optimal schedules for large projects. Many heuristics exist for the RCPSP, but it has proven difficult to decide in advance which heuristic will provide the best result, given a problem characterization in terms of parameters such as size and complexity. In this paper we discuss the use of genetic programming (GP) for heuristic selection, and compare it directly to alternative methods such as OLS regression and neural networks. The study indicates that the GP approach yields results which are an improvement on earlier methods. The GP solution also gives valuable information about project environments where a given heuristic is inappropriate. In addition, this approach has no problem evolving complex nonlinear functions to capture the relationship between problem parameters and heuristic performance. Thus the results given in this paper shed light on the logical domains of applicability of the various heuristics, while at the same time provide an improved heuristic selection process.

1 INTRODUCTION

Project scheduling problems with resource constraints and cash flows arise in many organizational settings, ranging from construction planning to research and development. The resource constrained project scheduling problem (RCPSP) investigates the scheduling of a collection of activities that are linked by precedence constraints and multiple resource restrictions. When significant levels of cash flows are present over the duration of the project, which occur in the form of expenses and payments for completed work, the development of project schedules which maximize the Net Present Value (NPV) of the cash flows is of considerable practical importance [2]. This is a difficult combinatorial optimization problem which precludes the development of optimal schedules for most projects encountered in practice [5].

Given the intractable nature of the problem, many heuristic procedures have been developed to obtain good feasible solutions [1, 3, 4, 9, 10, 13, 14]. However, different heuristics perform well under different conditions, i.e., no single heuristic dominates in any problem environment. The selection of an inappropriate heuristic may lead to poor suboptimal schedules resulting in significant monetary losses. Padman [8], and Zhu and Padman [11, 15] investigated the use of artificial neural networks on a randomly generated set of resource-constrained projects to address the problem of selecting the most appropriate heuristic for a given instance of the RCPSP. They concluded that the neural network procedure is better at selecting a category of heuristics rather than a single best heuristic. The results of the neural network study, while encouraging, indicated that the modular system of neural networks is successful in selecting the three best heuristics from a set of 16 heuristics in only about 46% of the test cases considered. In this paper, we investigate the feasibility of applying a genetic programming approach to the same data set to derive the functional forms that represent the relationship between the heuristics and project parameters.

For large projects, this study can lead to considerable savings in the time and effort required to determine an appropriate procedure for obtaining a feasible schedule, as well as cost savings from an improved schedule. In addition, these procedures can be embedded within a decision support framework to provide enhanced decision making capabilities for implementing large projects. The next section describes the problem in greater detail, provides a brief summary of the experimental data used in the study and the insights from the neural network implementation of Zhu and Padman [11, 15]. Section 3 describes the genetic programming approach. Section 4 discusses the results and conclusions of the experimental study and considers future extensions.

2 DESCRIPTION OF THE PROBLEM AND DATA

The resource-constrained project scheduling problem (RCPSP) with cash flows can be formulated mathematically as a mixed integer nonlinear program [13], [8]. Given a project with m activities and n events, with activity duration $d_k, (k = 1, \ldots, m)$, and cash flows $F_i, (i = 1, \ldots, n)$ at the events discounted at the rate α, the formulation is as shown below. R_{z_s} represents the resource requirements for the set Z_s of activities in progress at time $s(s = 1, \ldots, q)$, where q is the duration of the entire project. R_L is the resource limit vector that is assumed to remain constant. The variable T_i indicates the scheduled time of each event i. The objective is to obtain the optimal time for scheduling each task such that the project NPV is maximized subject to the constraints on precedence (1) and resources (2).

Maximize

$$\sum_{i=1}^{n} F_i e^{-\alpha T_i}$$

Subject to

$$T_j - T_i \geq d_k \quad k = 1, \ldots, m \tag{18.1}$$

$$R_{Z_s} \leq R_L \quad s = 1, \ldots, q \tag{18.2}$$

Figure 1 depicts an example of a project network with durations represented on the arcs and cash flows at the nodes. Each activity requires three types of resources, as shown in the figure. The resource limit vector is [12, 12, 12]. The optimal schedule is also indicated in the figure. As the resource usage of each activity increases or the resource limit decreases, it is significantly more difficult to solve this problem to optimality. In particular, for practical problems with hundreds, if not thousands of activities, it is computationally impractical to attempt optimal solutions. Hence heuristic procedures have found extensive application in project-oriented manufacturing and service industries.

Given the existence of a large number of heuristics for the RCPSP with cash flows and a large amount of data on the comparative performance of these heuristics, choosing a heuristic or category of heuristics that is appropriate for a specific problem instance is not straightforward. Padman [8], and Zhu and Padman [11, 15] investigated the use of neural networks to address the

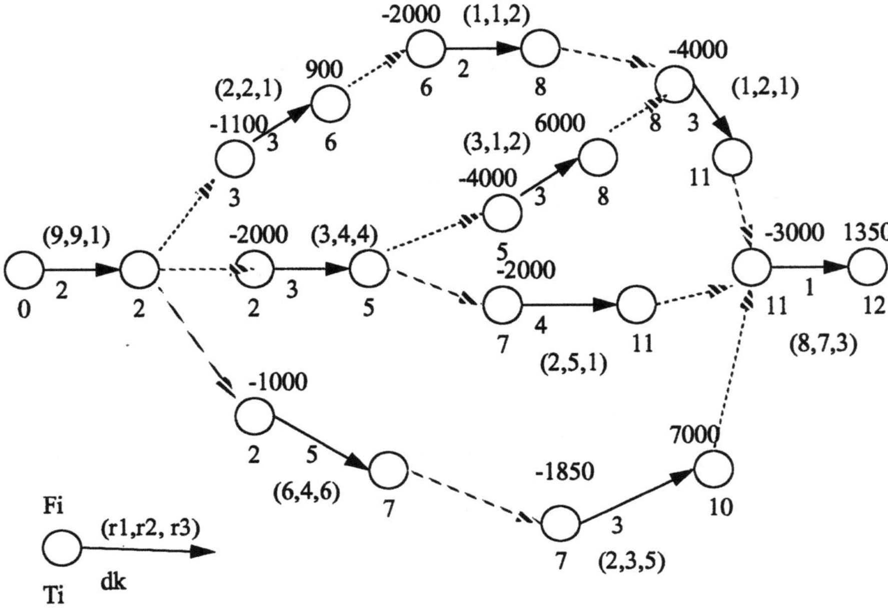

Figure 1 Example Project Network

problem of determining the most critical project parameters and to induce the relationship between these parameters and heuristic performance.

In order to study the effects of project parameters on heuristic performance, they investigated various optimization-based heuristics and heuristics derived from the critical path method that have been proposed in the literature [4, 7, 13, 8]. The sixteen heuristic rules examined in this study are the ones discussed in [10], which are listed in Table 1.

The data set used in this paper consists of 1440 randomly generated project networks [10]. These 1440 problems are generated from six experimental parameters including project size (size), project network structure (shape), resource constrainedness (AUF), interest-rate (CC), profit margin (PM), and frequency of progress payments (FP). All six parameters were each varied over two or three different levels in constructing the project networks. The size of the project is characterized by the number of activities, where all project networks consisting of 48 activities are considered small size problems while the medium size problems have 110 activities. The degree of resource constrainedness, characterized by AUF (average utilization factor) [7], is set at three levels: low level

OCS/LAN	Activities are selected for scheduling in ascending order of opportunity cost of scheduling. Activities with zero tardiness penalties are scheduled according to LAN.
OCR/LAN	Activities are scheduled in ascending order of the opportunity cost of resources, with zero tardiness penalty activities scheduled according to LAN.
NOC/LAN	Activities are scheduled in increasing order of net opportunity cost which balances the cost of delaying some activities against scheduling others. Zero tardiness penalty activities are scheduled according to LAN.
LFT/LAN	Priority is given to the activities with the minimum latest finish time as determined by critical path analysis. LAN is used to break ties between the activities.
MS/LAN	Select activity with the smallest slack, where the critical path is continuously revised as each activity is scheduled. LAN is used to break ties.
TS/LAN	Select activity with the maximum difference between current early finish time and optimal finish time as given by the relaxation model. LAN is used to break ties.
DUAL/TS	Activities with the highest dual price is selected for scheduling. Ties are broken using TS rule.
TS/DUAL	Select activities with maximum difference between current early finish time and optimal finish time as given by the relaxation model. Maximum dual price is used to break ties.
RAND-50	Random selection of eligible activities is chosen for scheduling, with the best of 50 replications being reported.
MTP/LAN	Activities are selected from the schedule queue in descending order of their tardiness penalties. Activities with zero tardiness penalties are scheduled according to LAN.
CFW/LAN	Select activity with highest cash flow weight for scheduling. LAN is used to break ties.
LTP/LAN	Activities are scheduled in ascending order of tardiness penalties. Activi- ties with zero tardiness penalties are scheduled according to LAN.
CFW-CFW	Use highest cash flow weight activities for scheduling. Activities with zero tardiness penalties are then scheduled using highest cash flow weights.
CFW-OCC	Use cash flow weights for the scheduling from the first queue, and opportunity cost of cash flows for scheduling from the second queue.
MTP/ET	Same as MTP/LAN, except that activities with zero tardiness penalties are scheduled according to early time rule.
TSCFW	Schedule activities based on highest cash flow weight.

Table 1 Heuristic Priority Rules and Descriptions

of 1.0; medium level of 1.5; and high level of 2.0. Project network shape also varies over three levels from balanced, to skewed to the right and skewed to the left. Three cash flow parameters are utilized to describe the financial characteristics of the project. The first, interest rate varies over two levels from a low of 10% to a high of 20% annually; the second, frequency of progress payments, also set at two levels where a low level indicates that on average every seventh activity yields a positive progress payment upon its completion while a high level indicates that on average every third activity yields a positive progress payment; and the third, the profit margin reflected by the final payment for the completed project computed as the sum of the cash expenditures for the project plus a profit percentage, is set at two levels, 30% and 50%. A full factorial experiment was conducted which resulted in 144 different scheduling environments leading to 1440 problems with 10 replicates in each environment.

More than 40 project characteristics that have been discussed in the literature to describe the RCPSP were identified by Zhu and Padman [11, 15]. Several of these were eliminated due to multicollinearity effects, resulting in the selection of a set of 30 variables. Two input preprocessing strategies using principal component analysis and feature extraction networks were employed to reduce input dimensionality from thirty to six.

The output layer of the neural network consisted of the normalized Net Present Values (NPV) from the sixteen heuristics described above [9, 10] which included nine revised optimization-guided heuristics and seven others from the literature. Some of these were proposed based on the same type of information such as dual price and opportunity cost. Two output prepocessing strategies, using expert opinion and cluster analysis, were used to reduce output dimensionality and to categorize the heuristic procedures. Extensive experimentation was conducted with single generalization as well as modular neural networks, differing network topologies, learning parameters, and data encoding schemes. When presented with the project parameter information on a new instance of the RCPSP, the neural network system produces a set of sixteen output values in the range of [0,1], representing the heuristics. The best heuristic for the given instance is then the one with the output value closest to 1.

The results indicated the ability of the single generalization neural network to successfully categorize heuristics into distinct groups and select a group rather than a single best heuristic. It is easy to implement and use when compared to the several levels used in the modular system. The modular system, however, had better success in identifying the most appropriate heuristic. The neural network system also performed marginally better than multiple regression and significantly better than discriminant analysis models.

3 GENETIC PROGRAMMING

3.1 Background

An ideal solution to the heuristic selection problem would be a function which takes as arguments the parameters describing the scheduling problem, and which returns an indication of the heuristic which is best. In addition to this predictive ability, a desirable property of such a function is descriptive power. Many regression equations, because of their simple structure, shed light on the relationships between variables, and can thus be considered descriptive. The drawback of regression is that one must have a firm idea, in advance, of the general form of the relationships (linear, quadratic, etc.). Conversely, in the case of a neural network, the resulting "function" can be arbitrarily complex but is essentially a black box; there is little if any descriptive power.

A technique which in principle can overcome the limitations of both regression and neural nets is genetic programming. This is an extension of the well-known fixed-string-length genetic algorithm. Genetic programming, pioneered by Koza [6], applies the crossover mechanism to functions represented as tree structures. Nodes in a tree correspond to either elementary operations (addition, multiplication, comparison, etc.), independent variables, or constants. Given values of the independent variables, a tree can be evaluated, with the result being taken as the corresponding dependent variable.

Given a suitably rich set of elementary operations, trees of arbitrary complexity may be constructed. The point of genetic programming, though, is that a process akin to natural selection should, over time, evolve functions which are "fit" in their environment. In the case of heuristic selection for the NPV scheduling problem, "fitness" means that the function will accurately select the best heuristic.

In practice, genetic programming starts with a set, called the initial generation, of trees randomly constructed from operations, variables, and constants. The only constraint on tree generation is that the resultant trees be syntactically correct, so that they may be evaluated.

To construct each member of this initial set, the following procedure is used. First, at level 0, a special node with 16 children is constructed. This node is really a dummy, there only to keep track of the 16 children. Each of these children (at level 1) is initially a single node, chosen at random from the set of operators $\{+, -, \times, /\}$. Then each of these 16 operator nodes is extended by choosing two level 2 nodes, which become its operands. These operand nodes are again chosen at random, but this time the choice set consists of operators, the independent variables X_1 through X_6, and a randomly chosen real number between 0 and 1 (for a total of 11 possibilities). Level 2 is then complete. However, if any of the level 2 nodes is an operator, then operands for it are chosen. This procedure proceeds recursively until no lowest level nodes are operators. Each tree so generated essentially provides 16 real-valued functions each of which may have as independent variables any subset of $\{X_1, \ldots, X_6\}$.

In our application, each tree returns a vector of real numbers which we wish to be as close as possible to the NPVs achieved by the heuristics. The fitness of each tree is then judged according to a problem-specific fitness measure; in the present case the measure indicates whether or not the index of the largest component is in the set of indices of the three best performing heuristics. A new generation is then constructed, using the fittest members of the initial generation.

Members of the new generation are formed either as direct copies of earlier, relatively highly fit trees, or by a crossover process. Crossover starts with a pair of trees, and swaps a randomly selected subtree from the first with a randomly selected subtree from the second. Since the trees selected to participate in crossover are themselves among the best of their generation, the new trees formed by crossover stand a chance of being even more fit than their parents. Naturally, they may easily be worse, but as the evolutionary process continues over time, succeeding generations as a whole, and the best individual trees in particular, improve.

The functions produced by genetic programming are often internally complex and redundant. They can easily be simplified by symbolic algebra programs, however, and the results interpreted like any other mathematical expression. Thus genetic programming applied to the problem of heuristic selection explores a very large space of functions over the set of scheduling parameters, and returns functions which can be examined for insights into the capabilities of the heuristics themselves.

3.2 Application to Heuristic Selection

We applied the genetic programming approach to a randomly generated subset of the 1440 benchmark problems. This subset (698 problems) was used to evolve trees which best approximated the normalized NPVs of the 16 heuristics. The remaining 742 problems were used to test the results.

The genetic programming process was begun by randomly constructing 500 trees, using the arithmetic operators $+, -, *, /$. Terminal nodes consisted of (1) the six variables which were used to generate the scheduling problems, and (2) randomly generated constants in the range $[0, 1]$. A fragment of a typical tree is shown in Figure 2. Each tree was evaluated for fitness, and a new generation of 500 trees produced. This process continued for 100 generations, and the best tree obtained from the overall process was retained for testing and analysis. Because of the high degree of randomness in the whole process, more than 40 runs were made in order to give some confidence that the best, or close to best, possible selection functions were obtained.

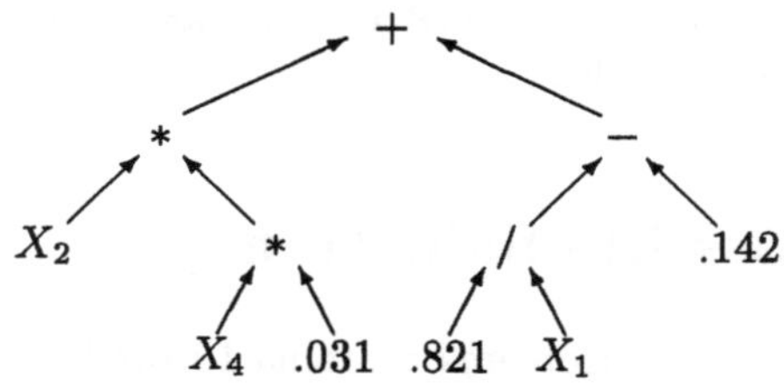

Figure 2 A Typical GP Tree

Since in many instances several heuristics produced NPVs which were close to the best, the following scoring rule was used to judge each tree's performance. The heuristics were ranked according to NPV (larger NPVs being better), and a tree scored a "hit" if it selected as best one of the top three heuristics. The tree was compared to the ranked heuristics for all 698 test problems (usually a different ranking for each problem), keeping a running total of accumulated hits. The total number of hits became the absolute fitness measure for the evolutionary process.

To produce the next generation, each tree was first given the relative fitness measure

$$R_i = \frac{A_i}{\sum_i A_i},$$

where A_i is the absolute fitness for the ith tree, and R_i is the relative fitness. Trees were then selected for reproduction by choosing pairs randomly, with the probability of a tree being selected equal to its relative fitness. Each pair thus chosen either became part of the new generation immediately (that is, without modification), or was subjected to the crossover operation described earlier. In all our experiments, the probability of crossover was set to 90%. No mutation operation was used in our experiments.

4 RESULTS

In this section we first examine the overall difficulty of the heuristic selection problem by considering an upper bound on the performance of any selection procedure. Then some sample GP outputs are presented and discussed, and finally the best results obtained by GP are compared with previous work using regression and neural networks.

4.1 Best Possible Selection

The data set of scheduling problems contains 10 randomly generated replications for each of the 144 combinations of the six coded input parameters. The selection task is to choose the best performing (largest NPV) heuristic. If we simplistically view the data as the entire universe of scheduling problems, it is straightforward to find the heuristic which is the top performer in each group of 10 problems. This leads to an upper bound, given the data, for a selection procedure. Varying this approach by selecting, for each group, the heuristic most often appearing in the top three gives an upper bound corresponding to the fitness measure used in the GP approach.

For the data at hand, it turns out that *any* selection procedure can do no better than 68% when trying to choose one of the top three heuristics. This gives us a benchmark against which to compare regression, neural networks and GP. It also gives an indication of the difficulty of the selection process; it is an inherently difficult problem.

4.2 Some Sample Output

Figures 3 and 4 present the output from two runs of the GP program. While these were not the best of all runs, we include them to illustrate several points.

	X_1	X_2	X_3	X_4	X_5
Y_2			0.025		0.025
Y_3	0.013				0.012
Y_4					0.006
Y_5	0.005				
Y_7	0.003			0.004	

Table 2 Fraction of Y_i in Top Three When $X_j = -1$.

First, both of the sample outputs have the same fitness. Each was able to identify one of the top three heuristics 345/698 = 49% of the time. Yet the two solutions differ considerably in the functions they propose for estimating the normalized NPV of each heuristic. This is a natural consequence of two facts: each run starts with a randomly generated set of trees, and each solution is attempting only to find one of the top three heuristics.

Perhaps the most interesting feature the two solutions have in common is simplicity of the prediction function for many of the heuristics, in particlar Y_2, Y_3, Y_4, Y_5, and Y_7. These are essentially of the form $Y_i = X_j$ for some pair i, j. In effect, GP has identified a single factor as most important in determining whether a particular heuristic is a viable candidate under certain conditions. Keeping in mind that the X_i can take on only the values -1, 0, and 1, the effect is to eliminate the possibility that a heuristic can be chosen as best when the X_i associated with it is -1.

That this is an effective strategy for GP can easily be checked by counting the number of times a given heuristic is actually among the best when the X variable associated with it by the GP solution is -1. Table 2 shows the fraction of times certain heuristics were among the top three when the "determining" X variable (as suggested by GP) was at the -1 level. These data suggest that the GP solution is generally quite correct in picking out situations where a given heuristic is inappropriate.

The two sample outputs lead to a second observation, namely that highly non-linear selection functions may be necessary to give good performance. The functions presented for Y_6, Y_{11}, Y_{12}, Y_{14} and Y_{15} in the first sample, and Y_1 and Y_{13}–Y_{15} in the second, include many (or all) of the X_i in complex relationships. (Note that the functions presented in Figures 3 and 4 are just as they appeared in the program output; algebraic simplification is obviously possible

```
Corfu 12-28-94no3.out  Fitness = 345

Y1 =  0.199561

Y2 =  X6

Y3 =  X6 + 0.0162969

Y4 =  X6

Y5 =  X1

Y6 =  (X5 / X2) + ((X4 - ((((X5 + ((0.757164 * X3) + 0.0162969))
      + X3) + 0.089877) * X4)) * X6)

Y7 =  X5

Y8 =  (0.757164 * X3) * X1

Y9 =  X5

Y10 = 0.201178 + 0.089877

Y11 = X1 - (X1 / X5)

Y12 = ((X4 / 0.538591) + X4) + X2

Y13 = X6 /0.578478

Y14 = ((((0.764733) - (X6)) * (0.614795)) - (X2)) - (X6)

Y15 = (((X3 * (X1 + X2)) - X1) - ((((X6 * X2) + X3) + (X4 + X6))
      + (X2))) * 0.101688

Y16 = (X6) + (0.0162969)
```

Figure 3 Sample Output 1

```
Corfu 1-6-95no5.out    Fitness = 345

Y1 =  (0.0827967 / (0.882168 - ((X1 / X5) - ((X6 * 0.122288) /
      0.247353)))) * (X4)
Y2 =  X3
Y3 =  X1
Y4 =  0.267708
Y5 =  X1
Y6 =  0.684133 - X4
Y7 =  X1
Y8 = -X4
Y9 =  (X6 * X1) / (0.247353)
Y10 = X2
Y11 = 0.208686
Y12 = 0.693594
y13 = ((((X5 - X5) + (X2 / X3)) + (X1 + ((0.267708 + X6) - X4))) /
      X4) / (X4 / 0.29722)
Y14 = (0.778375 / X4) / (0.640095)
Y15 = (0.332408 / (((0.208686 - (X4 - (X6 / X5))) * (X1)) -
      (X4 - X1))) + (X3)
Y16 = 0.268715
```

Figure 4 Sample Output 2

in some cases.) It would be difficult to anticipate or specify this complexity in a regression analysis of the data, but the GP has no problem evolving functions of this sort.

4.3 Comparison With Regression and Neural Nets

We tested the best selection function obtained by using GP on the "training" subset of problems. On the test set it was able to find a top-three heuristic 53% of the time. This compares favorably with 49% for standard multiple linear regression, and just over 46% for the neural network approach. Taken in the context of the upper bound of 68% given earlier, this represents an 8% relative improvement by GP over regression. While not earth-shattering, it nonetheless indicates the utility of the GP approach, since for large projects, and over time, potential dollar savings are sizable.

It is somewhat surprising, however, that GP did not perform better. Intuitively, one would imagine that it would not be too difficult to evolve simple functions which could precisely identify the best heuristic. After all, we only require a discrete mapping from the set of 144 possible parameter combinations to an index of the best heuristic. On reflection, though, it is clear (from the mediocre showing of linear regression) that the nonlinearity of this mapping makes finding an adequate closed-form expression extremely difficult.

We note once again that the best performing GP solutions were in many ways relatively simple. In observing evolving solutions, we often saw a pattern of simple (and poor) functions leading to very complex (and somewhat better) solutions. Almost always, however, the latter evolved further into much simplified, and much better performing, results. It is not immediately obvious why this process occurred, but we surmise that further compexity, to match the needs of the problem, might evolve in further generations. Unfortunately, the trials reported on here already required substantial computing horsepower (roughly 6 hours per run on a reasonably fast HP workstation). Since we wanted to explore a variety of (randomly generated) initial generations, extended runs have not yet been tried.

	1st	2nd	3rd	Percent
CFW-OCC	177	215	243	44%
LPT-LAN	294	129	107	37%
MPT-ET	98	152	138	27%

Table 3 Best performance, static choice.

4.4 Comparison With Static Choice Approaches

The data used in this study clearly indicate that some heuristics are more likely to be good performers. The point of heuristic selection, however, is that there is no single heuristic which dominates for every combination of input parameters. To illustrate how good a static choice can be, Table 3 shows the performance of three of the best overall performers. The column labels refer to first-, second- or third-place finishes (that is, highest, second highest or third highest NPV). The table cells list the number of such finishes, out of 1440 trials.

It is clear that CFW-OCC is an excellent heuristic (44% in the top three), but improvement is possible if we can choose from amongst the larger collection of heuristics, tailoring our choice to the problem parameters. For instance, GP was able to find one of the top three 53% of the time.

Since even the heuristics are time-consuming to run (not to mention exact procedures), it is unlikely that one would have the freedom to try more than one or two. However, if it were possible to run all three listed in the table, quite good results can be had. After eliminating duplicate "hits," the procedure of choosing the maximum NPV of the three yields 68% in the top three. (Note that this is *not* the same as the "best possible selection" given in §4.1; that result was predicated on choosing a *single* heuristic.) Running multiple heuristics for a given problem may or may not be an affordable luxury.

5 CONCLUSIONS

We have presented the results of some tests of the genetic programming technique applied to a difficult problem, that of heuristic selection in resource-constrained project scheduling. The genetic programming approach worked quite well, yielding results which improved on those previously obtained using other methods. The tests we performed reinforce the notion that evolutionary approaches to computation in general, and optimization and selection in particular, can bear significant fruit.

The GP approach was tried because we felt it would overcome some of the drawbacks of both fixed-functional-form methods (e.g., regression) and black-box techniques (e.g., neural networks). GP did develop a variety of arbitrarily complex selection functions, but at the same time was able to give a direct indication of which heuristics were likely to be unsuitable for some problem structures. Thus the results given here shed light on the logical domains of applicability of the various heuristics, while at the same time provide an improved heuristic selection process.

REFERENCES

[1] S. Baroum and J.H. Patterson, "A Heuristic Algorithm for Maximizing the Net Present Value of Cash Flows in Resource Constrained Project Schedule," Working Paper, 1989, Indiana University.

[2] R. Bey, R.H. Doersch and J.H. Patterson, "The Net Present Value Criterion: Its Impact on Project Scheduling," Project Management Quarterly, 1981, Vol.12, No.2, p. 35-45.

[3] E.W. Davis, "Project Scheduling under Resource Constraints – Historical Review and Categorization of Procedures," AIIE Transactions, 1973, Vol.5, No.4, p. 297-313.

[4] E.W. Davis and J.H. Patterson, "A Comparison of Heuristics and Optimal Solutions in Resource-Constrained Project Scheduling," Management Science, 1975, Vol.21, No.8, p. 944-955.

[5] M.R. Garey and D.S. Johnson, Computers and Intractability: A Guide to the Theory of NP-Completeness, 1979, W.H. Freeman and Co., New York.

[6] J.R. Koza, Genetic Programming: On the Programming of Computers by Natural Selection, 1992, The MIT Press, Cambridge, MA.

[7] I.S. Kurtulus and E.W. Davis, "Multi-Project Scheduling: Categorization of Heuristic Rules Performance," Management Science, 1982, Vol.28, No. 2, p. 161-172.

[8] R. Padman, "Choosing Solvers in Decision Support Systems: A Neural Network Application in Resource-Constrained Project Scheduling," Recent Developments in Decision Support Systems, 1993, Springer-Verlag, Berlin, p. 559-574.

[9] R. Padman and D. Smith-Daniels, "Early-Tardy Cost Trade-offs in Resource-Constrained Projects with Cash Flows: An Optimization-Guided Heuristic Approach," European Journal of Operational Research, 1993, Vol. 64, p. 295-311.

[10] R. Padman, D.E. Smith-Daniels and V. L. Smith-Daniels, "Heuristic Scheduling of Resource-Constrained Projects with Cash Flows: An Optimization-Based Approach," forthcoming in Naval Research Logistics Quarterly.

[11] R. Padman and D. Zhu, "On Integrating Knowledge Source Using Problem Spaces: An Application in Project Scheduling," forthcoming in Annals of OR.

[12] J.H. Patterson, "Project Scheduling: The Effects of Problem Structure On Heuristic Performance," Naval Research Logistics Quarterly, 1976, Vol.23, No.1, p. 95-122.

[13] R.A. Russell, "A Comparison of Heuristics for Scheduling Projects with Cash Flows and Resource Restrictions," Management Science, 1986, Vol. 32, No. 10, p. 1291-1300.

[14] K.K. Yang, B. Talbot and J.H. Patterson, "Scheduling a Project to Maximize its Net Present Value: An Integer Programming Approach," Proceedings of the Second International Workshop on Project Management and Scheduling, 1990, Compiegne, France, p. 384-397.

[15] D. Zhu and R. Padman, "Neural Networks for Heuristic Selection: An Application in Resource-Constrained Project Scheduling" in Stephen Nash and Ariela Sofer (eds.), The Impact of Emerging Technologies on Computer Science and Operations Research, 1995, Klewer Academic Publishers, Boston.

19

COUPLING A GREEDY ROUTE CONSTRUCTION HEURISTIC WITH A GENETIC ALGORITHM FOR THE VEHICLE ROUTING PROBLEM WITH TIME WINDOWS

Jean-Yves Potvin and François Guertin

Centre de recherche sur les transports et
Département d'informatique
et de recherche opérationnelle,
Université de Montréal,
Montréal, Québec, Canada, H3C 3J7

ABSTRACT

Vehicle routing algorithms can be divided into three broad classes: route construction heuristics that "build" routes through the insertion of new customers, route improvement heuristics that modify the location of customers within the existing routes through exchange procedures, and composite heuristics that mix route construction and route improvement procedures. In this paper, a greedy route construction heuristic for the vehicle routing problem with time windows is described. This heuristic inserts customers one by one into the routes using a fixed a priori ordering of the customers. Then, a genetic algorithm is proposed to identify the ordering that produces the best routes.

1 INTRODUCTION

The Vehicle Routing Problem with Time Windows (VRPTW) can be stated as follows: given a central depot, a homogeneous fleet of vehicles and a set of customers with known demands (e.g., known quantities of goods to be delivered), find a set of closed routes, originating and ending at the depot, that service all customers at minimum cost. The routes must also satisfy capacity and time window constraints. In the first case, the total quantity of goods to be delivered cannot exceed the capacity of the vehicle. In the second case, a time window or time interval is specified at each customer location in order to constrain the vehicle's service time. In the hard time window case, a vehicle cannot arrive at a customer location after its time window's upper bound. However, the vehicle can arrive at a customer location before the lower bound. In this case, a waiting time is incurred. Accordingly, the routing costs include the distance (or travel time) and the waiting time. Typically, the first objective is to service all customers with the minimum number of routes and, for the same number of routes, to minimize the total route time.

Problem-solving approaches for this problem can be divided into three broad classes: route construction heuristics that build routes through the insertion of new customers, route improvement heuristics that modify the location of customers within the existing routes through exchange procedures, and composite heuristics that mix route construction and route improvement procedures [4]. In this paper, a route construction heuristic for the VRPTW is proposed. This heuristic inserts the customers one by one into the evolving routes using a fixed a priori ordering of the customers. This heuristic is "greedy", since it looks for the best insertion place for each customer. It is also "myopic" because the insertion of the current customer is performed without any consideration for the remaining unrouted customers.

A distinctive feature of our problem-solving approach is the use of a genetic algorithm to search for the best ordering of customers during the insertion phase. This hybrid problem-solving approach has already been proposed for other combinatorial optimization problems with side constraints [11]. Under this scheme, the genetic algorithm does not explicitly consider the side constraints, because they are satisfied through the construction of the solution by the insertion heuristic. That is, the genetic algorithm only searches the permutation space of good orderings of customers.

In the literature, three approaches based on genetic algorithms have recently been proposed for the VRPTW. In [13], a genetic algorithm is used to find good clusters or groups of customers, within a "cluster-first route-second" problem-solving strategy. The clusters correspond to groups of customers that naturally fit together in the same route (due to geographical proximity). Then, the final sequence of customers within each cluster is produced with a classical insertion procedure. This algorithm also applies local search heuristics during problem-solving (i.e., customers are moved from one route to another or from one location to another within the same route). Accordingly, this is not a pure route construction heuristic, but rather a composite heuristic that mixes route construction and route improvement procedures. Another genetic approach is proposed in [9]. Here, parent solutions are merged with specialized crossover and repair operators in such a way that offspring solutions are likely to be feasible. Local search heuristics are then applied to the offspring for further improvement.

The genetic algorithm in [2] is based on a methodology that is similar to the one proposed in this paper. The authors also use a genetic algorithm to find a good ordering of customers for an insertion heuristic. Specialized genetic operators exploit a global precedence relationship among customers to create new orderings. For example, it is generally desirable to insert customer i before customer j during the greedy insertion phase, if the time window at customer i occurs before the time window at customer j. Accordingly, this relationship is exploited by the genetic operators to push customers with early time windows in front of the resulting orderings. However, the new orderings are often quite similar due to this bias, and the genetic algorithm is often limited to a fraction of the search space.

In the following, our problem-solving approach to the VRPTW is described. Section 2 first presents the greedy insertion heuristic. Then, the evolution paradigm is briefly introduced in Section 3. In Section 4, our genetic algorithm is described. Finally, computational results on Solomon's test problems [12] are reported in Section 5.

2 THE INSERTION HEURISTIC

Our greedy insertion heuristic is based on Solomon's work [12]. The routes are created one at a time, and each route is initialized with a "seed" customer. That is, the initial route only services a single customer (i.e., the vehicle leaves the depot, services the seed customer and comes back to the depot). Then, the other customers are inserted in this route, until it is full with respect to the time window or capacity constraints. At this point, the method initializes a new route which is filled with the remaining unrouted customers. Routes are sequentially created and filled in this way, until all customers are serviced. Note that the last route is rarely full, and often contains only a few customers.

The method uses an insertion cost measure $c_1(i, u, j)$ to guide the insertion of the next unrouted customer u between two consecutive customers i and j in the current route. Let $(i_0, i_1, i_2, \ldots, i_m)$ be the current route with i_0 and i_m

standing for the depot, and let u be the next unrouted customer to be inserted in the route (according to the a priori ordering). The best feasible insertion place for customer u is computed as follows:

$$c_1^*(i(u), u, j(u)) = min_{p=1,\dots,m}[c_1(i_{p-1}, u, i_p)]$$

where

$$c_1(i,u,j) = \begin{cases} \alpha_1(d_{iu} + d_{uj} - d_{ij}) + \alpha_2(b_{uj} - b_j) & \text{if } u \text{ can be feasibly inserted between customers } i \text{ and } j \\ \infty & \text{otherwise.} \end{cases}$$

d_{ij} is the distance (in time units) between i and j.
b_j is the current service time at customer j.
b_{uj} is the new service time at customer j due to the insertion of customer u before j.
α_1 is the weight for the additional distance (detour).
α_2 is the weight for the service delay at customer j.
$\alpha_1 \geq 0, \alpha_2 \geq 0, \alpha_1 + \alpha_2 = 1.$

Then, customer u is inserted between $i(u)$ and $j(u)$. The notation $i(u)$, $j(u)$ is used because the best feasible insertion place is a function of the unrouted customer u. Note also that the insertion cost c_1 of customer u is a weighted sum of two components: the additional distance (in time units) and the service delay at customer j, due to insertion of customer u before j. The two components are weighted by parameters α_1 and α_2. Of course, the best feasible insertion place for customer u minimizes this weighted sum (see Figure 1).

As mentioned before, the route construction heuristic inserts the customers one by one into the routes, using an a priori ordering. For example, if five customers are to be serviced, and the ordering is 5, 4, 3, 2, 1, then customer 5 is used as the seed customer for the first route. Then, customer 4 is inserted in this route, followed by customer 3. At this point, we assume that there is no feasible insertion place for customers 2 and 1. Consequently, customer 2 is used as a seed customer to initialize a second route, and customer 1 is inserted in this route. Obviously, different solutions can be produced by modifying the insertion order of the customers.

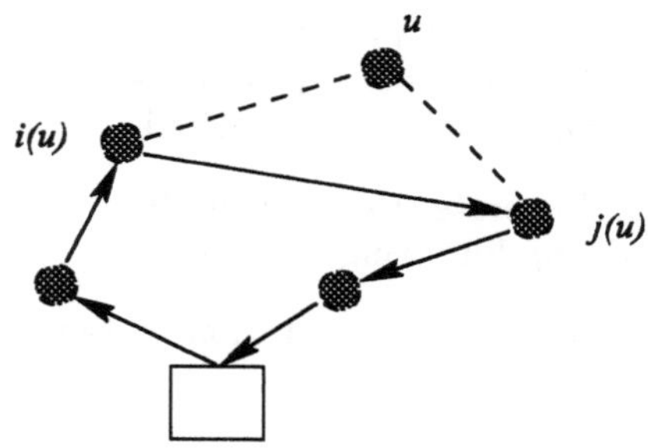

Figure 1 The best insertion place for customer u

In [2], a similar insertion heuristic is described. However, the routes are built in parallel, rather than one by one. That is, many routes are initialized at once with different seed customers, and the remaining unrouted customers are inserted into any one of those routes. Since the number of routes that must be constructed to service all customers is not known in advance, many different runs must be performed with an increasing number of routes until a feasible solution is found. This is a major drawback of the parallel approach. This difficulty does not arise with a sequential approach, because the routes are created one by one, as needed, until all customers are serviced.

Since the route construction heuristic inserts the customers based on a fixed ordering, a genetic algorithm is used to find the best possible ordering. In Section 3, a simple genetic algorithm is briefly introduced. Then, a specific implementation for our ordering problem is described.

3 A SIMPLE GENETIC ALGORITHM

In order to apply a genetic algorithm to a given problem, solutions to this problem must first be encoded as bit strings (i.e., chromosomes). Then, an evaluation function that returns the fitness or quality of each chromosome (solution) is defined. Using these two components, a simple genetic algorithm can be summarized as follows.

Step 1. *Initialization.* Create an initial random population of chromosomes and evaluate each chromosome. Set the current population to this initial population.

Step 2. *Reproduction.* Select two parent chromosomes from the current population. The selection process is stochastic, and a chromosome with high fitness is more likely to be selected.

Step 3. *Recombination.* Generate two offspring from the two parent chromosomes by exchanging bit strings (crossover).

Step 4. *Mutation.* Apply a random mutation to each offspring (with a small probability).

Step 5. Repeat Steps 2, 3 and 4, until the number of offspring in the new population is the same as the number of chromosomes in the old population.

Step 6. Evaluate each offspring. Set the current population to the new population of offspring and go back to Step 2.

This procedure is repeated for a fixed number of generations, or until no more improvement is observed. The best chromosome generated during the search is the final result of the genetic algorithm. Through this process, it is expected that an initial population of randomly generated chromosomes with low fitness values will improve as parents are replaced by better offspring.

The main components of the genetic search are now briefly described. First, the encoding transforms a solution to a problem into a bit string. For example, if the problem is to maximize a numerical function f whose domain is a set of integers, then each chromosome can represent an integer in base-2 notation. In this case, the fitness of each chromosome is $f(x)$, where x is the integer encoded in the chromosome. The reproduction phase favors the best chromosomes through a bias in the selection process. Namely, chromosomes with high fitness are more likely to be selected as parents. Once the parents have been selected, the recombination phase creates two new offspring through the exchange of bit strings taken from two parent chromosomes. Here, the underlying assumption is that a better chromosome can be produced by combining bit patterns taken from two good chromosomes. An example of a one-point crossover is shown in Figure 2 for two parent bit strings of length 5. Here, the cross point is (randomly) chosen between the second and third bit. As we can see, the end parts of both parents are swapped to create two new offspring.

Finally, the mutation phase is aimed at maintaining a minimum level of diversity in the population through random modifications to the bit values. Namely, the mutation operator changes the bit value from 0 to 1 or from 1 to 0 on each

1	1	|	0	0	0	(parent 1)
0	0	|	1	1	1	(parent 2)
1	1		1	1	1	(offspring 1)
0	0		0	0	0	(offspring 2)

Figure 2 One-point crossover on two bit strings

offspring, with a small probability at each position. A good level of diversity in the population is crucial to the effectiveness of the search. If the population is composed of similar chromosomes, the search cannot make any progress because the offspring look like their parents.

In the next section, an extension of this simple genetic algorithm is proposed to solve our problem.

4 THE ORDERING PROBLEM

Here, the genetic algorithm must find a good ordering of customers for the route construction heuristic. Accordingly, our implementation departs from the classical scheme in two important ways:

(a) chromosomes are sequences of integers rather than sequences of bits (where each integer represents a particular customer).

(b) specialized crossover and mutation operators are designed to produce new orderings (or permutations) from old ones.

Given these characteristics, the main components of the genetic algorithm will now be introduced.

4.1 Representation

As mentioned before, an ordering is encoded as a sequence of integers. Assuming that five customers must be serviced, the ordering 1 2 3 4 5 means that customer 1 will be the first to be inserted in a route, followed by customer 2, etc.

4.2 Reproduction

The fitness of each chromosome corresponds to the quality of the solution produced by the insertion heuristic, using the a priori ordering of customers on the chromosome. In this application, solution quality is based firstly on the number of routes, and secondly, on total route time. Hence, a solution with fewer routes is always better then a solution with more routes, independently of the total route time. For the same number of routes, the total route time is minimized.

The rank of each chromosome in the population is used to evaluate its fitness [15]. In the example below, the population is composed of four chromosomes, encoding four different orderings. The value of the solution produced by the route construction heuristic is shown on the same line. In this example, chromosome (3) gets rank 1 because it produces the minimum number of routes, while chromosomes (2), (1) and (4) get ranks 2, 3 and 4, respectively.

Chromosome		Number of Routes	Route Time	Rank
(1)	1 2 3 4 5	2	1612.0	3
(2)	5 4 3 2 1	2	1588.1	2
(3)	3 2 4 1 5	1	1660.0	1
(4)	2 4 3 5 1	2	1680.0	4

With these ranks, the fitness of each chromosome is computed as follows:

$$fitness_i = Max - \left[(Max - Min) \times \frac{(i-1)}{(n-1)}\right],$$

where i is the rank of the chromosome, and n is the number of chromosomes in the population. Hence, the best ranked chromosome gets fitness value Max and the worst chromosome gets fitness value Min. In the current implementation, Max and Min are set to 1.2 and 0.8, respectively. Then, a fitness proportional selection scheme is applied to these values. Namely, the selection probability for a chromosome of rank i is:

$$p_i = \frac{fitness_i}{\sum_{j=1}^{n} fitness_i} = \frac{fitness_i}{n}$$

It is worth noting that the summation over all fitness values in a population of size n is equal to n, because $Min + Max = 2$, and the average fitness is 1.

Since n different selections (with replacement) must be performed in order to select n parents and generate n offspring, the expected number of selections E_i for a chromosome of rank i is:

$$E_i = n \times p_i = fitness_i.$$

Hence, the fitness of each chromosome is also equal to its expected number of selections. For example, the best chromosome with fitness $Max = 1.2$, is expected to be selected 1.2 times on average over n different trials, while the worst chromosome is expected to be selected 0.8 times on average. As we can see, the gap between the best and the worst chromosome is not very large. Hence, the selection process is just slightly biased in favor of the best chromosomes. This is a way to favor exploration of new regions of the search space (note that we can favor exploration during the selection process, due to a conservative generation replacement scheme; see Section 4.5).

In order to reduce the variance associated with pure proportional selection (where each chromosome can be selected between 0 and n times over n different trials), stochastic universal selection or SUS is applied to the fitness values. This approach guarantees that the number of selections for any given chromosome is at least the floor, and at most the ceiling, of its expected number of selections [1].

4.3 Recombination

The permutation operator introduced in this section is an intermediate approach between "general purpose" permutation operators like OX [3, 8], and the heuristically-biased MX1 and MX2 operators [2]. In the first case, the crossover operator generates new permutations through manipulations that do not exploit any heuristic information about the problem. Since the time window constraints have a strong impact on the structure of the final solutions, such operators are unlikely to yield very good results. Conversely, the MX1 and MX2 operators are strongly biased by the precedence relationship among the customers in regard to their time window, which can prevent exploration of useful regions of the search space. In the following, the crossover operator MX1 and the new crossover 1X will be described.

Merge cross 1 (MX1)

Here, two parent chromosomes are processed position by position. Starting at position 1, the customers on the parent chromosomes are compared, and the one with the earliest time window is introduced in the offspring. Then, the procedure is repeated at the next position, until all positions are done.

In Figure 3, the precedence relationship among the customers in regard to their time window is 1 2 3 4 5. Starting at position 1, customers 1 and 4 are compared. Since the time window at customer 1 occurs before the time window at customer 4, customer 1 is introduced in the offspring (Figure 3a). Now, in order to produce a valid ordering, customers 1 and 4 are swapped on parent 2. Then, customers 3 and 5 are compared at position 2, and customer 3 is introduced in the offspring (Figure 3b). Customers 5 and 3 are swapped on parent 1, etc. The resulting offspring is shown in Figure 3e.

In practice, many offspring do not depart much from the precedence relationship established among the customers in regard to their time window (i.e., 1 2 3 4 5, in this example). Accordingly, premature convergence is frequently observed.

A complete description of the MX2 crossover operator, which is closely related to MX1, may be found in [2].

Permutation-based one-point crossover (1X)

This crossover is similar to the classical one-point crossover. Accordingly, a cross point is randomly chosen on both chromosomes, and the end parts are exchanged to create two new offspring. In the process, some customers are duplicated on each offspring, while other customers are left apart. The duplicates are eliminated, and the missing customers take the available positions. When many customers are left apart, the customer with the earliest time window is inserted at the first available position, the customer with the second earliest time window is inserted at the second available position, etc. Accordingly, this operator is not as strongly biased as MX1 and MX2 in favor of customers with early time windows.

In Figure 4, a cross point is randomly chosen between the second and third position on each parent chromosome. Then, the end parts are exchanged to create two new offspring (in the figure, only one offspring is shown). In the process, customers 1 and 5 are duplicated on one offspring and customers 3

(a)

parent 1	:	1	5	2	3	4
parent 2	:	4	3	1	2	5
offspring	:	1	-	-	-	-

(b)

parent 1	:	1	5	2	3	4
parent 2	:	1	3	4	2	5
offspring	:	1	3	-	-	-

(c)

parent 1	:	1	3	2	5	4
parent 2	:	1	3	4	2	5
offspring	:	1	3	2	-	-

(d)

parent 1	:	1	3	2	5	4
parent 2	:	1	3	2	4	5
offspring	:	1	3	2	4	-

(e)

parent 1	:	1	3	2	4	5
parent 2	:	1	3	2	4	5
offspring	:	1	3	2	4	5

Figure 3 MX1 crossover

and 4 are left apart. Hence, customers 1 and 5 are removed and replaced by customers 3 and 4, respectively. Note that customer 3 is inserted at the first available position, assuming that the precedence relationship among the customers in regard to their time window is 1 2 3 4 5.

parent 1	:	1	5 |	2	3	4
parent 2	:	4	3 |	1	2	5
		1	5	1	2	5
		1	5	-	2	-
offspring	:	1	5	3	2	4

Figure 4 1X crossover

4.4 Mutation

Mutation operators are useful to maintain a certain level of diversity in the population. In the following, four different order-based mutation operators are described.

Remove-and-Reinsert (RAR)

Here, a customer is randomly selected on a chromosome and is moved at another position. The customers located between the two positions are shifted accordingly. The absolute difference between the new position and the old position must be less than or equal to some upper bound. In the experiments of Section 5, this upper bound was set to 7.

In Figure 5, the RAR operator is illustrated. In this example, customer 3 is moved from position 2 to position 4.

chromosome:	1	$\underline{3}$	2	4	5
chromosome:	1	2	4	$\underline{3}$	5

Figure 5 RAR mutation

Swap

Here, two customers are randomly selected on a chromosome and exchange their respective position. This mutation operator is also used in [2]. As for RAR, the absolute difference between the position of the two customers must be less than or equal to some upper bound. In the experiments of Section 5, this upper bound was set to 7.

In Figure 6, the Swap mutation operator is illustrated. In this example, customers 3 and 4 are swapped.

chromosome:	1	$\underline{3}$	2	$\underline{4}$	5
chromosome:	1	$\underline{4}$	2	$\underline{3}$	5

Figure 6 Swap mutation

Last-will-be-first (LwbF)

This operator randomly selects a cut point on the chromosome. Then, the end part is moved in front of the chromosome. In Figure 7, the cut point is randomly chosen between the second and third position, and the substring 2 3 4 is moved in front of the substring 1 5. This operator does not look "natural" at first glance. However, it is very useful to modify the order of construction of the routes. Namely, the routes that were constructed last will now be constructed first after the mutation, and vice-versa. Often, the solution produced by the insertion heuristic with the new ordering will be slightly modified, because the last route is rarely "full", and the cut point does not necessarily occur at an exact boundary between two routes (i.e. between customers i and $i+1$, where customer i is the last customer inserted in route m and customer $i+1$ is the first customer inserted in route $m+1$).

It is worth noting that this mutation operator greatly modifies the chromosome to which it is applied, even if it does not modify much the solution produced by the route construction heuristic. Hence, if the initial ordering was a good one, the new ordering is likely to be good also. However, since the chromosome is greatly modified, this mutation operator, in conjunction with the crossover operator 1X, allows the genetic algorithm to explore new regions of the search space.

chromosome:	1	5 |	2	3	4
chromosome:	2	3	4	1	5

Figure 7 LwbF mutation

Last-will-be-First+ (LwbF+)

This operator is an extension of the previous mutation operator. Namely, all cut points along the chromosome are tried and the best one is selected at the end. If the length of the chromosome is L, then $L-1$ different cut points are tried, and $L-1$ new orderings are created. Hence, this mutation selects the best ordering in a "neighborhood" of the current ordering, and is more computationally expensive than the three previous operators.

4.5 Generation Replacement

If the size of the population is n, then n parents are selected and n offspring are created. Then, the next generation is composed of the n best chromosomes among the pool of $2n$ chromosomes (i.e., n parents + n offspring). Due to this conservative replacement scheme, the best chromosome at generation $m+1$ is always better than or equal to the best chromosome at generation m [5].

5 COMPUTATIONAL RESULTS

The genetic algorithm of Section 4 was applied to Solomon's test problems, which are all 100-customer Euclidean problems [12]. In these problems, the travel times are equal to the corresponding Euclidean distances. The geographical data were either randomly generated using a uniform distribution (problem sets R1 and R2), clustered (problem sets C1 and C2) or mixed with both randomly distributed and clustered customers (problem sets RC1 and RC2). Problem sets R1, C1 and RC1 have a narrow scheduling horizon. Hence, only a few customers can be serviced by the same vehicle. Conversely, problem sets R2, C2 and RC2 have a large scheduling horizon, and more customers can be serviced by the same vehicle.

In these problems, there is a fixed unload time at each customer. The sum of unload times over all customers is equal to 9,000 for each problem of type C, and 1,000 for the other problems. Accordingly, the total route time is equal to the sum of travel time (or distance), waiting time and unload time. More details about these problems may be found in [12].

5.1 Implementation details

The computational results of Section 5.2 were obtained as follows:

(a) The insertion order of the customers, based on Solomon's I1 heuristic [12], is used to seed the initial population. Namely, one chromosome encodes this ordering. Additional chromosomes are derived from this chromosome by applying the RAR and Swap mutation operators many times.

(b) All parent chromosomes undergo crossover (however, the best parents are preserved due to the conservative generation replacement scheme; see Section 4.5).

(c) All offspring undergo mutation (when mutation is used). A mutation operator is randomly selected when many mutation operators are available.

(d) The population size is equal to 150.

(e) The number of generations is equal to 50.

5.2 Experiments on Solomon's problems

The results on Solomon's problems are summarized in Tables 1a, 1b and 1c. The average number of routes, route time and computation time (in seconds, on a SPARC 10 workstation) are shown for each set of problems. Five independent runs were performed with each method. The first number in each cell corresponds to the best run (number of routes first, total route time second), the second number is the worst run, and the third number is the mean over the five runs.

In order to compare the crossover operators, three different implementations of the genetic algorithm with either 1X, OX, MX1 or MX2 (and no mutation operator) were tested. These implementations are referred to as MX (best of MX1 and MX2 on each set of problems), OX and 1X in Table 1. 1X-Mutation uses the RAR, Swap and LwbF mutation operators in addition to the crossover operator 1X. Finally, LwbF+ replaces LwbF in 1X-Mutation+.

Tables 1a, 1b and 1c show that the crossover operator 1X produces better solutions than OX, MX1 and MX2 on all sets of problems. The mutation operators RAR, Swap and LwbF also allow 1X to find better orderings at low computation cost. Finally, 1X-Mutation+ outperforms 1X-Mutation in most cases, but is much more computationally expensive. For example, on problem set R2, the average computation time of 1X-Mutation is only 44.5 seconds, while 1X-Mutation+ requires about 10 minutes.

Table 1a Results on problems of type R

Method	R1 (12 problems)			R2 (11 problems)		
	Number of Routes	Route Time	Comput. Time (sec.)	Number of Routes	Route Time	Comput. Time (sec.)
MX	13.5	2753.8	36.8	3.2	2515.4	49.5
	13.7	2758.1	39.7	3.2	2543.2	44.9
	13.6	2751.9	38.3	3.2	2531.1	46.0
OX	13.3	2658.2	34.8	3.2	2446.1	39.8
	13.4	2683.4	34.4	3.2	2474.0	42.5
	13.3	2671.8	34.5	3.2	2462.2	41.8
1X	13.2	2635.7	34.7	3.1	2427.1	42.0
	13.3	2660.4	35.4	3.2	2428.0	42.4
	13.2	2647.5	35.4	3.1	2435.1	42.6
1X-Mutation	12.9	2597.5	40.7	3.0	2381.2	44.4
	13.1	2628.1	37.6	3.1	2385.5	44.7
	13.0	2610.1	38.5	3.1	2395.1	44.5
1X-Mutation+	12.8	2585.0	560.7	2.9	2334.2	635.5
	13.0	2610.2	517.0	3.1	2375.9	599.1
	12.9	2596.0	529.8	3.0	2365.2	607.1

Table 1b Results on problems of type C

Method	C1 (9 problems)			C2 (8 problems)		
	Number of Routes	Route Time	Comput. Time (sec.)	Number of Routes	Route Time	Comput. Time (sec.)
MX	10.0	10098.6	22.7	3.1	9796.6	32.8
	10.0	10145.0	22.6	3.1	9833.4	35.7
	10.0	10121.1	22.6	3.1	9816.2	34.8
OX	10.0	10059.0	20.9	3.1	9770.9	33.2
	10.0	10078.1	20.7	3.1	9809.8	32.5
	10.0	10069.4	20.2	3.1	9791.2	33.0
1X	10.0	9978.3	22.4	3.0	9698.6	34.0
	10.0	10043.7	21.7	3.1	9732.8	34.5
	10.0	10006.2	21.2	3.1	9714.0	34.0
1X-Mutation	10.0	10037.6	23.1	3.0	9677.8	35.1
	10.0	10062.5	24.2	3.0	9712.9	35.9
	10.0	10048.8	23.8	3.0	9697.6	34.9
1X-Mutation+	10.0	9985.5	331.9	3.0	9648.4	511.7
	10.0	10002.1	322.8	3.0	9675.3	471.5
	10.0	9995.0	326.8	3.0	9664.7	481.1

Table 1c Results on problems of type RC

Method	RC1 (8 problems)			RC2 (8 problems)		
	Number of Routes	Route Time	Comput. Time (sec.)	Number of Routes	Route Time	Comput. Time (sec.)
MX	13.1	2773.6	33.2	3.5	2788.3	40.7
	13.5	2838.1	33.1	3.6	2848.6	38.3
	13.4	2827.5	32.8	3.6	2817.5	41.4
OX	13.1	2776.9	31.6	3.6	2770.9	38.3
	13.4	2779.4	30.8	3.6	2827.3	38.0
	13.3	2775.4	31.2	3.6	2798.5	38.4
1X	13.0	2714.8	32.6	3.5	2742.2	40.2
	13.3	2735.7	31.2	3.6	2781.7	38.2
	13.1	2722.4	31.9	3.6	2756.2	39.1
1X-Mutation	12.6	2706.2	35.7	3.4	2704.8	43.2
	12.9	2713.0	35.8	3.5	2708.9	43.0
	12.8	2704.7	35.4	3.5	2705.1	41.0
1X-Mutation+	12.6	2697.2	464.6	3.4	2664.4	567.8
	12.8	2711.1	502.6	3.5	2675.5	540.3
	12.7	2712.8	485.6	3.4	2675.7	549.0

5.3 Comparison with other heuristics

Table 2 compares the best solutions produced by 1X-Mutation and 1X-Mutation+ with the best solutions reported by other researchers on the 56 test problems of Solomon. The average number of routes is used for this purpose. It is not possible to compare the methods on the basis of our secondary objective, because it is not the same in all cases. Some authors use the route time, while others use the distance, or even a weighted sum of distance and route time. The other heuristics include the cyclic transfer algorithm (CTA) [14], the GIDEON sectoring algorithm (GID) [13], the greedy randomized adaptive search procedure (GRASP) [7], the GENEROUS routing system (GEN) [9] and the adaptive memory tabu search of Rochat and Taillard (RT) [10].

With respect to solution quality, this table shows that our problem-solving approach is outperformed by the most recent heuristics GENEROUS and RT, but compares favorably with CTA and GIDEON. Note that our method is based on simple problem-solving components (i.e., an insertion heuristic and a permutation-based genetic algorithm). Accordingly, it can be easily extended

Table 2 Average number of routes for different heuristics

Set	CTA	GID	GRASP	GEN	RT	1X-Mutation	1X-Mutation+
R1	13.0	12.8	12.6	12.6	12.3	12.9	12.8
R2	3.1	3.2	3.1	3.0	2.9	3.0	2.9
C1	10.0	10.0	10.0	10.0	10.0	10.0	10.0
C2	3.0	3.0	3.0	3.0	3.0	3.0	3.0
RC1	13.0	12.5	12.6	12.1	11.9	12.6	12.6
RC2	3.7	3.4	3.5	3.4	3.4	3.4	3.4

to problems with additional side constraints, like linehaul-backhaul precedence constraints [6]. In such a case, the insertion heuristic must be slightly modified to account for the new constraints when the feasibility of each insertion place is determined.

6 CONCLUSION

In this paper, a genetic approach to the VRPTW is described. The genetic algorithm is coupled with a greedy route construction heuristic to find a good insertion order for the customers. This approach has produced solutions of good quality on a standard set of problems. Since the method exploits simple problem-solving components, it is of general applicability and can be easily extended to problems with additional side constraints.

Acknowledgements

This research was supported by the Natural Sciences and Engineering Research Council of Canada (NSERC), and by the Quebec Fonds pour la Formation de Chercheurs et l'Aide à la Recherche (FCAR).

REFERENCES

[1] Baker J.E., "Reducing Bias and Inefficiency in the Selection Algorithm", Proceedings of the Second Int. Conf. on Genetic Algorithms, July 1987, Cambridge, MA, pp. 14-21.

[2] Blanton J.L. and R.L. Wainwright, "Multiple Vehicle Routing with Time and Capacity Constraints using Genetic Algorithms", Proceedings of the Fifth International Conference on Genetic Algorithms, July 1993, Champaign, IL, pp. 452-459.

[3] Davis L., "Applying Adaptive Algorithms to Epistactic Domains", Proceedings of the Int. Joint Conf. on Artificial Intelligence, August 1985, Los Angeles, CA, pp. 162-164.

[4] Desrosiers J., Y. Dumas., M.M. Solomon and F. Soumis, "Time Constrained Routing and Scheduling", Technical Report G-92-42, Groupe d'études et de recherche en analyse des décisions, Université de Montréal, Montréal, Canada, 1992.

[5] Eshelman L.J., "The CHC Adaptive Search Algorithm: How to Have Safe Search When Engaging in Nontraditional Genetic Recombination", Foundations of Genetic Algorithms, G.J.E. Rawlins Eds, Morgan Kaufmann, 1991, San Mateo, CA, pp. 265-283.

[6] Gélinas S., M. Desrochers, J. Desrosiers and M.M. Solomon, "A New Branching Strategy for Time Constrained Routing Problems with Application to Backhauling", Technical Report G-92-13, Groupe d'études et de recherche en analyse des décisions, Université de Montréal, Montréal, Canada, 1994.

[7] Kontoravdis G. and J. Bard, "A GRASP for the Vehicle Routing Problem with Time Windows", ORSA Journal on Computing 7, 1995, pp. 10-23.

[8] Oliver I.M., D.J. Smith and J.R.C. Holland, "A Study of Permutation Crossover Operators on the Traveling Salesman Problem", Proceedings of the Second Int. Conf. on Genetic Algorithms, July 1987, Cambridge, MA, pp. 224-230.

[9] Potvin J.Y. and S. Bengio, "A Genetic Approach to the Vehicle Routing Problem with Time Windows", Technical Report CRT-953, Centre de recherche sur les transports, Université de Montréal, Montréal, Canada, 1993.

[10] Rochat Y. and E. Taillard, "Probabilistic Diversification and Intensification in Local Search for Vehicle Routing", Technical Report CRT-95-13, Centre de recherche sur les transports, Université de Montréal, Montréal, Canada, 1995.

[11] Smith D., "Bin Packing with Adaptive Search", in Proceedings of the First Int. Conf. on Genetic Algorithms and their Applications, July 1985, Pittsburgh, PA, pp. 202-207.

[12] Solomon M.M., "Algorithms for the Vehicle Routing and Scheduling Problems with Time Window Constraints", Operations Research 35, 1987, pp. 254-265.

[13] Thangiah S.R., "Vehicle Routing with Time Windows using Genetic Algorithms", Technical Report SRU-CpSc-TR-93-23, Computer Science Department, Slippery Rock University, Slippery Rock, PA, 1993.

[14] Thompson P. and H. Psaraftis, "Cyclic Transfer Algorithms for Multivehicle Routing and Scheduling Problems", Operations Research 41, 1993, pp. 935-946.

[15] Whitley D., "The Genitor Algorithm and Selection Pressure: Why Rank-Based Allocation of Reproductive Trials is Best", Proceedings of the Third Int. Conf. on Genetic Algorithms, June 1989, Fairfax, VA, pp. 116-121.